Peter W. Zehna, Ph.D., teaches math and statistics at the Naval Postgraduate School in Monterey, California. In addition to his experience as a consultant and research participant for various defense agencies involved in modeling random phenomena, Dr. Zehna has been employed as a statistician.

STATISTICS BY CALCULATOR

Solving Statistics Problems With the Programmable Calculator

Peter W. Zehna

Donald R. Barr

A SPECTRUM BOOK

Prentice-Hall, Inc., Englewood Cliffs, New Jersey 07632

Library of Congress Cataloging in Publication Data

Zehna, Peter W.
Statistics by calculator.

(A Spectrum Book)—Verso of t.p.
Includes index.
1. Statistics—Problems, exercises, etc.—Data processing. 2. TI Programmable 59 (Calculating—machine) I. Barr, Donald Roy. II. Title.
QA276.4.Z43 1982 519.5′076 82–9810
AACR2

ISBN 0-13-844829-9

ISBN 0-13-844811-6 {PBK.}

This book is available at a special discount when ordered in large quantities. Contact Prentice-Hall, Inc., General Publishing Division, Special Sales, Englewood Cliffs, N.J. 07632.

A SPECTRUM BOOK

Printed in the United States of America

10 9 8 7 6 5 4 3 2 1

PRENTICE-HALL INTERNATIONAL, INC., *London*
PRENTICE-HALL OF AUSTRALIA PTY. LIMITED, *Sydney*
PRENTICE-HALL OF CANADA, INC., *Toronto*
PRENTICE-HALL OF INDIA PRIVATE LIMITED, *New Delhi*
PRENTICE-HALL OF JAPAN, INC., *Tokyo*
PRENTICE-HALL OF SOUTHEAST ASIA PTE. LTD., *Singapore*
WHITEHALL BOOKS LIMITED, WELLINGTON, *New Zealand*

Contents

3

Data Processing 40

4

Estimation 70

5

Hypothesis Testing 110

6

Bivariate Populations 128

7

Proportions 168

Preface

The computer age has arrived. That is not a novel announcement, and it may even seem late. But never before has the calculating power of the modern computer been within reach of so many potential users as it is today.

Part of that accessibility has come about with the modern hand-held, programmable calculators. Such devices, with their permanent recording capability and their capacity for effecting subroutines within a program, go beyond mere calculators and are truly miniature computers. As such, they offer the opportunity to exploit the potential of the computer as a labor-saving device, as well as a means for the rapid, consistently correct evaluation of various aspects of formulas in all of the mathematical sciences. While many feel that the computer is an untapped source of improving mathematics education, it has been somewhat hampered by its size, its expense, and its availability to individual users, as well as by the training required in special programming languages. All that is changing very rapidly, and we are now in a position to begin exploiting these ideas in realistic ways.

For about 2 years before beginning work on this book, we experimented with the use of hand-held programmable calculators in the classroom at the Naval Postgraduate School. We succeeded in exploiting such devices, not merely as peripheral tools for evaluating a numerical answer at the end of a problem, but as integral parts of the learning process. We began by asking ourselves repeatedly what tasks we want our students to be able to perform as a result of studying our courses, particularly in the areas of probability and statistics. In many instances, the answer was *not* to be able to derive the results in these subjects, an approach that we traditionally used for years. Rather, we wanted our students to become intelligent *users* of the results; that is, they should develop skills in modeling problems, in recognizing modeled problems, and in evaluating solutions, in either case. Actually computing the results and, certainly, being able to derive them algebraically, were incidental and of secondary importance at best for many curricula.

The ongoing results have been more than rewarding, and we continue to experiment with a view toward the rapid and dynamic changes presently occurring in the microcomputer industry. Of the many things we have learned, one is perhaps the most significant: the proper use of these devices enables those individuals who, for one reason or another, are mathematically handicapped to rise to a respectable level of competence—sometimes equal, in many respects, to those without such a handicap. We have seen too many such cases to attribute the results to chance. There are many reasons why such competence is possible, not the least of which are the attitudinal changes brought about when a past history of failure to obtain correct answers is replaced by virtually 100% success.

One special aspect of this approach is that students can repeatedly process a formula—or a number of formulas—in a model situation, both to see the effects of parameter changes and immediately to reinforce what they are learning. That type of reinforcement we have found to be invaluable, even for students having a good grasp of fundamental concepts. Since most textbooks are not written with that sort of "practice" in mind, we provide problem material that we feel is suited not only to mastering introductory statistics, but also to achieving that kind of reinforcement.

This volume is devoted to statistics. An earlier volume on probability is available in the series. In both cases, the material serves mainly as a supplement to other textbooks. As such, no attempt is made to derive statistical formulas, certainly not in any complete sense. Instead, statistical matters are summarized and exemplified, with some attention devoted to the experimental situations where those results are typically applied. Then, of course, the reinforcing problems with answers, keyed to programs we have written, are provided for the necessary practice. With an instructor to supply derivations and backgrounds, the book might very well be used as a classroom textbook.

We have tried to be complete, in the sense of screening standard introductory textbooks, including the several we use, for a list of topics that we perceive to be typical in such courses. Certainly some topics in statistics, perhaps amenable even to computer implementation, are not included here, because they are not commonly found in introductory courses.

While we have used both Hewlett–Packard and Texas Instruments calculators in our course work, this volume, and most of our work in the probability and statistics areas, is keyed to the TI–59 calculator, primarily because of its cost. This is no minor consideration when each member of an entire class needs to have one.

Since programming, per se, is not one of our goals, only a minimal amount of information in the manuals accompanying the calculator is necessary for utilizing these materials. The programs, keyed to various chapters of the book and tested by various classes, are available in an Appendix for permanently recording onto magnetic cards, which are supplied with the calculator. For more extensive programming information, the book *Calculator Clout*, by Maurice D. Weir,* is highly recommended.

Calculus should be a prerequisite to this book, because many aspects of the material require such a setting for a framework. Like so many elementary courses in the subject, however, the calculus does not really play a formal role, and so the material can be handled with ease based on only a rudimentary knowledge of the area under a continuous curve being computed by an integral. Again, the volume *Calculus by Calculator*, by Maurice D. Weir, in this same series, provides a more than adequate background for such an understanding and furnishes a wealth of problem settings for illustrating the use of the TI–59 calculator, as well.

The authors are grateful to the usual variety of people who make such a work as this possible. Globally, we must thank the many students who suffered through early versions of the programs, revisions of which came only through their sufferings. A special note of gratitude belongs to Mrs. Bea S. Green for her unlimited tolerance in typing the manuscript. Special credit also goes to Mssrs. George Nelson and Ed Stanton who supplied a large number of the problems for each chapter. Over 350 problems and solutions are provided from a variety of applied areas. Even so, there never seem to be enough problems, and one area or another will always be slighted. By keeping the notation and coverage within the mainstream of current offerings, however, the tools mastered here may be used for almost any other book on the subject. As such, it may be useful to beginner and professional alike. The authors are neither programmers nor computer specialists (which attests to the versatility and simplicity of the TI–59 programming language). The only claims regarding the programs are simply that they seem to work as tested and that they have been verified by scores of students and some colleagues. Otherwise, no warranty is made regarding their accuracy in all situations that might arise. Naturally, the authors would certainly be grateful if any errata were brought to their attention.

*Maurice D. Weir, *Calculator Clout: Programming Methods for Your Programmable Calculator* (Englewood Cliffs, N.J.: Prentice-Hall, Inc., 1981).

1
About the Book

Section 1.1: The Aim of the Book

There are certain things that this book is not. It is not a book about the theory of statistics, nor is it very much concerned with the conceptual development of statistics on intuitive grounds. Those topics may be found in dozens of competing text-books written at a variety of levels of mathematical maturity for various audiences. Except for the purest in theory, they all demonstrate the applications of statistical analysis applied to real or fictitious data.

They also share one other thing in common: you must arithmetically process certain laws and formulas to obtain numerical answers in problem situations used to illustrate those applications. Such problems not only enrich the subject matter, but they also provide further insight into the models by giving you the opportunity to observe the effects of parameter changes on the governing formulas. In addition, these problems offer you the means not only of observing a variety of applications, but also of practice, which is absolutely essential in some cases to clarify and amplify the concepts.

Unfortunately, the numerical evaluation of these formulas also of-
 fers many opportunities for committing arithmetic blunders. Someone

who encounters nothing but the frustration of failing to reach a correct answer soon loses sight of the concept being illustrated. Such failures, repeated often enough, may, and often do, create mental blocks. Yet all of this frustration is unnecessary with the widespread availability and portability of the modern electronic hand-held calculator. The proper exploitation of such devices can virtually eliminate computational drudgery. The time and energy you save you can better use to gain a deeper understanding of the subject matter and a better appreciation of points of the problems material.

Our experience shows that students suffer no loss by avoiding such arithmetic chores. Indeed, it is debatable whether such computation—per se—really contributes to their better understanding of the subject in the first place. In this regard, you have to make a careful distinction between computation, as a necessary vehicle for illustrating mathematical material, and the actual calculating skills necessary to carry out those computations. Take a very simple analogy: to verify a checking balance, you must add a sum of numbers. How you arrive at that sum makes no difference; you may count on your fingers, group by 5's or 10's, memorize addition facts, and, yes, press a button on a calculator.

The purpose of this book is to fill a large part of the need for computation in the area of statistics by reducing the required skills to a few simple procedures on a programmable hand-held calculator. To do so, we:

1. take most of the common situations encountered in a typical first course in statistics,
2. illustrate each by examples, and
3. provide a wide variety of problems taken from a broad spectrum of fields, along with programs keyed to such problems for automated solution.

In this way, the book may serve for student and professional alike as a supplement to either a textbook or a reference book.

Although the basic problems are surprisingly similar from one such book to another, the notation may differ. So we have adopted a notation and terminology that we perceive to be the most common, and we point out differences that you may expect to see in other sources when we think such differences are important.

Section 1.2: Your Training and Background

What sort of mathematical training do you need for this book? Technically, a course in calculus ought to be a pre-requisite for some of the material, but from a practical point of view it is not essential. All you need is the notion of the integral as a means of calculating the area under a curve—and we even summarize that when needed. In this regard, the book *Calculus by Calculator,* by Maurice D. Weir, is highly recommended. The skills acquired there are more than adequate for handling whatever calculus notions we need here. Curiously enough, most of the functions we commonly use in this subject, including our superstar, the normal curve, cannot be integrated in closed form anyway. For this very reason, such matters can be, and often are, included in noncalculus introductory courses in various curricula. Our basic goal is simply to provide an understanding of such basic material through computational experience.

It is also assumed that the reader is already familiar with the fundamental notions of probability. If not, such a background is available from the book *Probability by Calculator,* written by Zehna for this series. Indeed, for those readers who have prepared from that source, a great deal of what is said in this chapter may be omitted or at best skimmed for those parts peculiar to statistics. Obviously, those matters related to the hardware are the same regardless of the subject matter being studied and are duplicated here only for the sake of those readers who are not already familiar with them. For them, these details need to be studied carefully, for they govern the mechanics of implementing the magnetic card programs.

Section 1.3: The Calculator

To implement our plan, the TI–59 electronic programmable calculator from Texas Instruments is our choice, primarily due to its relatively low cost versus its capability, as well as to its notable feature of permanently recording programs on magnetic cards. Permanent storage enables you to construct a library of formulas, ready for processing at any time when you insert the magnetic card into the calculator and follow the simple input steps. You might possibly effect individual problem solutions by first keying in the program steps, which we do include for that contingency; yet most of our programs are so long that such a procedure is inefficient, and you only subject yourself to the risk of error. It is the recording feature, then, that we wish to exploit, so to this end, a set of programs is keyed to various formulas in each chapter of the book, for permanent recording on the magnetic cards furnished with the calculator. Once recorded, you need never consult the programs again, unless you have a particular interest in programming per se.

A prerequisite for using this book, then, is a TI–59 calculator. With the purchase of that calculator, you will receive a supply of blank magnetic cards, a programming manual entitled *Personal Programming* (PPM), a master module along with its manual entitled *Master Library* (MLM), among other peripherals. Before using this book, you should acquaint yourself with the mechanics of operating the machine through the guided tour of Chapter II, and possibly through some more details in

Chapter V, of PPM. For a greater in-depth study, *Calculator Clout* is again highly recommended.

You will also need the Applied Statistics Module, a solid state software module available from Texas Instruments, Inc. You will receive with the purchase of such a module a manual (to be referred to here as ASM for *Applied Statistics* manual), which covers the necessary instructions (page 1–3) for replacing the Master Module in the back of your calculator with the Applied Statistics Module. Also, you will find instructions on page 1–4 of ASM for running a diagnostic test to see that all is well with the module.

The Applied Statistics Module contains many useful programs and in particular contains several standard probability distributions which are vital to statistical inference. Even so, there are several standard techniques—confidence intervals, for example—that are not addressed by the programs in the module, and it is for this reason that magnetic card programs have been prepared for this book. Such programs may be found in the appendix, and almost all of them use one or more programs in the module as subroutines. In this sense then, our programs may be viewed as supplements to the Applied Statistics Module, designed once again to address the common problem situations found in most introductory statistics courses.

For our particular applications, you need not be thoroughly familiar with programming the calculator. Naturally, as with most things, the more you understand such matters, the better you will understand how our programs work, if the "how" aspect is of interest to you. But an extensive programming ability is not essential, since the programs for the book's applications are already devised and listed in the appendix.

In writing the programs for each chapter, care has been taken to make them each as independent as possible. So, if you have any interest in rearranging programs to suit your individual needs, you may do so with minimum difficulty. Accordingly, we have avoided direct addressing as much as possible. There are limits to the number of labels available to name programs, however, particularly among the primary labels A through E and A′ through E′. So be on guard not to use the same label or name for 2 different programs on the same magnetic card. For each of our programs, numbered according to the applicable chapter or section, we have listed the labels used and their locations to facilitate any such changes.

Once you have recorded the programs for a given chapter or section on the required magnetic card, you may solve any of the problems for that chapter or section simply by consulting the user instructions governing those programs. These instructions are conveniently located throughout this book near the point where they are used.

As you will discover, your calculator has a dual keyboard, with most keys serving double duty. To activate these secondary functions, which appear in gold lettering above the keys, first press the key [2nd] (as discussed in PPM). Like the writers of the TI–59 manuals, we found it redundant and space-consuming to include that key description every time a gold key use is described. Consequently, in our user instructions, gold keys like [π] appear alone. It is understood that the function will not be executed unless you press [2nd] first.

Answers for all the problems are provided with the problem statement. Although for numerical calculations, all programs are written to display the maximum number of decimal places allowed by the calculator, you might want to work with fewer. For one thing, rarely, if ever, do you need accuracy of more than 4 decimal places in any statistics application. Further, displaying more places is distracting to some people. Perhaps most important, displayed answers of many decimal places may even be misleading, because the calculator must ultimately round values, since the internal capacity is 13 significant digits. For example, algebraically, there should be no difference between the result $(1 - {}^{1}/_{35} \cdot 35)$ and $(1 - {}^{35}/_{35})$. Both should be 0. But on the TI–59 with full display, the result of the first calculation is shown as 1×10^{-12}. The difficulty is the rounding error in computing the reciprocal ${}^{1}/_{35}$. (Such matters are discussed in detail in Appendix C of PPM.)

Any very small answers, like the one in the example, need to be viewed with suspicion. So, as discussed in PPM, you may select any number of decimal places for the display using the [Fix] key. Fixing does not affect the internal accuracy of the calculator, and it avoids most rounding problems. Yet such fixing does affect some library programs, as well as the reading or writing of any magnetic card. You must use the keys [INV] [Fix] or [Fix] [0] to return the calculator to its original status prior to either of these contingencies. Usually, the answers in this book report no more than 4 decimal places.

Section 1.4: The Programs

As you encounter each chapter in this book, you will find the programs for it in the appendix. Chapter VII of PPM provides general guidelines for recording programs onto magnetic cards. After entering the learn mode with a press of [LRN], carefully key in those programs step by step. Following each numbered step (given in 3 digits), you will find a key code given as a 2-digit number (usually its row-column location on the keyboard), followed by its mnemonic name (typically the key name itself). A complete listing of these codes is found on pages V–49, V–50, and V–68 of PPM. It is important that your program be identical with the one listed step by step.

Despite the facts that you should make no modifications and that, for most keystrokes, the procedure is straightforward, you may run into some unique problems when translating the *printed* code into keyboard strokes. Here are some helpful guidelines:

Both [STO] and [RCL], very commonly used instructions, are

called *merged codes* (see IV–17 and V–51 of PPM). Once they are pressed, the next requirement is an address, which is a 2-digit register number (xx) that is placed on one line. For example, the key sequence [STO] [3] [2] would appear in the printout for, say, Steps 124 and 125, as:

STEP	KEYCODE	NAME
124	42	STO
125	32	32

Now let's say that, when you are checking before recording, you discover that you inadvertently pressed the wrong key for the address and saw the display:

124	42
125	62

In such a case, you may *not* simply change 62 to 32. If you were to go to Step 125 and press [3] [2], thinking you were overwriting the mistaken 62, you would discover (by using [SST] and [BST] to step through the program) the display:

124	42
125	03
126	02

The reason is that 3 and 2 are separate keystrokes, and the calculator has no way of knowing that they were to be merged when they are entered beginning at Step 125. Whatever you intended to store in R_{32} ("R" stands for register) will be stored instead in R_{03}, register number 3. The correct procedure is to go back to Step 124 and press [STO] once again. Then you may enter the merged address [3] [2] as separate keystrokes. Now you should see:

124	42
125	32

in the display as required. The same remarks apply to the keys [SUM], [Prd], [Exc], [Op], and [Pgm], as well as [RCL].

You should know that code 92, listed as [RTN] in a program, is effected by the double *merged* keystroke sequence [INV] [SBR]. No key is labeled RTN on the calculator.

The *set flag instruction* (the gold key above [RST]) requires a *single* keystroke for a register address. Thus the printed display:

126	86	STF
127	01	01

is accomplished by the keystrokes [St flg] [1] and *not* by [St flg] [0] [1], which would produce:

126	86	STF
127	00	00
128	01	1

The same remarks apply to the keys [If flg] and [Dsz]. By the way, if you have made a mistaken entry, it may be easily removed with the gold key [Del] above [BST].

Indirect addresses are designated on the printed sheet by asterisks. Thus the printed display:

078	73	RC*
079	20	20

refers to an indirect recall instruction for register 20. This coding is effected by the keystroke sequence [RCL] [Ind] [2] [0], in that order. The instructions for it are discussed on pages IV–84 to IV–88 and listed on page V–68 in PPM. Once again, since this is a merged code, you must precede any corrections in addresses by the complete keystroke sequence.

By contrast, [GTO], another merged code, also requires either a single label address, like [A] or [sin], or a 3-digit address, which must be keyed in *as 3 digits.* For example, the printed instructions:

021	61	GTO
022	01	01
023	23	23

is an instruction intended to send the program to Step 123. It is properly executed by the keystroke sequence [GTO] [1] [2] [3], with the 3-digit address 123 printed as the merged 2-step code 01 23. The unwary user seeing 01 at Step 022, tends to key this set of instructions as [GTO] [0] [1] [2] [3]. But the calculator accepts only the first 3 keystrokes as the address and prints:

021	61	GTO
022	00	00
023	12	12
024	03	3

which will send the program to Step 012 instead of to the intended 123. Again, do not attempt to correct this error at Step 022. You must press [GTO] at Step 021 first to signal merging, then ignore the first 0 you see and enter in order the next 3 digits. For example, to send a program to Step 009, the program will appear as:

021	61	GTO
022	00	00
023	09	09

But you enter this instruction with the sequence [GTO] [0] [0] [9]. These same remarks apply to the test keys [If flg], [$x=t$] (shown as EQ in print), [$x \geq t$] (shown as GE in print), and [Dsz] when they are followed by 3-digit addresses.

While we have tried to avoid direct addressing as much as possible, the foregoing warnings suggest that we have not done so completely. If you undertake any modifications of these listed programs, be sure that you verify all direct address statements and modify them if changes have taken place. After you are satisfied that the program steps have been entered correctly into your calculator, you are ready to record them onto a blank magnetic card. The procedure is simple enough. If the number of steps does not exceed 240, take the following steps:

1. Make sure your calculator is out of any fixed decimal format. You can do so by pressing [INV] [Fix].
2. Be sure you are in standard partitioning, which is the way we wrote all the programs. If you press [Op] [1] [6], you should see 479.59 in the display. If you do not, then follow this with [6] [Op] [1] [7], and you will.
3. Make certain that your batteries have a good charge, so as to handle the extra power needed for the recording mechanism.
4. If the total number of steps does not exceed 240, simply press [1] [Write], at which time the display goes blank.
5. Now grasp a magnetic card right side up (so that you can read the name "Texas Instruments"), and insert the card into the slot provided on the right side of the calculator case. Let the drive motor pull the card through.
6. If all went well, you should see a 1 in the display, indicating that Side 1 of the card is now recorded. If the calculator flashes any number, stop the flashing with [CLR] and repeat the procedure, for recording has not been effected.

If the program has more than 240 steps, you must use Side 2 of the blank card to record Steps 240 through 479:

1. First key the sequence [2] [Write], and the display will go blank again.
2. Now insert the magnetic card upside down, and let the drive motor pull it through.
3. If recording went properly, then you should see a 2 in the display.

Since the card is now permanently recorded, handle it with the precautions discussed on pages VII–7 and VII–8 of PPM. See the same source for advice on marking the cards. Our users have found that very little marking, other than for identification, is useful, since the companion user instructions are essential for running the programs anyway. But that is a matter of personal taste.

Once the calculator is turned off, it loses all program steps and data storage. But the programs remain permanently recorded on the cards and may be recovered simply by following the read instructions as outlined on page VII–5 of PPM. The programs are then ready once more to work for you in solving the problems to which they are keyed. With just a little practice, you begin to use them quite automatically, and you can depend on them to produce consistent results in application after application.

2
Sampling

Section 2.1: Random Variables

Statistics, in contrast to probability theory, is mainly concerned with analyzing data for meanings, patterns, inference about an underlying model. Such analysis takes place within the framework of probability, however, both in terms of its formulation and its assessment. The key concept needed for most modern formulations is that of a random variable. You should have encountered that term in your study of probability theory in one form or another. From a pragmatic point of view, it suffices to think of a random variable, X, as a place-holder for an experiment which has been described in such a way that the outcome of one performance or *trial* of that experiment can be represented by a unique real number.

For example, X may stand for the *time* it takes for an electronic device to fail, in which case X may be replaced by any positive real number. Or, perhaps X is a code for whether or not an interviewee agrees (1) or disagrees (0) with a certain issue. Again, X may stand for the score an individual makes on a certain examination, in which case X may be replaced, typically, by a number from 0 to 100 score points. In all cases, we need to know the *set* of possible replacements (real numbers) for X. This set is called the *range* of X and is denoted R_X. It is typical among probabilists to denote a particular outcome of the experiment, a point in R_X, by lower case x, if capital X is used for the random variable.

It is not enough to know simply the range of a random variable X; to describe a random experiment (one whose outcome cannot be predicted with certainty), it is necessary also to describe the probability distribution of X. In most applications, such distributions fall into two categories, discrete and continuous. The discrete case arises when the range of X is finite or countably infinite (like $\{1,2,3,...\}$), in which case the distribution is characterized by a mass function $p(x)$, whose value at x is the probability that X will be replaced by x when the experiment is performed. We denote this $p(x) = \Pr(X = x)$ and such mass functions have the two properties $p(x) \geq 0$ and $\sum_{x \epsilon R_X} p(x) = 1$.* When the range of X is uncountable, we usually model X by defining a probability density function as a real-valued function $f(x)$ with properties $f(x) \geq 0$ and $\int_{\infty}^{\infty} f(x)dx = 1$. In that case, $f(x)$ is not a probability at all; indeed, $\Pr(X = x) \equiv 0$ for all choices of x. Instead, we define probabilities for intervals by calculating the area under the density curve, such as

$$\Pr(a < X < b) = \int_a^b f(x)dx$$

This is the probability that, when the experiment is performed, X will be replaced by a number between a and b.

In both cases, there is defined the notion of the *cumulative distribution function* (CDF) as a function $P(x)$ whose value at x is the cumulated probability up to and including x, that is, $P(x) = \Pr(X \leq x)$. This is calculated as $\sum_{t \leq x} p(t)$ when X is discrete and $\int_{\infty}^{x} f(t)dt$ when X is continuous. The *complementary* cumulative distribution function (CCDF) also has a lot of use and is defined as a function $Q(x)$ whose value at x is given by $1 - P(x)$. That is $Q(x) = \Pr(X > x) = 1 - \Pr(X \leq x) = 1 - P(x)$. Hence, it is always the case that

$$P(x) + Q(x) = 1 \tag{2.1}$$

whether X is discrete or continuous. We will have examples shortly.

Of considerable interest in analyzing distinguishing features of any random variables are certain moments. In particular, the *mean* of X, often denoted μ, is also called the *expected value* of X, denoted $E(X)$, and is given by

* Capital Greek sigma, Σ, stands for sum, so $\sum_{x \epsilon R_X} p(x)$ means add $p(x)$ for every x in (ϵ) the range R_X.

$$\mu = E(X) = \begin{cases} \sum_{x \in R_X} xp(x) & \text{if } X \text{ is discrete} \\ \int_{-\infty}^{\infty} xf(x)dx & \text{if } X \text{ is continuous} \end{cases} \tag{2.2}$$

As such, the mean is the center of gravity of the distribution of X viewed as a mass system. It is a measure of "central location" for the probability distribution and is often used as a reference point for discussing the possible values of X.

In general, if X is a random variable, so is $g(X)$ where $g(x)$ is an ordinary function of real variable x. As such, $g(X)$ possesses its own distribution with associated characteristics. But, thanks to a basic theorem in probability theory, it is not necessary to determine this distribution (which may be a difficult task to say the least) in order to compute moments. In fact the expected value of $g(X)$ may be found using the distribution of X itself by the formula,

$$E[g(X)] = \begin{cases} \sum_{x \in R_X} g(x)p(x) & \text{if } X \text{ is discrete} \\ \int_{-\infty}^{\infty} g(x)f(x)dx & \text{if } X \text{ is continuous} \end{cases} \tag{2.3}$$

As a special application of this result (sometimes called the law of the unconscious—blind, wise—statistician), we may compute higher order moments of X. In particular, $E(X^2)$ is the second moment (about 0), and $E[(X - \mu)^2]$ is the second moment about μ. The latter is of special significance in probability and statistics and is called the *variance* of X. It is denoted σ^2 or $V(X)$ and is related to the former by the relationship

$$\sigma^2 = V(X) = E[(X - \mu)^2] = E(X^2) - \mu^2 \tag{2.4}$$

where $E(X^2)$ is either $\sum_{x \in R_X} x^2 p(x)$ or $\int_{-\infty}^{\infty} x^2 f(x)dx$

The positive square root of the variance is denoted σ and is called the *standard deviation* of X. Both σ and σ^2 are measures of variability and give relative measures of how "spread out" from μ the distribution of X is.* Indeed, a result known as Chebyshev's inequality states that at most $1/k^2$ of the probability distribution will be found outside a band of x-values extending from $k\sigma$ below μ to $k\sigma$ above μ in *any* distribution. That is,

$$\Pr(|X - \mu| \geq k\sigma) \leq 1/k^2 \tag{2.5}$$

* The standard deviation has the advantage that it is measured in the same units as X, whereas σ^2 is in square units,

Even for $k=3$, most (at least 89%) of the probability mass is *within* 3σ units of μ, that is, $\Pr(\mu - 3\sigma < X < \mu + 3\sigma) \geq .89$.

Another characteristic of X that we are very much concerned with in statistical applications is the percentile of its distribution. For any proportion α, we define x_α to be the $100(1 - \alpha)$th *percentile* (or quantile) of X as basically that value of X which divides the probability distribution into two parts with $1 - \alpha$ *below* x_α and α *above* x_α. For discrete distributions, the definition is ambiguous and awkward to make precise. When X is continuous with a strictly increasing CDF (the case we will be almost entirely concerned with in this book), the definition reduces nicely to the requirement

$$P(x_\alpha) = 1 - \alpha \qquad \text{or} \qquad Q(x_\alpha) = \alpha \tag{2.6}$$

Using standard superscript notation, -1, for the inverse of a function, we may rewrite Equation 2.6 as

$$x_\alpha = P^{-1}(1 - \alpha) \qquad \text{or} \qquad x_\alpha = Q^{-1}(\alpha) \tag{2.7}$$

In particular, $x_{.5}$ is called the *median* (sometimes denoted md).

Again, examples of these concepts are forthcoming. We attempt here only to summarize the pertinent facts as a brief review of basic probability concepts to set a notation.

Section 2.2: Random Sample

Whenever the experiment described by X is repeated, making it possible to record the successive outcomes or values of X, we need a notation to account for this compound experiment. Since we are planning to replace X with one of its possible values in R_X at each trial, we will continue to use X as a symbol for a variable to be replaced by a unique element of R_X when the experiment is performed. But now we add a subscript to account for the sequence of trials, so that X_1 denotes the first trial, X_2 denotes the second trial, and so on. For a fixed n, the collection $X_1, X_2, \ldots, X_n$ is a *sample* of X of size n. As repeated copies of X, the random variables $X_1, X_2, \ldots, X_n$ all have the same probability distribution as X, and so are said to be *identically distributed.* This being the case, each sample random variable X_i has precisely the same characteristics (mean, variance, percentiles) as X. In this context, X itself is referred to as the *population* random variable, and we speak of drawing a sample of size n from the ($X-$) population. We also refer to such population characteristics as μ and σ^2 as *population* moments—the *population mean* and *population variance* in particular.

After the n trials of the $X-$ experiment are actually performed, by agreement, each X_i is replaced by exactly one value from the common range R_X; let this value be denoted, logically, by x_i, the lower-case version of the random variable X_i. Then the collection $x_1, x_2, \ldots, x_n$ is just a set of n real numbers. As such, we may refer to them as the sample *values* or *data*, although often you will find them ambiguously referred to as the *sample*.

Taken collectively, the sample random variables $X_1, X_2, \ldots, X_n$ have a joint probability distribution defined by a discrete mass function (when X is discrete) which evaluates the probability that, simultaneously, X_1 will be replaced by x_1, X_2 will be replaced by x_2, and so on. That is,

$$p(x_1, x_2, \ldots, x_n) = \Pr(X_1 = x_1, X_2 = x_2, \ldots, X_n = x_n)$$

for each choice of possible *values* $x_1, x_2, \ldots, x_n$. The notion of *independent* random variables, wherein the replacement of X_i in no way affects the replacement of X_j when $j \neq i$, is made precise by requiring that this probability factor into the product of probabilities, each given by the common (population) mass function $p_1(x)$. That is,

$$p(x_1, x_2, \ldots, x_n) = p_1(x_1)\, p_1(x_2) \cdots p_1(x_n) \tag{2.8}$$

defines independence.

In the continuous case, we have a slightly more complicated problem, because the population density function $f(x)$ is not a probability. But, by analogy, we say $X_1, X_2, \ldots, X_n$ are *independent* in this case if

$$f(x_1, x_2, \ldots, x_n) = f_1(x_1)\, f_1(x_2) \cdots f_1(x_n) \tag{2.9}$$

where $f(x_1, \ldots, x_n)$ is then a non-negative function whose multiple (n-fold) integral is unity. Such a function is called a *joint density function.* In general, such functions are not particularly easy to manage, and we are content to confine our attention to the special case when independence applies.

When the two notions of independence and identical distributions are put together, we have what the mathematical statistician means by a random sample. That is, a set of sample random variables $X_1, X_2, \ldots, X_n$ of a population random variable X, is said to be a *random sample* if the sample variables are both independent and identically distributed (as X). The reason this point should be clarified is that this concept is at the very heart of statistical analysis. As we will soon see, almost every statistical procedure we will examine requires, for its validity, that the random variables under consideration be independent and identically distributed, that is, a random sample. This brings up an interesting and difficult matter. It is virtually impossible to tell from a given set of data $x_1, x_2, \ldots, x_n$ whether or not they constitute the values of a random sample $X_1, X_2, \ldots, X_n$ from a *given* population random variable X. The numbers in the data simply will not reveal this. Moreover, the technical requirements of Equation 2.8 or 2.9 are not easy to implement in practice. It is safe to say that it is unlikely that any sample values obtained in real experiments are truly random in the sense we are discussing. And yet the validity of our procedures depends on this being so. About the best we can do is exercise extreme caution in drawing our samples, so that it is not patently obvious that one sample value will affect another and that we are sampling from the same population each time.

For example, if we draw items off a shelf, such as milk cartons in a grocery store, to check some sort of status (like defective), it is reasonably clear that if we do not replace the item each time, then such outcomes are not independent. More on that later. If we wanted to infer something about the general population of junior high school students based on scores on a given test, it might not be appropriate to sample only the junior high school students in Monterey, California. Even with precaution on obtaining "independent scores," it would be risky to suppose that the probability distribution of such scores is the same as those that would be obtained from the general population. Certainly we ought to remove such "obvious" inequities, if we can control them. Removing the less obvious ones is not easy, and we often just assume that samples are random if there is nothing stated to the contrary, or we have no reason to believe otherwise. This is especially true of textbook problems, and this book is no exception. Nevertheless, we wish to make an issue of the point, for any user of statistics, even the casual problem solver, ought to be aware that his or her conclusions can be valid only to the extent that necessary assumptions are satisfied, and randomness is a key one.

Section 2.3: Simulation

There are no simple, easy replacements for the required mathematical definition of random sample, but there are ways to approximate that mathematical ideal in practice and we investigate an important one in this section.

Suppose we define a population of "grades" given in all the courses taught at a given university in a given quarter. To make things numerical in order to fit the framework of random variables, let us agree to identify each letter grade by its grade point equivalent. At this university A = 4.0, B = 3.0, C = 2.0, D = 1.0, and F = 0.0. There were 20,000 students who earned grades that quarter, and registrar records show that the proportions of A, B, C, D, F, respectively, were .13, .21, .43, .14, .09. Thus, *as a model,* we may assume that a "grade" is the value of a random variable X whose probability distribution is the discrete distribution specified by

x:	4.0	3.0	2.0	1.0	0.0
$p(x)$:	.13	.21	.43	.14	.09

For this population: $\mu = \sum_x xp(x) = 2.15$, $E(X^2) = \sum_x x^2p(x) = 5.83$, and so $\sigma^2 = 1.2075$ or $\sigma = 1.10$.

We digress momentarily to address the problem of caluclating population means and variances for simple discrete distributions like this. The book *Probability by Calculator* in this series provides a brief program for effecting such calculations, and that program may certainly be used here, since it was deliberately made independent of the module located in the calculator. But now that you have the statistics module in the calculator, one of its programs may be "fooled" into calculating these parameters for you. That program is ST–03, a program designed to evaluate various aspects of data analysis, topics we will be discussing later. For now, you may use Steps 1–4 of the User Instructions on page 3–6 of ASM for the grouped univariate data program. The only modification you need for the present application is to input $p_i = p(x_i)$ in place of f_i at Step 4a. This will input (x_i, p_i) pairs and store them in successive pairs of registers starting with (R_{32}, R_{33}). Without repartitioning (Step 3), this will handle $N = 14$ pairs. For N up to 34, you may repartition the calculator at Step 3 in accordance with the capabilities expressed in Table 3.2 on page 3–4. (Actually, you will seldom need that step.)

That much inputs the probability distribution. For our university example, the sequence would be:*

STEP	ENTER	PRESS	DISPLAY
1		[Pgm]03	*xxx*
2		[E′]	1.
3		(Unnecessary)	
4a	0.13	[B]	0.13
4b	4	[A]	1.
4a	0.21	[B]	0.21
4b	3	[A]	2.
4a	0.43	[B]	0.43
4b	2	[A]	3.
4a	0.14	[B]	0.14
4b	1	[A]	4.
4a	0.09	[B]	0.09
4b	0	[A]	5.

Notice that a running count of the number of pairs entered is displayed at the conclusion of each press of [A], concluding with $5 = N$ in our present example. As another minor check on your entry, you should find 1 in register R_{03}, since this will contain Σp_i. To find the parameters μ and σ, all you need do now is recall R_{04} to find $\mu = 2.15$ for our example. You may also find μ by pressing the gold key [$\bar{x}$] located above [6],

* Consistent with ASM, we will usually denote most keys from now on by enclosing them in square brackets. Numeral keys are an exception.

followed by [$x \leftrightarrows t$], for it is stored in the T-register. As for σ^2, this can be calculated with a special internally-wired program coded Op 11. You can effect that program by pressing the gold key [Op] (above [9]) followed by 11 and then [$x \leftrightarrows t$], for σ^2 is located in the T-register also. In this case, that procedure yields $\sigma^2 = 1.2075$ and, of course, if you then press the [$\sqrt{x}$] key, you may verify that $\sigma = 1.10$ to 2 decimal places in our example.

To return to our original problem, suppose now we wanted actually to draw a random sample of 4 "grades" from this population of grades. Then we need to replace 4 random variables X_1, X_2, X_3, X_4 (remember, they are place-holders), each one by exactly one of the numbers 0, 1, 2, 3, 4, so that the probability that a replacement is 0 will be .09, that it is a 1 will be .14, that it is a 2 will be .43, and so on. Moreover, we want to effect these replacements in such a way that X_1, X_2, X_3, X_4 may be considered independent. One method to accomplish this would be to take 100 identical chips (since our probabilities are rounded to hundredths) and mark 13 of them with a 4 (for A), 21 of them with a 3, 43 of them with a 2, 14 of them with a 1, and 9 of them with a 0. Now place the chips in a bowl, mix thoroughly, and draw out a chip. Whatever number is marked on that chip, replace X_1 with that number—call it x_1. Now what was the probability beforehand that $x_1 = 0$? As a model, the probability of getting any one number is $^1/_{100}$ and 9 of them have 0 on them, so, adding point probabilities, $\Pr(X_1 = 0) = {}^9/_{100} = .09$. Similarly, $\Pr(X_1 = 1) = .14$, $\Pr(X_1 = 2) = .43$, and so on.

Let us move on to X_2. If we just proceed to draw another chip, then, with only 99 left, $\Pr(X_2 = 0)$ cannot be .09 no matter what X_1 was replaced by. So the first thing we should do is place the first chip back in the bowl. This will guarantee that the count for computing probabilities for X_2 will be the *same* as for X_1—that is, that X_1 and X_2 are identically distributed. We should shake the bowl again, mixing the chips thoroughly, so that the replacement of the first chip does not affect the second draw. This ensures that X_1 and X_2 are independent. Proceeding this way for X_3 and X_4 will then assure us that the resulting data x_1, x_2, x_3, x_4 constitute values of a random sample.

Thanks to modern electronics, we can replace the chips, the bowl, and the mixing very easily and still accomplish the same results through what are called *random number generators*. Program ST–02 in the TI statistics module is such a generator, and it is discussed on pages 2–6 through 2–8 in ASM. The basis for the generator is the simplest continuous density function we can conceive: namely, the function whose value is 1 over the unit interval, $f(x) = 1$ if $0 < x < 1$ (0, otherwise). If U is a random variable having this density function, then finding a value of U is what we mean by "picking" a number from 0 to 1 "at random." The reason is that the probability of such a number's falling in a given interval between 0 and 1 is just the length of the interval and hence does not depend on where that interval is located. Thus, $\Pr(.1 < U < .3) = \Pr(.2 < U < .4) = \Pr(.8 < U < 1) = .2$. For this reason U, and its simple density, is called *uniform*. Generally, U is *uniform* over the interval from A to B, if its density is given by $f(x) = \frac{1}{B-A}$ for $A < x < B$ (0, otherwise). The value u of U is referred to simply as a *random number* (from A to B).

As for actually choosing a random number from 0 to 1 using the random number generator, we have only to press [Pgm] 02 and initialize by pressing [E′] as per Steps 1 and 2 of the user instructions for ST–02 on page 2–7 of ASM. The next step (Step 3) is to enter what is called a *seed,* which is any number between 0 and 199,017 inclusive, and press [E]. (This corresponds to shaking a bowl or shuffling a deck of cards and should be arbitrarily selected for each application.) For the sake of illustration, let us take Seed = 49. Now, moving to Step 13, the subroutine [D.MS] (the gold key above [2] will in fact produce a value u between 0 and 1 to 5 decimal places. If you followed the procedure here, you should see 0.48483 in the display register. In the process of generating a random number, the generator produces internally a new seed for a second application (it is stored in R_{09}), so it is not necessary to go back to Step 3 again. This explains the ASM remark to repeat Step 13 as needed. So a second press of [SBR] [D.MS] produces 0.09563, which may be viewed as a second value of u selected independently of the first; that is, 0.48483 and 0.09563 represent the values of a random sample of size 2 from the uniform (0,1) distribution.

If we wanted to generate samples from the uniform (A, B) distribution, we have only to go to Steps 4–6 of ST–02. For example, if $A = 10$ and $B = 67$, then enter $x_{min} = 10$ at Step 4 and press [A], followed with $x_{max} = 67$ at Step 5 and press [B]. A press of [C] will now generate the value 16.84912 and a second press yields 24.67066. A third press gives us 28.44406 and a fourth press 63.87868. These may be viewed as values of a random sample of size 4 from this uniform (10,67) distribution. The density for this distribution is just the function $f(x) = 1/57$ for $10 < x < 67$ (0, otherwise) and for this distribution, $\Pr(10 < U < 11) = \Pr(11 < U < 12) = \Pr(12 < U < 13) = \ldots = \Pr(66 < U < 67) = 1/57$, for example.

The latter fact allows us to modify the random number generator easily to handle our "100 chips in the bowl" problem of drawing grades against a given distribution. We want each "chip" to appear with probability $1/100$ so one way we could do this is to generate a random number from 0 to 100 using the random number generator and then *truncate* (delete any fractional part) each observation so that all numbers between 0 and 1 will be called 0, all numbers between 1 and 2 will be called 1, and so on. The integer function [Int], the gold key above the right-hand parenthesis [)], will accomplish that. Accordingly, the probability of obtaining any integer k this way is

$$\Pr(k < U < k + 1) = (k + 1 - k)/100 = 1/100$$

as required, for $k = 0,1,2,...,99$.* (We could accomplish the same goal by using the uniform from 1 to 101 or 53 to 153, for that matter.) Now let us assign the labels 00 through 08 to $x = 0$, labels 09 through 22 to $x = 1$, labels 23 through 65 to $x = 2$, labels 66 through 86 to $x = 3$, and finally labels 87 through 99 to $x = 4$. Then every time we generate a value u between 0 and 100 using Step 6 of ST–02 and truncate, we obtain a label R. If we compare R to our table of assignments and record x, this procedure imitates or *simulates* the experiment of actually observing a value of X with the required probability.

Let us examine how this works for our example. Initialize ST–02 and input a seed of 21 for illustrative purposes. Now input $A = 0$ and $B = 100$ at Steps 4 and 5. The first press of [C] should yield 6.631, and if you follow this with [Int] you will find $R_1 = 6$ for the first label generated. From our table of assignments, $R = 6$ corresponds to $x = 0$, so let $x_1 = 0$. Repeating, press [C] [Int] and find $R_2 = 84$, so that $x_2 = 3$. A third application yields $R_3 = 86$, so $x_3 = 3$ again, and then $R_4 = 57$, so that $x_4 = 2$. Thus, our *observed* random sample of X of size $n = 4$ provides the data 0,3,3,2. But it is important to see that, because of the way we generated labels, $\Pr(X_1 = 0) = .09$, since the probability of getting any particular one of the labels 00 through 08 is precisely .01 and so the probability of one among the nine labels is, by mutual exclusiveness, $(.01) + (.01) + ... + (.01) = .09$. Thus, the probability, before we started, that X_1 would be replaced by 0 is exactly .09, as required. Similarly for the other possible values of each X_i, and, finally, the built-in independence of the generator output (by internally changing the seed each time) guarantees that the X_i's are independent.** In short, 0,3,3,2 may be viewed as values of a random sample of size 4.

In general then, if the range of possible value of the population random variable consists of N values,

$$R_X = \{e_1, e_2,...,e_N\} \text{ and } p_i = \Pr(X = e_i) \text{ for } i = 1,2,...,N$$

we can simulate a random sample of X. Choose K labels (let us agree to number labels starting with 0, since so many computer applications use this scheme) in such a way as to reflect the lowest rounding values of the probabilities. Thus, if probabilities are measured to the nearest thousandth, let $K = 1{,}000$, to the nearest hundredth, $K = 100$; if the p's are given in fractions, let K be the least common denominator. Next, assign labels to each e_i by any scheme that ensures each p_i times K is the number of labels assigned to e_i. Use $A = 0$ and $B = K$ in ST–02 to generate random numbers using Step 6 and truncate the result to produce a random label R. Record the value e of X that corresponds to R as x_1, and repeat as often as desired to generate a random sample.

* For more details on these and related matters, the reader is advised to consult *Probability by Calculator* in this series.

** Actually this is not quite true of any physical device, but it is close enough for practical purposes.

EXAMPLE 2.1

Sampling

Rainfall X (in inches) in Gotham City, USA, for the month of August appears to be a random variable, and past records reveal the following relative frequencies, which are assumed to be probabilities. Simulate the rainfall for August of the next 3 years.

x:	9	10	11	12	13	14
$p(x)$:	.10	.24	.30	.20	.12	.04

Solution Since probabilities are measured in hundredths, select $K = 100$ and use the following assignment scheme.

LABELS	VALUE OF X
00–09	9
10–33	10
34–63	11
64–83	12
84–95	13
96–99	14

Using a seed of 101 in ST–02, a generated random sample of 3 (labels are shown in parentheses) would be 12(83), 11(41), and 10(15). (Remember to ignore fractional parts of random labels; do not round.)

EXAMPLE 2.2

When 4 unbiased coins are tossed, the frequency with which heads ($x = 1$) occurs is usually modeled as follows.

x:	0	1	2	3	4
$p(x)$:	1/16	1/4	3/8	1/4	1/16

Simulate the experiment of tossing 4 coins 5 times and recording the total number of heads each time.

Solution What is required are the values of a random sample of size 5, namely, replacements for X_1, X_2, X_3, X_4, X_5 where each X_i has the above distribution and they are mutually independent. Since the probabilities are presented as fractions with a lowest common denominator of 16, let $K = 16$, and assign x-values according to the scheme:

LABELS	VALUES OF X
0	0
1–4	1
5–10	2
11–14	3
15	4

Using a seed of 225, the results are (with labels in parentheses):

4(15)   4(15)   2(9)   2(10)   3(11)

Please observe that values of X are repeated, which mimics the *replacement* of the "chips in the bowl" in our imaginary generator.

Problems

In the following problems the seed used is shown for the purpose of allowing you to verify answers. Normally, you would provide your own seed.

1. A local corner drugstore stocks 5 of the town's morning papers for sale each morning. The number of papers sold on a given day is a random variable with the distribution:

x:	0	1	2	3	4	5
$p(x)$:	.01	.01	.04	.03	.67	.24

Find the mean and variance of X and simulate a week's sales for the corner drugstore. Compare the average sales with the expected value.

Ans: Using program ST–03, $\mu = 4.06$, $\sigma^2 = .6764$. Using label assignments and a seed of 0,

x:	0	1	2	3	4	5
Labels:	00	01	02–05	06–08	09–75	76–99

the generated labels are 50, 41, 17, 67, 86, 11, 59, yielding corresponding sample sales of 4,4,4,4,5,4,4 for a week. (Note the preponderance of 4's, reflecting the probability distribution.) The total sales for the week are $4(6) + 5 = 29$ and the average is thus $^{29}/_{7} = 4.14$, slightly higher than the expected value.

2. Repeat Problem 1 with a seed of 455 to reflect a different week's sales. Observe yet another average, differing from both the average of the sales in the problem, and μ.

Ans: Labels are 5, 19, 82, 92, 1, 92, 18, yielding sales of 2, 4, 5, 5, 1, 5, 4, for an average of 3.71 (bad week).

3. A certain rabbit farm in Iowa has 15% black rabbits, 40% brown rabbits and 45% white rabbits. Use the random number generator to indicate the color of the next 5 rabbits born during any specific period on this farm.

Ans: (seed=0)   Brown (50)   Brown (41)   Brown (17)
White (67)   White (86)

4. There is a 25% chance that a certain company will sell 10,000 shares of stock to raise needed capital, a 40% chance of selling 20,000 shares, and a 35% chance of selling 30,000 shares. The price per share is \$1.

a. Use simulation to find the average capital raised in 5 trials.

Ans: $\bar{x} = \$22{,}000$ (seed = 0)

b. Compare this to the expected value of capital raised.

Ans: $\mu = \$21{,}000$

5. At a certain sports event 45% of the spectators present are alumni of the school, 8% are faculty, 23% are students and 24% are "others." Simulate the type of the first 10 spectators to exit the building.

Ans: (seed = 0) Faculty (50) Alumnus (41) Alumnus (17) Student (67) Other (86) Alumnus (11) Student (59) Alumnus (0) Student (53) Other (94)

Section 2.4: Simulating Continuous Distributions

When X is continuous, the foregoing method of simulation no longer applies. Indeed, there would be no way even to allow enough labels for the possible values of X in R_X since they now lie on a continuum of values. Fortunately, there are other ways, one due to a theoretical result called the Probability Integral Transformation. This theorem states that if X is *any* continuous random variable having CDF $P(x)$, then the random variable $U = P(X)$ (note the change from lowercase to uppercase) always has a uniform distribution on the interval (0,1). Hence, if we can solve the equation $u = P(x)$ for x in terms of u to arrive at $x = P^{-1}(u)$ (we may have to "solve" such an equation by table), we can express the random variable X as a function of the uniform random variable U.

But, since we can generate the value u with a random number generator, we may view $x = P^{-1}(u)$ as a generated or simulated value of X. Moreover, repeated applications of a generator like ST–02 ensure that the successive values of x may be viewed as values of a random sample. Incidentally, $P^{-1}(u)$, as a number on the x-axis, would just be a *percentile.* More specifically, if we let $\alpha = 1 - u$ (a traditional notation), then $P^{-1}(u)$ would be x_α, the $100(1 - \alpha)$th percentile as previously defined (see Figure 2.1).

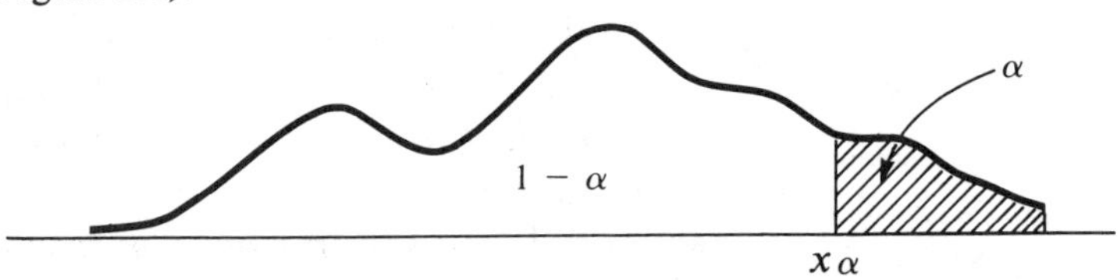

Figure 2.1. The 100 (1 − α)th percentile

Thus, $x_{.05}$ is the 95th percentile and represents a point that divides the area under the density curve into 2 parts with 95% to the left and 5% to the right of the point. As previously noted, the 50th percentile, $x_{.50}$, is referred to as the *median* of the distribution. Similarly, $x_{.75}$ and $x_{.25}$ are often called the first and third *quartiles*, respectively.

EXAMPLE 2.3

Sampling

The function $f(x) = \lambda e^{-\lambda x}\ (x > 0)$ defines, for each $\lambda > 0$, a density function called *exponential.* Such a density is often used as a model for time to failure of certain kinds of equipment. Then the parameter λ is called a *failure rate,* and it is related to the mean of this family by $\mu = {}^1/_\lambda$. These facts are easily verified by elementary integral calculus, as is the formula for the CDF given by $P(x) = 1 - e^{-\lambda x}\ (x > 0)$. If we let $u = 1 - e^{-\lambda x}$, then we may solve for x to find $x = -ln(1 - u)/\lambda$. In other words, $P^{-1}(u) = -ln(1 - u)/\lambda$. We can then simulate a random sample of size n using the random number generator, first to generate values of u, and then to substitute in the formula for $P^{-1}(u)$ to obtain values of x. For example, taking $\lambda = .001$ and using a seed of 635, 5 successive values of u are: .02962, 0.22340, 0.75139, 0.84628, 0.47193. Substitution in $x = -ln(1 - u)/.001$ yields rounded x-values of 30, 253, 1392, 1873, and 639. These may be viewed then as a random sample of 5 observations from the exponential density.* If λ is measured in units per hour (1 every 1,000 hours of operation), then the x-values may be viewed as time in *hours* before that particular unit placed on test failed. Thus, the first one failed much sooner than expected—after only 30 hours—and the third one survived nearly 1,900 hours before it failed. The mean time to failure is, of course, 1,000 hours (the reciprocal of λ).

What made this application particularly easy is the fact that the exponential CDF admits an easy formula to work with. Even so, we have programmed the results as a subroutine in our first program for this chapter called ZS–2 (its chapter reference), whose user instructions are located at the end of this section and again in the appendix.** This will automate the process of generating samples, as well as allow you to calculate other items of interest connected with the CDF of the exponential distribution using the instructions under code E (for exponential). Illustrations, including the implementation of Example 2.3, are provided following the user instructions as a guide to the use of the program.

The existence of such a simple closed expression for the CDF is certainly not true of all continuous cases, even many that are used in practice, as the next example shows.

EXAMPLE 2.4

The density function

$$\phi(z) = \frac{1}{\sqrt{2\pi}}\ exp\left[-\left(\frac{1}{2}\right)z^2\right]$$

* Actually, the formula $x = -ln\ u/.001$ would do as well, for the random variable 1 − U has the same uniform (0,1) distribution as U.

** For this and succeeding chapters, we recommend that you record the chapter program on a magnetic card (following the instructions of chapter 1) at the start. Such a card will then cover all the problems in that chapter.

is called the (standard) normal density function and is a key model used in statistical analysis. More generally, the function

$$f(x) = \frac{1}{\sigma \sqrt{2\pi}} \, exp \left[\frac{-(x-\mu)^2}{2\sigma^2} \right]$$

defines a similar function with the property that μ is the mean and σ^2 is the variance of a random variable having this density. (The standard case then corresponds to $\mu = 0$ and $\sigma^2 = 1$). For this particular density function, it is impossible to write down the formula for the CDF for any choice of parameters. Yet, so important is this family, that the CDF for the standard case has been extensively tabulated with great accuracy.

Both the statistics and master modules of the TI–59 calculator contain the equivalent of a normal table. In the statistics module, it may be found as ST–19, discussed on pages 6–4 through 6–7 of ASM. In Figures 6.4a and 6.4b on page 6–4 of that manual, you will see the graphs of several normal densities appearing always as a bell-shaped curve.

The program itself only yields values of the CDF and CCDF for the standard case, and the method of finding these is discussed in the user instructions on page 6–5. After selecting the program as usual, you have only to enter z and press [B] for $P(z)$ or [C] for $Q(z)$. [$A(z)$ has a special application we will discuss later.*] We are rarely interested in $\phi(z)$ (label A), since it is not a probability. For example, the area under the standard normal density curve to the left of 2.69 (see figure) is, according to this calculation,

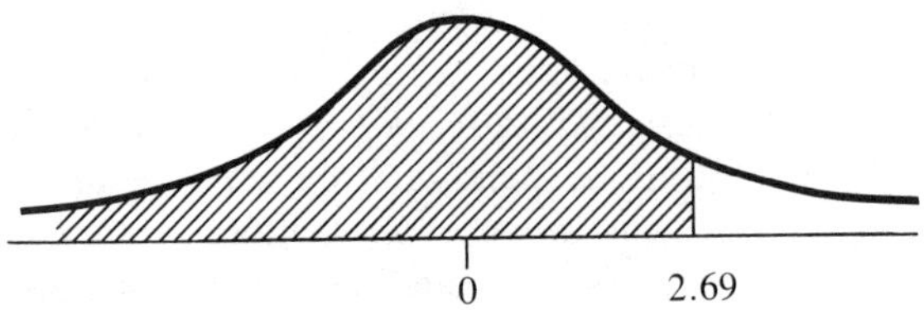

$P(2.69) = .9964$ to 4 decimal places. The area to the right of 1.73 would be found by simply entering 1.73 and pressing [C] to yield $Q(1.73) = .0418$. The program will handle negative values of z as well, so that $Q(-1.73) = .9582$, which is the same as $1 - P(-1.73)$, of course where $P(-1.73) = .0418$ may be found by pressing [B]. Incidentally, in the context of percentiles, 2.69 would be the 99.64th percentile and could thus be denoted $z_{.0036}$. Similarly, $z_{.0418} = 1.73$ is the 95.82nd percentile, and -1.73 is the 4.18th percentile.

When X is normal, but not standard, a simple linear relationship exists between it and the standard Z, namely $X = \sigma Z + \mu$ where $\mu =$

* You should correct your manual so that it reads $A(z) = 1 - 2Q(|z|)$, for it is intended that this be the area between $-|z|$ and $|z|$ under the curve. If $z < 0$, then $1 - 2Q(z) < 0$ and that cannot be right. Also, Step 5 should be corrected then to read $\Pr(|Z| \leq |z|)$.

$E(X)$ and $\sigma^2 = V(X)$. This allows us to find the value of $P(x)$ and $Q(x)$ in terms of standard P_Z and Q_Z, say, using the formulas

$$P(x) = P_Z\left(\frac{x-\mu}{\sigma}\right) \quad \text{and} \quad Q(x) = Q_Z\left(\frac{x-\mu}{\sigma}\right)$$

Put another way, $P(x) = P_Z(z)$ where $z = (x - \mu)/\sigma$. The latter formula is sometimes referred to as *standardizing* the value of X. For example, if X is normal with mean $\mu = 50$ and standard deviation $\sigma = 10$, then $P(x)$ may be found by entering ST–19 with the value $z = (x - 50)/10$. For $x = 25$, $P(x) = P_Z(-2.5) = .0062$, using ST–19, label B. For the x-distribution, the 0.62th percentile is 25, that is, $x_{.9938} = 25$. In general, $x_\alpha = \sigma z_\alpha + \mu$.

The program makes no provision for calculating percentile values directly, that is, for finding x_α when α is given. But there is a relatively simple polynomial approximation to the function $P^{-1}(z)$ which allows us to compute any z_α for given α and, from this, x_α using the linear relationship $x_\alpha = \sigma z_\alpha + \mu$. Such a subroutine is included in the program ZS–2 along with the exponential case. As you see from the user instructions for Code N (for normal), your first step is to enter the two parameters μ and σ. Then, for any choice of α, pressing [C'] outputs the standard percentile, z_α, while [D'] outputs x_α, the percentile of the required distribution. For convenience, the program has also been arranged to allow you to compute $P(x)$, $Q(x)$ and $\Pr(x_1 < X < x_2)$ (a common requirement) without the bother of standardizing X if you so desire. The implementation is illustrated in the examples following the user instructions.

As for simulating a random sample from a normal distribution, we are apparently hampered by the fact that $P(x)$, hence $P^{-1}(u)$ admits no simple expression. But, since $P^{-1}(u)$ is just another symbol for z_{1-u}, a percentile, we can generate samples at will using subroutine D' in program ZS–2. As previously remarked, if $P(X)$ is uniformly distributed, then it is a theoretical fact that so is $Q(X) = 1 - P(X)$. Hence, if we wish, we may calculate z_u for any generated value of u and interpret this as a random choice of Z so that $x = z_u\sigma + \mu$ is a random selection of x. To proceed then, we have only to generate u using program ST–02 as with the exponential case, then press [D'] to generate x. We illustrate.

EXAMPLE 2.5

Generate a random sample of size 5 from a normal distribution having mean $\mu = 50$ and standard deviation $\sigma = 10$.

Solution

Step 1. Read in a magnetic card having program ZS–2, and intialize the normal routine by storing $\mu = 50$ and $\sigma = 10$ as required by Step N1 of that program.

Step 2. Enter ST–02 with a seed of 198 (for illustrative purposes) and generate random $u = 0.67625$ with [SBR] [D.MS]. Now exit ST–02 by pressing [RST].

Step 3. Next, press [D′] to see a displayed x-value of 45. This may be taken as the first generated sample value x_1.

Step 4. Repeated presses of [D.MS] in ST–02, followed by [RST] then [D′] in ZS–2, yield the next 4 values of X(u's are in parentheses).

57(.23903)   45(.69025)   45(.68062)   55(.32017)

Actually, we can easily program the routine outlined in Example 2.5 to automate the procedure and have done so under label B' in ZS–2 (see user instructions). The implementation using Example 2.5 is illustrated following the user instructions. The method we have used is another example of the use of the Probability Integral Transformation. There are other methods of generating random samples from a normal distribution —*normal deviates,* as they are called. One of these is incorporated by the TI–59 in the random number generator program ST–02 as you may have already observed. Referring to pages 2-6 and 2-7 of ASM, the formula for generating such normal deviates is recorded and implemented under Steps 7, 8, and 9 of the user instructions for ST–02. The transformation converting a pair of uniform outcomes u_1 and u_2 to a normal outcome (page 2-6) need not concern us here. By entering the required parameters μ and σ at Steps 7 and 8 (after the usual initialization, of course), as many normal observations as you wish may be generated by successive presses of [D]. The results, as usual, may be viewed as the values of a random sample for all practical purposes. We might compare this program with the one used previously for Example 2.5.

EXAMPLE 2.6

Generate a random sample of size 5 from a normal distribution having mean $\mu = 50$ and standard deviation $\sigma = 10$ using program ST–02.

Solution Enter program ST–02 and use a seed of 198 in order to compare with Example 2.5. Now enter 50 and press [A] followed by 10 with a press of [B] as per instructions. The first press of [D] yields 50.61 and 4 successive presses of [D] generate x_2, x_3, x_4, x_5 as successive values, 46.36, 65.09, 38.84, 49.30.

With this example, we conclude our development of simulation for continuous random variables. True, we have treated only two probability distributions, but they happen to be the standard ones encountered in most introductory texts and remains within our scope. Other examples in such texts are generally contrived to illustrate theory, rather than being motivated by practical applications. In any case, for any example for which the equation $u = P(x)$ can be solved explicitly for $x = P^{-1}(u)$, the method illustrated here may be employed.

Problems

6. Time to failure of a very sensitive component of an electronic device has been modeled as exponential with mean time to failure 1,500 hours. What is the probability that it will survive 2,000 hours? Generate a random sample of 6 times to failure.

Ans: Q(2,000) = .26. (seed = 469) 2,157 2,589 4,401 320 1,316 1,685.

7. Generate a random sample of size 4 from a normal distribution having mean $\mu = 110$ and $\sigma = \sqrt{63}$ in two ways using a seed of 11,224. Compare the two resulting averages with μ.

Ans: Using ZS–2 the 4 values are 102, 102, 102, 97, totalling 403 for an average value of 100.75. Using ST–02, the 4 values are 113, 114, 125, 106, totalling 458 with an average of 114.50.

8. The function $f(x) = 3x^2$, $0 < x < 1$ defines a density function on the unit interval. Generate a random sample of size 5 from this "population."

Ans: Since $\int 3x^2 dz = x^3 + c$, we can see $P(x) = x^3$, $0 < x < 1$. Hence $u = x^3$ implies $x = \sqrt[3]{u} = u^{1/3} = P^{-1}(u)$. Using a seed of 0, the first 5 values from ST–02 are .50242, .41088, .17351, .67797, and .86408. When these are substituted in $x = u^{1/3}$, the 5 x-values are .79, .74, .56, .88, and .95.

9. A machine produces bolts whose lengths are supposed to be normally distributed with mean $\mu = 4$ inches. The specification limits in the contract call for bolts to be scrapped if they are not between 3.95 and 4.05 inches. About how many from a lot of 1,000 would you expect to be scrapped, if the process is assumed to have $\sigma = .030$?

Ans: $Pr(X < 3.95) + Pr(X > 4.05) = .096$. Multiplying by 1,000, you should guess about 96.

10. Grade point averages for seniors at a certain university based on a 4.0 scale are assumed to be normally distributed with mean 2.0 and variance .65. Only 1% of the graduating class is to be admitted to an honor society. What is the minimum grade point average necessary for admission to this society? If 5 seniors were randomly selected, what sort or grade point averages would one expect to see?

Ans: $x_{.01} = 3.88$. Using ST–02 with a seed of 633, 5 random selections would be 2.37, 2.17, 2.17, 3.22, and 2.13.

Section 2.5: Bernoulli Trials

One special circumstance that arises a lot in practice is the case where the experiment has been dichotomized so that there are only two possible outcomes. Modelling such basic experiments by a population random variable X, attention may be focused on the two possible values of X which, for convenience, may be taken as 0 and 1. A classic prototype is that of tossing a coin and observing the upturned face upon landing. If "heads" is associated with 1 and "tails" with 0, then the experiment is one of recording 0 or 1. Again, 1 might stand for success and 0 for failure of an electronic component. (In some cases we might even find it convenient to let 1 stand for real life failure, as is often done in reliability studies, where the attention is focused on trying something out until it *fails.*) There are just a lot of similar circumstances that arise, like classifying an object as good-bad, a score as pass-fail, and so on. You will find such random variables often called *Bernoulli* in honor of a famous mathematician; others sometimes use the terminology *0–1 or attributes variables* or even *coin-tossing* experiments.

Whatever the name, dealing with such random variables is conceptually simple, because there are only 2 values in the range of X and their point probabilities must add to 1. Traditionally, $\Pr(X = 1)$, is denoted p (sometimes π), and so $\Pr(X = 0) = 1 - p$, and we often denote that value by q. Since p may be assigned any value between 0 and 1, it may be viewed as a parameter in the same sense as λ for the exponential, or μ and σ for the normal cases. Then $q = 1- p$ is not a separate parameter, but merely a convenient way of representing $1- p$. As for moments, when X is Bernoulli with parameter p, then

$$\begin{aligned} E(X) &= 1 \cdot p + 0 \cdot q = p \\ E(X^2) &= 1 \cdot p + 0 \cdot q = p \qquad (2.10) \\ \sigma^2 &= p - p^2 = p(1 - p) = pq \\ \sigma &= \sqrt{pq} \end{aligned}$$

Simulating a random sample $X_1, X_2, \ldots, X_n$ of Bernoulli random variables trials is a simple matter with the random number generator. We can rely on the tools of Section 2–3. We can choose labels $0, \ldots, K$ to correspond to the accuracy with which we wish to measure the probability p, and then simply assign the proportionate number of labels to 1 and the remaining ones to 0. A record of such a sample is then a string of 0's and 1's. But the typical reason for generating such a sample is to count the number of 1's in the string—the total number of heads, successes, good items—whatever it is that 1 stands for. Such a number can be conveniently represented by the sum

$$Y = \sum_{i=1}^{n} X_i$$

of the n independent, identically distributed Bernoulli random variables. This is so standard in practice that Y is also given a special name, *binomial.* The reason stems from the derivation of the probability distribution of Y.

First, the *range* of possible values of Y is clearly the set of integers $\{0,1,2,...,n\}$ since the value of Y merely counts the numbers of 1's in the n Bernoulli trials. If the value of Y is, say k, then there must be exactly k 1's and $(n - k)$ 0's in the string $x_1,x_2,...,x_n$. From our independence assumption, the probability of *any* such arrangement would be $p^k q^{n-k}$ by multiplication. But, how many such distinct arrangements are there? As many as there are ways of selecting k of the x_i's to be replaced by 1 (the others are understood to be replaced by 0). This is just the combination of n objects k at a time, denoted $C\binom{n}{k}$ [or $\binom{n}{k}$ in some sources*]. This quantity is defined mathematically as

$$\binom{n}{k} = \frac{n!}{k!\,(n-k)\,!}$$

where, for any positive integer m, $m! = m(m-1)\cdots 3\cdot 2\cdot 1$ and $0! = 1$. Consequently, if Y is binomial, and we denote its *mass function* by f (so as not to become confused with the parameter p),

$$f(k) = \binom{n}{k} p^k q^{n-k} \qquad k = 0,1,2,...,n \tag{2.11}$$

This formula may be recognized as the kth term in the *binomial* expansion

$$1 = (p + q)^n = \sum_{k=0}^{n} \binom{n}{k} p^k q^{n-k}$$

derived in elementary algebra.

Because both n and p are fixed in the formula for the binomial mass function, both are referred to as parameters, and so Equation 2.11 defines a 2-parameter family of probability distributions. With a little algebraic manipulation, it may be shown that the moments of binomial Y are related to its parameters by the formulas

$$\begin{aligned} \mu &= E(Y) = np \\ \sigma^2 &= V(Y) = npq \\ \sigma &= \sqrt{npq} \end{aligned} \tag{2.12}$$

This probability distribution is so important that it has been extensively tabulated for a variety of parameter choices and has been programmed

* For more details, consult your probability reference or see *Probability by Calculator* in this series.

in your statistics module as ST–20. The relevant discussion and instructions begin on page 6–8 of ASM. You see under the user instructions that the parameters n and p are entered by pressing [A] and [B] respectively, and you may enter them in either order, changing one or both of those entries at any time. Once those parameters are entered, pressing [A′] and [B′] will, respectively, output μ and σ in accordance with Equation 2.12. For any integer k in the set $\{0,1,2,...,n\}$ entered in R_D, pressing [C] will then cause $f(k) = \Pr(X=k)$ to be displayed. Consistent with our previous notation, [D] will output the CDF at k, namely,

$$P(k) = \sum_{j=0}^{k} \binom{n}{j} p^j q^{n-j}$$

while [E] will output the CCDF at k,

$$Q(k) = \sum_{j=k+1}^{n} \binom{n}{j} p^j q^{n-j}$$

You want to be very careful in translating binomial problems to solution when inequalities are involved for, unlike our continuous cases, it makes a difference here whether or not equality is included. For example, if a problem required the evaluation of the probability that the value of Y is *at least* k, then the answer is not $Q(k)$ but $Q(k - 1)$. We illustrate several circumstances.

EXAMPLE 2.7

Imagine the birth of a boy has probability $p = .51$. A couple is planning to have 4 children.

a. What is the probability that the couple will have no boys?
b. What is the probability that there will be at least three boys?
c. What is the probability that there will be at least three girls?
d. What is the expected number of boys? Variance?

Solution We may model the experiment as consisting of $n = 4$ independent Bernoulli trials with $p = .51$. This means identifying 1 with "boy." (We have a choice. We could just as well model 1 to stand for "girl," in which case $p = .49$. It makes no difference, as long as we are consistent.) Then (a) requires an evaluation of $f(0)$. Since the answer to (b) is $\Pr(Y \geq 3)$, we can answer this immediately with $Q(2) = \Pr(Y > 2)$. As for (c), there will be 3 or more girls precisely when there are 0 or 1 boys, so (c) may be answered by $P(1) = \Pr(Y \leq 1)$. The calculator solutions (rounded to 2 decimal places) are

STEP	ENTER	PRESS	DISPLAY
1		[Pgm]20	xxx
2	4	[A]	4.
3	0.51	[B]	0.51
6	0	[C]	$0.06 = \Pr(Y = 0)$
8	2	[E]	$0.33 = \Pr(Y > 3) = Q(2)$
7	1	[D]	$0.30 = \Pr(Y \leq 1) = P(1)$
4		[A′]	$2.04 = \mu$
5		[B′]	$0.99979998 = \sigma$
		[x^2]	$0.9996 = \sigma^2$

Another fact that might be useful in the solution of Example 2.7 is that $n - Y$ is also binomial with parameters n and q whenever Y is binomial with parameters n and p. Hence, if Y counts the number of boys, then $n - Y$ counts the number of girls. We could then answer (c) by merely changing the entry of p from 0.51 to 0.49 with label B, and then find $Q(2)$ by entering 2 and pressing [E]. Naturally pressing [A′] now yields $nq = 1.96$, the mean number of girls. So this has the same effect as redesignating what it is that 1 stands for.

There are limitations to the use of the program, even though the allowances are quite generous. Basically, the program is limited by the use of the [y^x] key to evaluate probabilities. For example, if $p = .5$, then n may not exceed 328, while if $p = .001$, values of n up to nearly 228,000 could be used. But, if $p = .999$, then the program will not function with a value of n as small as $n = 34$. In the next chapter, we will discover a way out of this small dilemma where the normal distribution may be used to approximate the binomial (as alluded to on page 6–8 of ASM). But you will find ST–20 quite adequate for solving the vast majority of binomial problems posed in the textbooks currently in use, although some may require more running time in this program than you may be used to observing.

One special type of situation in which the binomial distribution arises is in connection with a type of sampling often referred to as "drawing balls from an urn." We imagine an urn consisting of N balls of 1 color and $N - M$ of another. Drawing a ball from an urn under uniform conditions may then be viewed as a Bernoulli trial with parameter $p = M/N$. The value of X is then 1 precisely when the ball drawn is 1 of the M of 1 color, and is 0 when the ball is 1 of the $N - M$ of the other color. Now, in fact, the urn may be a lot or stockpile of items consisting of defective and nondefective items.

Provided we replace a ball and mix the contents of the urn well, we can model successive draws of a ball in an urn problem as independent repeated Bernoulli trials. Then the total number of balls of the type whose probability is M/N will have a binomial distribution. Answers to various questions that are posed in that context can then be effected with program ST–20 taking $p = M/N$ and n, the number of draws, as parameters. For obvious reasons, we call such a scheme *sampling with replacement.*

EXAMPLE 2.8

An urn contains 7 white and 10 black balls. Three are drawn with replacement from the urn. The probability that 2 of these will be white is given by

$$\Pr(Y = 2) = \binom{3}{2}(7/17)^2(10/17) = .30$$

while the probability of no more than 1 white ball would be, using ST–20 with $n = 3$, $p = 7/17$, $P(1) = .63$. The probability of drawing 2 black balls can be solved by observing that this is just the probability of 1 white or $\Pr(Y = 1) = .43$. Or, we can reparameterize the problem so that $p = 10/17$, and Y now stands for the total number of black balls in a sample of 3. Then, $\Pr(Y = 2) = .43$ using ST–20 with $n = 3$, $p = 10/17$.

Occasionally, M is not given directly, but is already expressed as a proportion M/N, making the problem even easier.

EXAMPLE 2.9

A manufacturer regularly ships items in lots of size 100 and has a record of 5% defectives in his shipment. How likely is it that, in sampling 5 items from the lot with replacement, no defectives will be observed?

Solution We model this as a binomial problem with $n = 5$ and $p = .05$, so that $\Pr(Y = 0) = .77$ is the answer. The probability of at least 1 defective would of course be .23 or $Q(0)$. The mean number of defectives would be .25, while the mean nondefectives number must be $4.75 = 5 - .25$.

If sampling were carried out without replacing the ball each time, then the binomial model may be a poor one. For one thing, the Bernoulli random variables cannot be identically distributed, since the proportion M/N is subject to change each time. Moreover, the random variables cannot be independent either, for, clearly, the probability that the value of X_2 will be 1 definitely depends on what the value of X_1 was. In these circumstances, *sampling without replacement,* the total number of outcomes of one type has a probability distribution that differs from the binomial and is known as the *Hypergeometric* distribution. This situation does not arise typically in statistical analysis where we deal almost exclusively with random samples, however. The interested reader is invited to examine *Probability by Calculator* in this series, where typical problems in that setting are treated. It should be clear that, if N is very large compared to n, it will not make a great difference whether we sample with or without replacement, and in that special case, the binomial model may serve adequately to evaluate probabilities. Some applications follow.

Problems

11. A fair coin is tossed 5 times.

a. What is the probability of exactly 3 heads?

Ans: $f(3) = .31$

b. Find the probability that at least 3 heads are tossed.

Ans: $Q(2) = .50$

c. What is the expected number of heads?

Ans: $\mu = 2.5$

d. What is the probability that heads will occur on each of the last 2 tosses?

Ans: Same as changing n from 5 to 2 and evaluating $f(2) = .25$, because of assumed independence.

12. What is the probability, correct to 2 decimal places, that in 20 tosses of a biased coin ($p = .1$), you will observe at most 2 heads?

Ans: $P(2) = .68$

13. Suppose that the probability of hitting a target in a single trial is .4. Assuming independent trials, at least how many trials are required so that the probability of at least 3 hits is at least .5?

Ans: $Q(2) \geq .5$ yields $n = 7$ by trial and error, using [A] and [E] in ST–20.

14. A projectile is fired at a target with probability .2 of a direct hit. In 7 independent firings, what is the probability of 7 hits?

Ans: .0000128

What is the probability of at least 1 hit?

Ans: $Q(0) = .79$

What is the expected number of hits?

Ans: $\mu = 1.4$

15. A population of 1,000 insects infests a rose garden. The gardener sprays the garden with insecticide. Suppose the probability is .995 that an insect will be killed.

a. What is the probability distribution of X, the number of surviving insects?

Ans: Binomial $\quad n = 1{,}000 \quad p = .005$

b. What is the expected number of insects surviving?

Ans: $\mu = 5$

c. Calculate the probability that at least 997 insects are killed.

Ans: $1{,}000 - Y \geq 997 \quad \text{iff} \quad Y \leq 3, \quad P(3) = .26$

16. Five men each fire 1 round, simultaneously, at the same target. The probability is .8 that each man hits the target.

a. What is the distribution of X, the number of hits on the target?

Ans: Binomial $n = 5$ $p = .8$

b. What is the probability that the target is hit at least twice?

Ans: $Q(1) = .9933$

c. What is the expected number of target hits?

Ans: $\mu = 4$

17. If X is binomial with parameters $n = 10$ and $p = .4$, find values for $\Pr(6 < X \leq 8)$, $\Pr(2 \leq X < 3)$, $\Pr(X > 4)$, and $\Pr(X < 8)$.

Respective Answers $P(8) - P(6) = .0531$
$f(2) = .1209$
$Q(4) = .3669$
$P(7) = .9877$

18. It is conjectured that half of the people in a certain community have exactly 2 cars. To test this claim, a plan is devised whereby the claim is accepted if, in a sample of 10 families, the total number having 2 cars is between 2 and 8 inclusive; otherwise, the claim will be rejected. What is the probability that the claim will be rejected, if in fact it is true?

Ans: Model: binomial $n = 10$ $p = .5$
$f(0) + f(1) + f(9) + f(10) = .0215$

19. Past history shows that only 6 out of 10 applicants for a certain position meet the qualifications specified. What is the expected number of successful applicants out of a group of 30 persons?

Ans: Binomial $\mu = 18$

20. An auditor is hired to inspect the financial accounts of a firm. Rather than inspect every individual record, a "random sample" of the records will be inspected. Assume the firm has many accounts, and 20 will be selected for inspection. Suppose that there is at least 1 error in every 15 of the firm's accounts. Let X be the number of erroneous accounts among the sampled ones.

a. What is the probability distribution of X?

Ans: With no value of N specified, we must assume X is binomial, $n = 20$, $p = 1/15$.

b. Calculate the probability that exactly 3 sampled accounts are erroneous.

Ans: $f(3) = .1045$

c. What is the expected number of erroneous accounts?

Ans: $\mu = 1.33$

21. Suppose one baby in 1,000 is born blind. A large city hospital has 2,000 babies born in a year.

a. What is the expected number of babies born blind in this hospital this year?

Ans: $\mu = 2$

b. Calculate the probability that at least 3 babies are born blind this year.

Ans: $Q(2) = .32$

c. Determine the variance of the number of babies *not* born blind this year.

Ans: $\sigma^2 = 1.998$ (*Note*: $n - Y$ is the number not born blind, and this random variable has the same variance as Y.)

22. Of the light bulbs made by a manufacturer, 5% are defective. These bulbs are sold in packages of 4.

a. Calculate the probability that all of the bulbs in a package are nondefective.

Ans: Binomial $n = 4$ $p = .95$ $f(4) = .8145$

b. You buy 3 packages of these bulbs. Let Y denote the number of packages that have no defective bulbs. Calculate $\Pr(Y = 2)$.

Ans: Y is binomial, $n = 3$, $p = .8145$ from part a. Then $f(2) = .3692$.

23. A keypunch operator has probability .005 of making an error in a single keystroke. In a program requiring 600 keystrokes,

a. What is the probability that the operator makes no errors?

Ans: Binomial $n = 600$ $p = .005$ $f(0) = .0494$

b. Calculate the probability of 2 or more errors.

Ans: $Q(1) = .8016$

c. What is the standard deviation of the number of errors?

Ans: $\sigma = 1.73$

24. The probability that a handblown glass is defective is .01. In 1 week a good worker turns out 200 glasses.

a. What is the probability distribution of the number of defective glasses produced by the worker in 1 week?

Ans: Binomial $n = 200$ $p = .01$

b. Compute the probability that he produces no more than 1 defective glass in a week.

Ans: $P(1) = .4046$

c. What is the expected number of defectives he will produce in a 1 week and the variance?

Ans: $\mu = 2 \qquad \sigma^2 = 1.98$

25. Suppose an examination has 20 multiple choice questions with 4 possible answers to each one. A student simply guesses the answer to each question.

a. What are his chances of getting 15 or more answers correct?

Ans: $Q(14) = .0000038$ where $n = 20$ and $p = .25$

b. What is the mean and standard deviation of the number of correct answers?

Ans: $\mu = 5 \qquad \sigma = 1.94$

ZS–2: User Instructions

STEP	PROCEDURE	ENTER	PRESS	DISPLAY
E	*Exponential Distribution*			
1	Initialize.		[RST]	xxx
2	Enter parameter.	λ	[STO] 16	λ
3	Compute $P(x)$ and $Q(x)$.	x	[E]	$P(x)$
	Note: λx must not exceed 227.		[$x \rightleftharpoons t$]	$Q(x)$
4	Calculate 100 $(1 - \alpha)$th percentile.	α	[E′]	x_α
	Note: Repeat steps 3 and 4 at will.			
5	Generate sample of size n.			
	a. Initialize random number generator.		[Pgm] 02 [E′]	0.
		Seed	[E]	Seed
	b. Exit ST–02.		[RST]	Seed
	c. Execute Step 2 (if necessary).			λ
	d. Generate x (repeat n times).		[B]	x
N	*Normal Distribution*			
1	Enter parameters.	μ	[STO] 17	μ
		σ	[STO] 18	σ
2	Compute $P(x)$ and $Q(x)$.	x	[C]	$P(x)$
			[$x \rightleftharpoons t$]	$Q(x)$

STEP	PROCEDURE	ENTER	PRESS	DISPLAY
3	Compute $\Pr(x_1 < X < x_2)$ or $1 - Pr(x_1 < X < x_2)$.			
	a. Enter x_1.	x_1	[D]	$Q(x_1)$
	b. Enter x_2 and compute.	x_2	[R/S]	$\Pr(x_1 < X < x_2)$
			[$x \rightleftarrows t$]	$\Pr(X < x_1) + \Pr(X > x_2)$
4	Calculate standard $100(1-\alpha)$th percentile.	α	[C′]	z_α
5	Calculate general $100(1-\alpha)$th percentile.	α	[D′]	x_α
	Note: Repeat N_2–N_5 as often as desired.			
6	Generate sample of size n.			
	a. Initialize random number generator.		[Pgm] 02 [E′]	0.
		Seed	[E]	Seed
	b. Exit ST–02.		[RST]	Seed
	c. Execute Step 1 (if necessary).			
	d. Generate x (repeat n times).		[B′]	x

Register Contents

Register	Contents	Register	Contents	Register	Contents
00	Used	10	Used by ST–02	20	$z = (x - \mu)/\sigma$
01	Used by ST–02	11	Used by ST–02	21	Used by ST–19 and ST–20
02	Used by ST–02	12	Used by ST–02	22	Used by ST–19 and ST–20
03	Used by ST–02	13		23	Used by ST–19 and ST–20
04	Used by ST–02	14		24	Used by ST–19 and ST–20
05	Used by ST–02	15		25	Used by ST–19 and ST–20
06	Used by ST–02	16	λ	26	Used by ST–19 and ST–20
07	$P(x)$	17	μ	27	
08	$P(x_2) - P(x_1)$	18	σ	28	
09	Seed	19	z_α	29	

EXAMPLES ZS–2

1. Let X have an exponential distribution with parameter $\lambda = .001$, and suppose X measures time to failure in hours.

   a. Calculate the probability that time to failure will exceed 1,200 hours.

   b. Compare the mean time to failure with the median time to failure.

   c. How many hours may we reasonably depend upon for survival of 90% of such items?

   d. Generate a random sample of 5 times to failure.

Solution

ZS STEP	ENTER	PRESS	DISPLAY	COMMENTS
E 1		[RST]	*xxx*	Initialize the exponential subroutine.
2	0.001	[STO] 16	0.001	Single parameter λ stored in R_{16}
	1200	[E]	.6988	Displays $P(1200) = \Pr(X \leq 1200)$
		[$x \rightleftharpoons t$]	.3012	Displays $Q(1200) = \Pr(X > 1200)$, which is the answer to (a)
4	0.50	[E′]	693.15	Calculates and displays the median $x_{.50}$ (in hours)
		[RCL] 17	1000.	Recall μ, the mean time to failure. This answers (b).
4	0.10	[E′]	105.36	Displays $x_{.90}$
5a		[Pgm] 02	*xxx*	Initialize the random number generator.
		[E′]	0.	
	635	[E]	635.	Enter seed = 635 for illustrative purposes.
5b		[RST]	635.	Exits statistics module to return to control of ZS–2 program
5c	(0.001)	([STO] 16)	(0.001)	Enter the parameter λ, if not already entered.
5d		[B]	30	Displays the first generated sample value, x_1
		[B]	253	The second simulated time to failure
		[B]	1392	Successive times to failure (rounded to whole hours) for a random sample of size 5.
		[B]	1873	
		[B]	639	

2. A standardized test is administered to incoming freshmen at a university. Scores, X, are assumed to be normally distributed and, based on thousands of past scores, it is assumed that $\mu = 100$ and $\sigma = 16$. For an incoming freshman chosen at random, what is the probability that the test score will be

   a. greater than 110?
   b. less than 90?
   c. between 75 and 125?
   d. If only the top 80% of incoming freshmen are to be admitted on the basis of this test, what would the minimum passing score be?

Solution

ZS STEP	ENTER	PRESS	DISPLAY	COMMENTS
N 1	100	[STO] 17	100.	Enter the mean value.
	16	[STO] 18	16.	Enter the second parameter σ.
2	110	[C]	.73	Displays $P(110) = \Pr(X \leq 110)$

		[$x \rightleftarrows t$]	.27	Displays $Q(110)$, the required probability.
		[RCL]20	0.625	Shows the standardized value for $x = 110$, namely, $z = (110 - 100)/16$
2	90	[C]	.27	Displays $P(90)$
3a	75	[D]	.94	Displays $Q(75)$, of minor interest
3b	125	[R/S]	.88	Calculates and displays $\Pr(75 < X < 125)$
		[$x \rightleftarrows t$]	.12	Displays $\Pr(X < 75) + \Pr(X > 125)$
5	.80	[D′]	87	Displays the 20th percentile for X, so that $\Pr(X > 87) = .80$

3. Example 2.5

Solution

		[Pgm] 02	*xxx*	Initialize random number generator.
N 6a		[E′]	0.	
	198	[E]	198.	Enter seed = 198 for illustrative purposes.
6b		[RST]	198.	Exit ST–02 to return to ZS–2.
6c	50	[STO] 17	50.	Enter the normal parameters,
	10	[STO] 18	10.	and store in appropriate registers.
6d		[B′]	45.43	Displays first generated sample value x_1.
		[B′]	57.09	Successive sample values
		[B′]	45.04	are generated and
		[B′]	45.31	displayed.
		[B′]	54.67	

3

Data Processing

Section 3.1: Sample Characteristics

Often in practice we do not know precisely what the distribution of a population random variable X is. But it may be possible to get some ideas about the distribution by observing a number of outcomes, that is, a sample, and examining various attributes of this sample. For example, we might compute the *sample average* and use this as a best guess or *estimate* of the unknown population mean μ. *Descriptive statistics* is a collection of techniques used to process data to describe the sample itself.

If we have a notion of the distribution family involved, we may wish to use sample data to estimate or make other inferences about the parameters of the distribution. This process is called *statistical inference.* For example, we may believe that the success or failure of a particular type of electronic component is well represented as a Bernoulli random variable, but we may not have a good idea as to the value of the success probability parameter (reliability), p. We might plan a series of tests with a sample of the components, to obtain a random sample of the Bernoulli population. How should the sample values be used to estimate unknown parameters such as p?

We begin our study of descriptive statistics by considering the sample analogs of the mean, μ, and standard deviation, σ, of a population.

Suppose we have obtained a set of data as a result of a random sample of size n of a population random variable X, resulting in data, $x_1, x_2, \ldots, x_n$. We first introduce the procedures for entering these data into your TI–59, and illustrate how the sample mean, denoted $\bar{x}$, and the sample standard deviation, s, are calculated. The sample mean is used as a measure of central tendency, a focal point or "center" of the sample data. It is the ordinary average of the sample values. The *sample mean* is defined by

$$\bar{x} = \frac{\sum_{i=1}^{n} x_i}{n} \tag{3.1}$$

Similarly, the sample standard deviation, denoted s, is a measure of how spread out the data values in the sample are in relation to the sample mean. We will return shortly to more discussion of measuring dispersion; for immediate reference, the *sample standard deviation* is defined by

$$s = \sqrt{\frac{\sum_{i=1}^{n} (x_i - \bar{x})^2}{n-1}} \tag{3.2}$$

The appearance of $n - 1$ instead of n will be explained in Chapter 4.

We now give an example set of data and discuss entry of the data and computation of the sample mean and sample standard deviation. There are two ways to enter and summarize univariate (single-variable) data. The simplest method is to use the [Σ+] key following each data entry. This key has the effect of summing an entered number into R_{01}, summing its square into R_{02} and recording in R_{03} the number of times a value has been entered. In that case, the gold key [$\bar{x}$] may be used to calculate the sample mean of the entries, and the key sequence [INV][$\bar{x}$] may be used to find the sample standard deviation of the entries. These two sample moments are the bread and butter of statistical applications.

Try this entry scheme with the following data: 12,10,8,9,16. First, make sure registers R_{01}—R_{06} are clear by pressing [Pgm]01[SBR][CLR], then [RST], (or simply [CMs]). Enter each of the numbers and follow it with [Σ+]. Each time you will see a count of how many numbers you have entered and, at the conclusion, you should see 5 in the display register (R_D). Now press [$\bar{x}$] to see the mean, 11, in R_D, then [INV][$\bar{x}$] to verify a (rounded) standard deviation of 3.16. If you successively recall R_{01}, R_{02}, and R_{03}, you should verify the sum of 55, sum of squares of 645 and the number of entries, 5, in those respective registers as a check on your entry.

Incidentally, you may delete any data entry with the sequence [INV][Σ+]. For instance, to delete the last entry of 16, key in 16 [INV] [Σ+], and 4 then appears in R_D. Recalling registers 01, 02, and 03 now reveals respective values of 39, 389, and 4. Pressing [$\bar{x}$] and [INV][$\bar{x}$] will yield adjusted values of 9.75 and 1.71, respectively. As a final note, the above sequence used to clear registers R_{01} through R_{06} also clears the *T*-register (R_T) without disturbing any other memory storage while [CMs] does not. These are the registers used by [Σ+] (more about R_{04}, R_{05}, and R_{06} later), and so they may be cleared for a new problem without affecting anything else you might have in program memory.

Warning: Be sure to press [Pgm]00 or [RST]* to get back into the calculator memory after using any module program such as ST–01.

There is practically no limit to the number of observations that may be entered.

Another method of entering univariate data is to use program ST–03 of the Statistics module. It is slower, is more complicated, and there is a limit (generous) to the number of observations it can handle. But, the program is useful for certain aspects of data analysis beyond computing sample means and standard deviations. This method of entering data is discussed in Chapter III of ASM beginning on page 3–1. With this method, not only are the data summarized as indicated on page 3–1 for registers R_{03} through R_{18}, but a record of the data entered is preserved as well. Basically, the data are stored one-by-one starting with register 31, up through register 99, for a total capability of 69 entries. Meanwhile, Registers R_{00} through R_{29} are used to process the data in what is referred to as the Intermediate Data Base (IDB). It is this processing which makes the program so much slower than the use of [Σ+] for each entry, for you see that not only sums and sums of squares are computed and stored, but many other quantities as well. We will rarely use some of these quantities, but nevertheless we wish to use ST–03 for data entry in conjunction with other programs. Observe that when data are entered using ST–03, the sum of the *x*'s is stored in R_{04} and the sum of the squares in R_{05}. This makes it necessary to press [$x \rightleftarrows t$] after both [$\bar{x}$] and [INV][$\bar{x}$] to display the sample mean and standard deviation, for they are stored in the T-register by program ST–03.

To illustrate, take the same example you just worked and follow the directions on page 3–5 of ASM for KEYBOARD ENTRY of data, entering the values 12, 10, 8, 9, 16. With only 5 entries, you may ignore the repartitioning instructions at Step 3. Notice how much longer it takes for the count to appear after entering a datum.

Warning: Be sure to wait for that count to appear in the display before entering the next data value!

Now press [$\bar{x}$] then [$x \rightleftarrows t$] to find the value 11 as before and [INV] [$\bar{x}$] followed by [$x \rightleftarrows t$] to read 3.16. This time you will find R_{01} and R_{02} empty, while R_{04} and R_{05} contain the respective values 55 and 645, as explained above. Notice that R_{12} and R_{13} contain, respectively, the minimum and maximum values entered. This information can be useful. For

* [RST] also lowers all flags and should be used with caution. If you have nothing in program memory to worry about, a simple press of [CMs] will suffice as initialization as already noted.

example, label *C* in program ZS–3 computes the distance between the largest and smallest values, called the *range* of the sample, after you have used ST–03 for data entry. (The range is 8 for our example data.)

On page 3–4 of ASM, Table 3.2 shows you the various limits on the number of data values that may be entered in various partitioned states of the calculator. You may plan ahead and repartition accordingly at Step 3 of ST–03. Remember, each time you partition up to allow for more data, you give away program step capacity. Whether or not that is a problem depends on the use to which you will be putting the program memory. Program ZS–3, for example, only has 171 program steps, so that its use will not be encumbered by partitioning up to 9[Op]17.

If you neglect to plan partitioning ahead of time and are working in the standard partitioning (which will be the case most of the time), you may still repartition at Step 4. As you can see from Table 3.2, standard partitioning allows only 29 data values. If you attempt to enter a 30th value, the calculator will flash the attempted value in R_D to indicate that the allowable data registers are filled. Press [CLR] or [CE] at this point, and the flashing value in R_D is cleared. Now you may repartition by pressing *n*[Op]17 as required to allow for additional data entry, subject to the restrictions just discussed. Then continue entering data as instructed in Step 4, starting with the 30th value again.

When the Raw Data Base is filled, either because you could only allow so much repartitioning or you have more than 69 data values, you have 2 options open to you. If you have no further need for the individual sample values, follow Instruction 5b by pressing [C] and then proceed to enter additional data. On the other hand, if you need a record of the data you have entered, or may have occasion to work additional problems with these same data, you should record them on magnetic cards. As outlined on page 3–4 of ASM, you may record R_{30}—R_{59}, stored in Bank of 3 of the calculator memory, by pressing 3 [2nd] [WRITE] and inserting a blank card. Similarly, banks 2 and 1 may be recorded, giving a permanent record of the block of registers R_{60}—R_{89}, and R_{90}—R_{99}, respectively. You should not record Bank 4, since it is reserved for the IDB which is not complete until all the data are entered. You may use additional cards as needed to record the banks, if they are refilled by your data entry.

For example, suppose you had 60 data values to enter, you wanted to work several problems with them, and one of the prepared magnetic cards you wish to use has 390 program steps recorded on Banks 1 and 2. At Step 3, you may repartition the calculator with 7[Op]17, allowing for 39 data values and a maximum of 400 program steps. That is as high as you can go in the partitioning, for if you used 8[Op]17, only 320 program steps are allowed, and program steps numbered 320 through 389 of your program would be lost. So you enter the first 39 data values, which will be stored in registers R_{31}—R_{69}. To record these you will have to use Bank 3 (for R_{30}—the pointer—through R_{59}) and Bank 2 (for R_{60}—R_{69}) on 2 edges of 1 blank magnetic card. Now press [C], then enter data values starting with the 40th and proceeding through the 60th. These will now be stored in registers R_{31} through R_{51}, writing over what was previously there. At this point, these remaining 21 values in Bank 3 may be recorded on a second blank magnetic card using only 1 edge of that card.

All of this time, the IDB is being compiled with the final results stored in R_{00} through R_{29} (Bank 4). These may be recorded (on 1 edge of a card), if desired, as described in Footnote 5 on page 3–4 of ASM. These procedures cover keyboard entry. For future analyses of these same data, the data need not be entered by keyboard, but rather can be read off your recorded magnetic cards. Then the IDB will have to be recompiled using routine [D] in ST–03 for each card entry, as described in the user instructions for ST–03. Note that, if the data are stored in their proper registers and the location of the last data value, say the *n*th, (please note that this is not the *n* for repartitioning) is within the allowable range of the partitioning, the IDB may be compiled at any time by pressing [C], then [D], in ST–03, *provided* $31 + n$ is stored in R_T. In summary, to recompile the IDB, be sure that $31 + n$ is in R_T, 31 in R_{30}, and press [D].*

Please recall that the calculator will not normally read a card which was recorded with the calculator in a partitioning differing from its current state. (You may check current partitioning using [Op]16.) All of the magnetic cards you will prepare will be recorded in standard partitioning, so, unless they are read in when you first start your session, you may experience difficulty trying to read some of your cards, with so much repartitioning taking place. There are several alternatives to allow reading of such cards. The simplest is to read in the card by first entering $-n$ (versus n or 0) in R_D. The card will then be read regardless of the partitioned state of your calculator.

Let us return now to a discussion of measuring the dispersion or spread of a sample. We already discussed s as a measure of spread, in connection with the use of [INV] [$\bar{x}$], (with or without [$x \rightleftharpoons t$], depending on the mode of data entry). There are other measures of spread in common use. The concept we need is that of how spread out are the data from $\bar{x}$, our chosen measure of central tendency. Suppose, in one set of data, we have observations -1, 0, $+1$ and, in a second set, we have -5, 0, $+5$. Clearly, $\bar{x} = 0$ for both sets of data, but, intuitively, one would say the second set of data is more "spread out" than the first. Any measure of spread should reflect this.

The crudest measure of spread in common use is the range, R, defined before to be the distance between the smallest and largest sample values, that is, $R = x_{max} - x_{min}$. As a rough measure, the range gives some idea of spread, but, since it ignores all of the intermediate observations, you would not expect it to be as useful or informative as one that did not ignore such observations. For our trite examples, the values of R

* These matters will never arise in our applications.

are 2 and 10 respectively. If the data consisted of 10 values at -1, one at 0 and 10 at $+1$, we would still have $\bar{x} = 0$ and $R = 2$, as with the data -1, 0, 1. But now consider having 1 observation at each of -5 and $+5$ as before, 100 at 0, and 10 at each of .01 and $-.01$. Again, $\bar{x} = 0$ and $R = 10$, and yet, we would think surely this new set of data has less spread than -5, 0, $+5$, and indeed perhaps even less than -1, 0, $+1$, because of the data points crowded at $-.01$ and $+.01$. However, the range will not reveal this.

One way to reflect the spread in such patterns would be to consider *deviations* from $\bar{x}$ for each data point, say, $x_i - \bar{x}$ for the i-th data point. Some of these deviations will be positive, and some will be negative. Indeed, it is an inevitable fact that these deviations will always sum to 0. For

$$\sum_{i=1}^{n} (x_i - \bar{x}) = \sum_{i=1}^{n} x_i - \sum_{i=1}^{n} \bar{x} = n\bar{x} - n\bar{x} = 0$$

recalling that

$$\bar{x} = \frac{1}{n} \sum_{i=1}^{n} x_i \quad \left(\text{hence } \sum_{i=1}^{n} x_i = n\bar{x}\right)$$

and summing $\bar{x}$, n times, gives a sum equal to $n\bar{x}$. Consequently, adding deviations to measure spread will get us nowhere. On the other hand, it would seem natural that we should treat a deviation of -4 the same as $+4$, since the "distance" of each x_i from $\bar{x}$ in those 2 cases is the same. So we attempt to destroy the algebraic sign.

If we just ignore the sign, that would be equivalent to defining the absolute value of the deviations $|x_i - \bar{x}|$. We could add up all such unsigned deviations, and then average them out to determine an overall absolute deviation. Such a measure is called *mean* (for average), *absolute* (for taking the absolute value) *deviation*—or MAD for short. Thus, the (sample) mean absolute deviation for a data set $x_1, x_2, \ldots, x_n$ is defined by

$$\text{MAD} = \frac{1}{n} \sum_{i=1}^{n} |x_i - \bar{x}| \tag{3.3}$$

One of the reasons MAD does not enjoy more popularity is that the absolute value operator is not particularly easy to work with mathematically. As far as computing the value, however, the [$|x|$] key on the TI–59 may be exploited. Once data have been stored by ST–03, they can be successively recalled and processed through Equation 3.3. This is done for you in program ZS–3, along with calculation of the range. To compute MAD using ZS–3, it is essential that the data be stored in the calculator, which means you may have to repartition the calculator. However, the program steps in ZS–3 are so few in number that you may use up to a partition of 9[Op]17, thereby allowing for 59 data values.

A much more mathematically appealing way of destroying the sign of a deviation is to square it, producing, for the ith data value, $(x_i - \bar{x})^2$. This is partly because such an expression immediately admits an expansion to

$$(x_i - \bar{x})^2 = x_i^2 - 2\bar{x}x_i + \bar{x}^2$$

When we average these squared deviations, we produce a measure of spread called the *mean* (averaged) *squared deviation*, or MSD for short. Thus, for the data $x_1, x_2,..., x_n$, MSD is defined by

$$\text{MSD} = \frac{1}{n}\sum_{i=1}^{n}(x_i - \bar{x})^2 \tag{3.4}$$

Using the foregoing expansion, we can obtain the identities

$$\sum_{i=1}^{n}(x_i - \bar{x})^2 = \sum_{i=1}^{n}x_i^2 - 2\bar{x}\sum_{i=1}^{n}x_i + n\bar{x}^2 = \sum_{i=1}^{n}x_i^2 - 2n\bar{x}^2 + n\bar{x}^2$$
$$= \sum_{i=1}^{n}x_i^2 - n\bar{x}^2 = \sum_{i=1}^{n}x_i^2 - \frac{\left[\sum_{i=1}^{n}x_i\right]^2}{n}$$

In this way, we see that the numerator for MSD can be calculated directly from the cumulated sums Σx_i^2 and Σx_i, and $\bar{x}$ need not be subtracted from each x_i. No such simplification can be constructed for MAD. At any rate, the TI–59 is hard-wired to calculate MSD using [Op] 11. If data have been entered with [Σ+], the key sequence [Op]11 produces the MSD measure. If data have been entered with ST–03, then [Op]11 must be followed by [$x \rightleftharpoons t$] to produce MSD.

We mentioned earlier, in connection with the use of [Σ+] and [INV][$\bar{x}$], the sample standard deviation, s. The square of s is often called the *sample variance*

$$s^2 = \frac{1}{n-1}\sum_{i=1}^{n}(x_i - \bar{x})^2 \tag{3.5}$$

You see that the difference between MSD and s^2 is whether the sum of squared deviations is divided by n or by $n - 1$. It can be shown that MSD is *biased*, in that it tends to underestimate the population variance σ^2. Use of the smaller divisor in Equation 3.5 is precisely the adjustment needed to make s^2 an unbiased estimate of the population variance. That is, over a very long history of the use of Equation 3.5 to estimate an unknown variance parameter σ^2, the average of the estimates would be close to σ^2.* It might be well to remember the underlying relationship between MSD and s^2, namely,

$$(n)\ \text{MSD} = (n - 1)s^2 \qquad \text{or} \qquad \text{MSD} = \frac{n-1}{n} s^2$$

To compare MSD directly with s, we define *root mean squared deviation*, RMSD, as simply $\sqrt{\text{MSD}}$. Of course, then

$$\text{RMSD} = \sqrt{\frac{(n-1)}{n}}\ s$$

EXAMPLE 3.1

For the data set below, calculate the sample mean and the following measures of dispersion: s, s^2, R, MSD, RMSD.

4 17 −2 11 8 11 5 0 1

Solution All of the measures required can be obtained using the [Σ+] data entry, and since this is simpler than using ST–03, we choose this method. If the calculator has been in use, press [Pgm]01 [SBR] [CLR], to clear the necessary registers. Enter the data, using [Σ+]. (To enter −2, first press [2], then [+/−].) When the last value is entered, you should see 9 in R_D, which is the sample size, n, for this problem. Press [$\bar{x}$] to obtain $\bar{x}$ = 6.11; [INV][$\bar{x}$] yields s = 6.17, and if this is followed by [x^2], you will find s^2 = 38.11. By inspection of the data, $R = 17 - (-2) = 19$. Press [Op]11 to obtain MSD = 33.88, then [$\sqrt{x}$] to obtain RMSD = 5.82.

EXAMPLE 3.2

Compute MAD for the data of Example 3.1.

Solution Computation of MAD is accomplished with ZS–3, which requires data entry with ST–03. First, you should load ZS–3 into the calculator (be sure to record it for future use, if you have not done so already), then enter the data following the instructions for ST–03 on page 3–5 of ASM. With a sample size of 9, no repartitioning is necessary, so ignore Step 3. You thus use only Steps 1, 2, and 4 of ST–03 for this data entry. Now press [RST], then [B] to calculate MAD = 5.01, following the instructions for ZS–3. The range may be recalculated by pressing [C], giving R = 19, as before (this is Step 1 of ZS–3).

You will find that some books (very few) call our MSD "sample variance." This means that you should examine, in any discussion involving "sample variance," whether the author really means MSD or s^2. It is also common to use s and RMSD to measure spread, as already

* We will explore this in more detail in Chapter 4.

mentioned. The latter measures have the advantage of being in the same units in which the data values were measured. You will sometimes also find RMSD written simply RMS.

Returning to our two simple data sets -1, 0, $+1$, and -5, 0, $+5$, we see that the respective values of MAD, s, and RMSD are .67, 1, .82, and 3.33, 5, 4.08. Each in its own way measures data variability or spread. You see that the data set we felt was more variable is so, as measured by each of the three measures. If we go back to 10 values at -1, 10 at 0, and 10 at $+1$, we find MAD $=$ .67, $s =$ .83, and RMSD $=$ .82. Thus you see that s is the only calculation that was affected, and it is now more in line with RMSD, which tells us that the 30 data values are no more variable than when there were only 3. If we had 1 observation each of 5 and -5, 10 each of .01 and $-.01$, and 100 0's (as we discussed earlier), we find MAD $=$ .084, $s =$ .643, and RMSD $=$ .640. As predicted, all 3 measures verify that this data set is less variable than any of the others.

Problems

1. Find $\bar{x}$, s^2, MSD, MAD, and R for the data set -6, -2, 1, 11, -10.

   Ans: $\bar{x} = -1.2$ $s^2 = 63.7$ MSD $= 50.96$ MAD $= 5.76$ and $R = 21$

2. Add 5 to each data value in Problem 1, to obtain the set -1, 3, 6, 16, -5. Now calculate $\bar{x}$, s^2, MSD, MAD, and R, and compare with the results in Problem 1.

   Ans: $\bar{x} = 3.8$ $s^2 = 63.7$ MSD $= 50.96$ MAD $= 5.76$ and $R = 21$ Note that adding a constant, such as 5, to all data values changes the measure of location $\bar{x}$ by the constant added, but does not affect the measures of dispersion.

3. Multiply by 2 each data value in Problem 1 to obtain the set -12, -4, 2, 22, -20. Now calculate $\bar{x}$, s^2, MSD, MAD, and R, and compare with the results of Problem 1.

   Ans: $\bar{x} = -2.4$ $s^2 = 254.8$ MSD $= 203.84$ MAD $=$ 11.52 and $R = 42$ Note that multiplying each data value by a constant factor, such as 2, affects the mean, MAD and range by the same factor, but it multiplies the variance and MSD by the square of the factor.

4. Find the sample mean and sample standard deviation of the measurements: .50, .42, .34, .01, $-.31$, $-.41$, $-.61$.

   Ans: $\bar{x} = -.01$ and $s = 0.443$

5. Times to failure of 50 electronic tubes (in hours) are:

166	28	252	76	300	84	37	4	84	221
34	9	15	143	16	36	151	30	77	97
5	241	394	230	17	17	127	4	54	7
299	18	42	102	20	322	22	117	82	99
117	62	120	8	131	48	84	105	20	33

Compute all of the measures, including MAD, for these data.

Ans: $\bar{x} = 96.14$ $s = 95.40$ $s^2 = 9101.06$ MSD = 8919.04 RMSD = 94.44 $R = 390$ MAD = 72.448.

6. The data below are heights, in inches, of 10 randomly selected students at PG High. Compute $\bar{x}$ and s for these students.

53 56 61 60 56 58 55 63 58 59

Ans: $\bar{x} = 57.9$ inches $s = 2.998$ inches

7. a. Find the mean $\bar{x}_1$ of the data 4, 5, 7, 2.

Ans: 4.50

b. Find the mean $\bar{x}_2$ of the data 1, −2, 8, 4, 2, 3, 0, 5, 4.

Ans: 2.78

c. Find the average of $\bar{x}_1$ and $\bar{x}_2$.

Ans: 3.64

d. A different way of combining averages is to "weight" each one by the number of observations involved. Such a measure is called a *weighted* average of $\bar{x}_1$ and $\bar{x}_2$. It will be denoted $\bar{x}_c$ and is defined by

$$\bar{x}_c = \frac{n_1\bar{x}_1 + n_2\bar{x}_2}{n_1 + n_2} = \left(\frac{n_1}{n_1 + n_2}\right)\bar{x}_1 + \left(\frac{n_2}{n_1 + n_2}\right)\bar{x}_2$$

Find the mean $\bar{x}_c$ of combined data in parts a and b.

Ans: 3.31 (This is the same answer you would get if you averaged all 13 observations.)

8. Similar to 7d, a combined variance can be obtained by weighting and is called the *pooled* variance, s_p^2. It is defined by

$$s_p^2 = \frac{(n_1 - 1)\,s_1^2 + (n_2 - 1)\,s_2^2}{n_1 + n_2 - 2}$$

Work (a) through (d) for Problem 7, except in terms of the sample variances, s_1^2, s_2^2, and s_p^2.

Ans: $s_1^2 = 4.33$ $s_2^2 = 8.69$ $s_p^2 = 7.50$

9. In a study of sex differences in golf ability, the scores of 10 brother-sister pairs were observed, with the following results:

Brother:	92	84	93	91	93	90	86	89	91	88
Sister:	88	85	82	90	81	93	87	92	86	85

Compute the average and standard deviation of the differences between the brother and sister pairs.

Ans: $\bar{x} = 2.8$ and $s = 5.35$

10. The data below represent the weights of 10 packages of a drug, correct to the nearest thousandth of a gram. If the weights are too variable, the machine being used to package the drug will be replaced.

What is the standard deviation of the sampled weights? Note: You should report your answers correct to the nearest thousandth of a gram. Your TI–59 does its calculations with 13 digits; 10 digits are displayed. Thus, you will have sufficient accuracy for this problem. (Save the data base.)

9.981 10.001 9.995 9.991 10.010 9.880 9.999, 10.000 10.003 10.008

Ans: $s = .038$

11. Compute RMSD for the data of problem 10.

Ans: RMSD = .036

12. Compute the range of the data of problem 10.

Ans: $R = .130$

13. Compute MAD for the data of problem 10.

Ans: MAD = .023

14. The following scores were achieved on a statistics test by 28 students.

75	97	89	64	73	82
68	62	80	92	72	79
75	71	91	91	77	80
96	83	69	84	71	
81	86	80	65	88	

Calculate for these data all of the summary measures of this section.

Ans: $\bar{x} = 79.3$ $s = 9.69$ $s^2 = 93.86$ MSD = 90.50 RMSD = 9.51 $R = 35$ MAD = 7.87 $x_{min} = 62$ $x_{max} = 97$

Section 3.2: Grouping Data

When data are presented as a whole, particularly large data sets, it is difficult to see patterns of distribution in the numbers. Summary statistics, like $\bar{x}$ and s, certainly help, but even they can leave a lot to be desired, for such summaries tend to oversimplify the results. One alternative is to group the data into convenient classes, in order to see better how the values are distributed. Techniques for doing this generally fall into the category of an art rather than a science, so you will find that rules are not hard and fast, but vary from one writer to another.

Take the data for Problem 14, the scores achieved by 28 students.

75	97	89	64	73	82
68	62	80	92	72	79
75	71	91	91	77	80
96	83	69	84	71	
81	86	80	65	88	

Is it reasonable that such scores might have come from a normal distribution? It is hard to tell just looking at the conglomerate of data. To group these data, you first must decide how many total classes there should be. Right now, there are essentially 28, and they are not telling us much. Reporting $\bar{x} = 79.3$ reduces everything to a single class, and that is too much the other way. The rule of thumb is to strike for about 5 to 15 classes altogether, with an average of 10. From the solution to Problem 14, you know that scores range from 62 to 97 for $R = 35$. So if you want about 10 classes, you should let each class be $^{35}/_{10} = 3.5$ score units wide, or 4 in rounded units.

A second rule-of-thumb is to decide on precisely 1 class for each score to belong in. There should be no possibility that it can belong to 2 classes. That means suitably defining the boundaries of each class. One convention, when it is reasonable to assume that the underlying scale is continuous, as it is here, is to imagine that each score represents a continuous value, extending from half-a-unit below to half-a-unit above the measurement. Thus, a score of 77 really stands for an interval of values from 76.5 to 77.5. If the observation were 6.39 inches, we imagine that as representing 6.385 inches to 6.395 inches. With this convention in mind, we may begin the first class boundary at any convenient point that will at least fall below the minimum score of 62. Thus, 61.5 would do, as would 60.5. If we use 61.5, with a width of $w = 4$, then the first class interval will extend from 61.5 to 65.5, with a midpoint of $61.5 + 2 = 63.5$.

Another convention is to start at any score to be included in the left boundary and to extend up to, but not including, the upper boundary point. Thus, we could begin the first class at 61 and let it extend up to, but not including, 65. The second class would then run from 65 to 69, and it is understood that 65 is included, but not 69. Now the midpoint will be a whole number (only because $w = 4$ is even), but that need not be a prime consideration. The two different arrangements of class intervals under these two conventions would then appear as follows:

CONVENTION I		CONVENTION II	
LIMITS	MIDPOINT	LIMITS	MIDPOINT
61.5–65.5	63.5	61–65	63
65.5–69.5	67.5	65–69	67
69.5–73.5	71.5	69–73	71
73.5–77.5	75.5	73–77	75
77.5–81.5	79.5	77–81	79
81.5–85.5	83.5	81–85	83
85.5–89.5	87.5	85–89	87
89.5–93.5	91.5	89–93	91
93.5–97.5	95.5	93–97	95
		97–101	99

Either way, we obtain cells with the property that no score will ever be considered in 2 cells at the same time. And, the midpoints may then be thought of as representatives for all of the scores that belong to that cell. Grouping data, then, deliberately destroys the individual characteristics of a measurement, so as to see better the pattern of distribution. Whichever convention you follow, the next task is actually to tally the scores according to the classes to which they belong, obtaining a frequency count for each class. This may be done by hand for small data sets, adding a "1" mark, for individual tallies, with ~~1111~~ representing 5 tallies (a convenience). For the 2 conventions, those tallies for the example would be

CONVENTION I

LIMITS	MIDPOINT	TALLIES	FREQUENCY
61.5–65.5	63.5	111	3
65.5–69.5	67.5	11	2
69.5–73.5	71.5	1111	4
73.5–77.5	75.5	111	3
77.5–81.5	79.5	~~1111~~	5
81.5–85.5	83.5	111	3
85.5–89.5	87.5	111	3
89.5–93.5	91.5	111	3
93.5–97.5	95.5	11	2
			$n = 28$

CONVENTION II

LIMITS	MIDPOINT	TALLIES	FREQUENCY
61–65	63	11	2
65–69	67	11	2
69–73	71	1111	4
73–77	75	111	3
77–81	79	~~1111~~	5
81–85	83	1111	4
85–89	87	11	2
89–93	91	1111	4
93–97	95	1	1
97–101	99	1	1
			$n = 28$

The TI–59 statistics module program ST–07 has been devised to automate this tallying procedure for you. Program ST–09 is then used to display (print when attached to a printer) the class boundaries and frequency counts. We illustrate with this same example, using the second convention on limits, to show that the program is designed to handle that convention. That is, all values that fall exactly at an upper limit will be automatically placed in the *next* class, so you want to be sure you choose parameters in such a way as to cover the largest x_{max}, as well as the smallest observation.

If you will consult the user instructions on page 3–13 of ASM, you will see that you must have planned matters so that you know how many cells you want (10 here*), what the width of each cell is ($w = 4$ here), and where counting is to begin, x_{min}, the lowest class limit (61 here). After initializing ST–07 by pressing [E′], you should enter these parameters as per Steps 3–5. Now you are ready for keyboard entry just as though you were using ST–03. Indeed, this program utilizes ST–03 precisely to store the usual summary statistics. At the conclusion of data entry the [$\bar{x}$] key, as well as [Op] 11, will produce the same values as before, namely, $\bar{x} = 79.3$, $s = 9.69$, and MSD $= 90.5$ in R_T. (Clear any flashing display with [CE]—all is well.)

Now if you will turn to program ST–09, discussed on page 4–9 of ASM, you will notice that you have already effected Steps 1 and 2. So if you will select program ST–09 and initialize with [E′], you are ready to duplicate the foregoing frequency table. Successive presses of [B] and [C] yield counts and upper limits for each class. Thus, the first such presses give you 2 and 65, verifying two tallies in the interval 61–65. The second time yields 2 and 69, and so on to verify the entire tallying. You may also verify the values of $\bar{x}$ and s by pressing [A] and [A′] within this program.

You will notice that program ST–09 is called *Histogram Construction*. That is because the output, as you have just seen, is the frequency table, when suitably arranged. These results may now be graphed into a simple diagram called a histogram, as illustrated on page 4–8 of ASM. Starting with the data minimum (61 here) and using a cell width of w (4 here), frequencies are graphed as heights to produce rectangles over each class interval. This is where you get a visual interpretation of your data. For the example, the histogram is displayed in Figure 3.1. Now you can see that assuming normality for these data is not too unreasonable.

Figure 3.1. Histogram

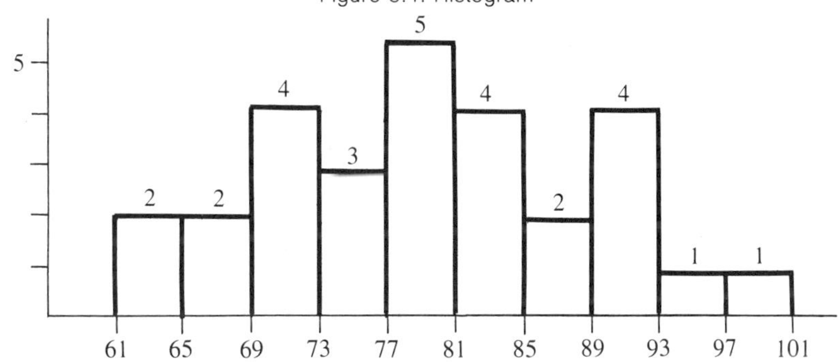

* Please observe that you will never be allowed more than 12 with this program.

If you have followed all of these steps, you will discover that the data are stored for you, starting with R_{31} as in ST–03. (Repartitioning may be necessary.) At this point you may wish to see a different grouping of the data. Or perhaps you had already entered raw data using ST–03, completed various calculations, and now wish to group the data. In either event, you certainly do not want to have to reenter the data (assuming you have not recorded the data on a magnetic card). Fortunately, you can avoid this, but you may have to make one slight modification in ST–07 to accommodate this revision. You will have to initialize again with [E′] (Step 2) in ST–07, because the IDB has to be recompiled with [D′]. This initialization routine clears registers R_{00} through R_{29} for this purpose, stores the pointer address 31 in R_{30} and transfers the last address ($31 + n$) to R_T. If, for any reason, that address is not in R_{30}, the transfer will not take place—essential in order to halt the IDB compilation at the proper time. The solution is simply to enter $31 + n$ manually into the T-register, and then proceed with Step 12 using [D′] for compiling.

In summary, the steps for using Program ST–07 after it has already been used, or if data have been entered with ST–03 and there is a possibility that R_{30} has been changed are as follows:

a. Complete Steps 1–6 in ST–07.
b. As a precaution, enter $31 + n$ and press [$x \leftrightarrows t$], store 31 in R_{30} and press [D′].

If you had recorded your data on a magnetic card using either ST–03 or ST–07, the foregoing steps are harmless, but redundant. In that case, just follow Steps 1–6, followed by Steps 10–13 in ST–07—the usual procedure for magnetic card entry.

Try this out on the same data to reproduce the frequency distribution shown previously for convention I. First press [Pgm]07 [E′] again to initialize. Now observe $59 = 31 + 28$ is in R_T, so the stopping point is where it belongs. Place 59 back in R_T by pressing [$x \leftrightarrows t$]. Enter 9 and press [B], then 61.5 [B′], followed by 4 [C]. This sets the parameters for convention I. Simply press [D′] to recompile and group the data already stored. At the conclusion, you should see $n = 28$ in R_D to signal that all the data have been reprocessed. Proceed to ST–09, press [E′], and, with [B] and [C], reproduce the frequency table.

When data have been grouped, for whatever purpose, the original data may be lost entirely, or may have not been presented to you in the first place. In any event, grouped data have the feature that each midpoint must stand for all observations in that class. To compute summary statistics, like $\bar{x}$ for such data, would require many unnecessary repititions. Fortunately, program ST–03 has been modified to accommodate grouped data, and the user instructions are discussed on page 3–6 of ASM. You see that you initialize this time with [E′], instead of [E].

Standard partitioning allows for 14 cells, and on page 3–4, you can see what you will need to do at Step 3 to accommodate a larger table (never in this book). Now, enter in order, each pair f_i, x_i (midpoint), with successive presses of [B] and [A]. If you will turn to page 3–2, you will see the layout of data storage, with the pointer address in R_{31} and data beginning in R_{32}.

Incidentally, both the ungrouped data and grouped data versions of ST–03 allow for deleting unwanted data. But the procedure is complicated, because the IDB has to be recompiled as well. This can take a lot of time and is subject to some unexpected error if not followed exactly. As a general rule, it is much more desirable to exercise care in entering the data at the outset. A little extra time and caution at the data entry point can save a lot of time and frustration later. Try to develop the habit of entering data slowly and deliberately, with as little distraction as possible.

Practice this with the grouped data under Convention II. First, press [Pgm]03 [E′], then start with $f_1 = 2$, [B], followed by 63 [A]. Successively enter the other 9 pairs this way, concluding with a running count of 10 in R_D. Now when you press [$\bar{x}$] and follow it with [$x \rightleftarrows t$], you will find a grouped average value of $\bar{x} = 79.6$, expectedly different from the ungrouped value of 79.3. Similarly, [INV] [$\bar{x}$] [$x \rightleftarrows t$] produces grouped $s = 9.66$, and [Op]11 [$x \rightleftarrows t$] yields grouped MSD = 89.96. Also, your magnetic card program ZS–3 has been arranged to produce the other grouped characteristics. If you will press [RST] to exit ST–03, then [B′], you should see the value of grouped MAD = 7.80 in the display. With $w = 4$, enter this, and press [C′] to see a range of 40, indicative of spanning all score possibilities from 61 through 101, the limits of the frequency table.

When frequencies are quite high, storing Πx^f in R_{11} may cause a flashing display, due to the limitations of the [y^x] key being used.* Such would have been the case, for instance, if the last frequency corresponding to $x_{10} = 99$ were 51 instead of simply 1. In a case such as this, the quantity cannot be calculated by the TI–59. Fortunately, this overflow does not affect any other part of the IDB compilation, so whenever you encounter this condition—a flashing count—just press [CE] to stop the flashing and continue.

Pay particular attention to Footnote 8 for ST–03 on page 3–6 of ASM. The calculator will not accept a 0 frequency. It will flash an overflow condition whenever an attempt is made to enter 0 with *B*. Of course a 0 frequency corresponds to not having any observations at all in that particular interval or cell. Hence, it is redundant to enter the midpoint of that cell as a data value, and such cells are simply best ignored. *By contrast, a 0 midpoint can and should be entered using* [*A*], if that situation occurs.

* $x \ln y$ must exceed -228.

15. Group the data of Problem 5 into 10 cells, starting at 0 with $w =$ 40, using Convention II for class limits, and graph the histogram. Would you say the data are normal? (*Note*: You will have to repartition the calculator by 9[Op]17 to accommodate the data.) The graph is asymmetric, indicating nonnormal data.

Ans:

Classes:	0–40	40–80	80–120	120–160	160–200	200–240	240–280	280–320	320–360	360–400
Frequencies:	20	6	10	5	1	2	2	2	1	1

16. Compute the grouped values of the various statistics called for in Problem 5.

Ans: $\bar{x} = 98.4$ $s = 95.6$ $s^2 = 9140.24$ MSD = 8957.44 RMSD = 94.64 $R = 400$ MAD = 71.936.

17. Group the following data, representing signal-to-noise ratios measured for a sample of telephone equipment. Use 5 class intervals with midpoints 115, 145, 175, 205, 235.

107 136 217 118 153 200 166 187 121 186 199 169 202 111 114 174 151 196 142 129 174 225 246 110

Ans:

Frequency	7	4	6	5	2
Class interval	100–130	130–160	160–190	190–220	220–250

18. For the ungrouped data of problem 17, calculate $\bar{x}$, s, MSD, MAD, and R.

Ans: $\bar{x} = 163.875$ $s = 40.49$ MSD = 1571.11 MAD = 34.22 R = 139

19. For the grouped data of problem 17, calculate $\bar{x}$, s, MSD, MAD, and R.

Ans: $\bar{x} = 163.75$ $s = 40.36$ MSD = 1560.94 MAD = 34.69 $R = 150$

20. The following data are a random sample of fees, in dollars, charged by doctors for a certain treatment.

120 86 87 75 100 120 100 80 110 105 95 90 100 85 95 85

Compute the mean and standard deviation, using ST–03 for data entry.

Ans: $\bar{x} = 95.8$ $s = 13.25$

21. For the data of Problem 20, compute MAD, R, and RMSD.

Ans: MAD = 10.5 $R = 45$ RMSD = 12.83

22. Group the data of Problem 20, using class intervals 70–80, 80–90, and so on.

Ans:

Frequency	1	5	3	4	1	2
Interval	70–80	80–90	90–100	100–110	110–120	120–130

23. Find the grouped data range for Problem 22.

Ans: 60

24. Find the grouped data mean and standard deviation for Problem 22.

Ans: $\bar{x} = 98.1 \qquad s = 14.93$

25. Find the grouped data MAD and RMSD for Problem 22.

Ans: MAD = 12.27 RMSD = 14.46

Section 3.3: Transformations

Data are sometimes "transformed" to simplify the values, to change the units, or to affect the variance in some way. In this section we will not discuss in detail the motivations for data transformation. Rather, we will concentrate on how you can use the TI–59 to accomplish such transformations. Technically, a "transformation of the data" is a mapping of the data values to new values, using some function $f(x)$. This function is often referred to as the transformation, and we say x is transformed into $x' = f(x)$.

A particularly simple transformation, and one encountered in many applications, is a linear transformation $x' = mx + b$, where m and b are constants. For example, the standardization of a normal random variable, X to $Z = (X - \mu)/\sigma$, may be written

$$Z = \frac{1}{\sigma} X + \left(- \frac{\mu}{\sigma} \right)$$

and so is a linear transformation. So is the inverse $X = \sigma Z + \mu$, expressing X in terms of Z. Other commonly used ones are quadratic (involving x^2) and logarithmic. Since we have so many function keys available to us on the TI–59, there is a great deal of latitude in the transformations we can use.

Although program ST–11 of the statistics module is arranged for such transformations, it is limited to certain logarithmic cases. So, we have replaced it with Step 5 of the ZS–3 program. It is not quite as straightforward as the other instructions you have been following, because it is open for you, the user, to enter any function of interest. This requires you to write an algorithm for $f(x)$, keeping in mind that each x will be placed in R_D so $f(x)$ can be computed. This can be easy or difficult, depending on the formula for $f(x)$.

Here is a simple case. Suppose we wanted to transform the data by a logarithmic (base e) transformation, $x' = ln\ x$. Since the TI–59 is hard-wired with a [*ln x*] key, we can use a simple 1-step instruction. If you consult the User Instructions for ZS–3, you will see that after initialization and possible repartitioning and exit from ST–03, you must press [GTO] [A′], followed by [LRN], to prepare the program for $f(x)$. Now just press [*ln x*] and [INV] [SBR], then [LRN] to go back out of the learn mode, and the calculator is prepared to make the transformation.

EXAMPLE 3.3

Transform the data below with the natural logarithmic function, then compute the sample mean, MSD, and standard deviation. Compare these with those for the original data.

75.9 23.7 42.0 115.1 66.8

Solution Load program ZS–3 and initialize as per Steps 5a–5c. Then enter the transformation as described above, using Steps 5d, 5e. At Step 5f, enter each value for x and press [A]. The calculator then transforms x to $x' = \ln x$ (that is what [A′] does from the keyboard), begins storing in R_{31} and compiling the IDB, as with ST–03. For the example, the transformed values are

4.33 3.17 3.74 4.75 4.20

You can find these by just pressing [D] (after [RST]) at the conclusion of data entry to see the successive values. With the data thus stored, [$\bar{x}$] [$x \rightleftharpoons t$] displays the mean of the transformed sample, $\bar{x}' = 4.04$. Similarly, [INV] [$\bar{x}$] [$x \rightleftharpoons t$] gives $s' = .605$, and [Op] 11 [$x \rightleftharpoons t$] gives MSD′ = .293. For the original data, $\bar{x} = 64.7$, $s = 34.88$, and MSD = 973.26. Notice that s' and MSD are much smaller than s and MSD for the original data. Logarithmic transformations are sometimes used for such effects on variability. Notice also that $\bar{x}' \neq \ln \bar{x}$.

EXAMPLE 3.4

Transform the data of Example 3.3 by the transformations

$$x' = x - 600 \qquad \text{and} \qquad y = \frac{(x - 69)}{3}$$

Solution The first transformation requires that we program the formula $x' = x - 600$. Remembering that x will be in R_D, we only need the simple steps [−] 600 [=], followed by [INV] [SBR], to complete the algorithm. Be sure always to initialize with Steps 5a–5c first. If you will enter these steps at Step 5d (writing over $\ln x$), (if your calculator is still on from the preceding case) and follow through, you should be able to verify that the data in Example 3.3 are all reduced by 600. Incidentally, [−], which clears all pending operations, can be avoided in the program, with a safer program utilizing parentheses and the fact that CE acts as a dummy variable in a program (see PPM, page V–15). The sequence of steps would be

[(] [CE] [−] 600 [)] [INV] [SBR]

Here, CE has the effect of bringing x from R_D into the parentheses for operation with minus. To transform the data by the linear transformation $y = (x - 69)/3$, a sequence of programming steps that would do the job would be

[(] [(] [CE] [−] 69 [)] [÷] 3 [)] [INV] [SBR]

Try this out on the data. For a linear function, the mean of the transformed data, $\bar{y}$, is the same as the linear function of the mean of the original data, that is, $\bar{y} = f(\bar{x})$. This is not true in general for other than linear functions, such as *ln x*.

If data are already stored in R_{31}, R_{32}, and so on, either through prior entry with ST–03 (ungrouped) or a magnetic card entry, you do not have to recall x item by item to transform the data. Step 5f has been arranged for automatically transforming x_i to $x_i' = f(x_i)$, and then storing the new data in R_{31}, R_{32},... . Then, statistics such as $\bar{x}'$, s', RMSD′ may be computed for the new data. If you are worried about losing the original raw data with this process, you may easily recover them by using the inverse of $f(x)$ at Step 5d. For example, if $f(x) = \ln x$, just program [INV] [*ln x*] *in A′*. If $x' = (x - 100)/50$, program the formula $50 \cdot x + 100$ in A'.

Problems

26. Daily consumption of coffee, in quarts, at a certain restaurant is as follows:

12 11 19 16 22 20 14 17 14 15 21 21 16 9 15

Using ST–03, enter the data and record the values of $\bar{x}$, s, RMSD, MAD, and R for these data.

Ans: $\bar{x} = 16.13$ $s = 3.89$ RMSD = 3.76 MAD = 3.09 $R = 13$

27. Transform the data of Problem 26, using label E of ZS–3, to new values $x' = (x - \bar{x})/s$. Compare the statistics for the transformed values. (Use rounded values given in Ans.)

Ans: $\bar{x}' = 0.00$ $s' = 1.00$ RMSD′ = .97 MAD′ = .80 $R' = 3.34$

28. Transform the data of Problem 27 back to the original data using the transformation $x' = x\,s + \bar{x}$. Use Step 6 of ZS–3 to verify that the retransformation was successful.

29. Perform the transformation $x' = -3\,x + 2$ on the data of Problem 26. Verify that $\bar{x}' = -3\bar{x} + 2$ and $(s^2)' = 9s^2$ or $s' = 3s$, confirming that in general, if $x' = mx + b$, then $\bar{x}' = m\bar{x} + b$, $s' = |m|s$; MSD$' = m^2$MSD; MAD$' = |m|$MAD.

Ans: $\bar{x}' = -46.4$ $s' = 11.67$ MSD$' = 127.04$ MAD$' = 9.28$ (There will be small differences due to rounding)

30. Transform the data of Problem 29 back to the original form and perform a square root transformation. How does MAD$'$ compare with MAD?

Ans: MAD$'$ = .387, much smaller than MAD

31. The monthly production of coal at a certain mine was, in tons, 137240, 137511, 137208, 137020, 137481. Enter these data with ST–03 and calculate $\bar{x}$, s, MSD, and MAD.

Ans: $\bar{x} = 137{,}292$ $s = 204.59$ MSD = 33485.2 MAD = 163.2

32. In pre-calculator days, data like that of Problem 31 were "coded" for ease of computation by linear transformations. In that problem, 137,000 would be subtracted from each observation. Do this, and use the general results stated in Problem 29 for calculating the answers to Problem 31. Hint: $x' = x - 137{,}000$, so that $x = x' + 137{,}000$.

Ans: $\bar{x}' = 292$, so $\bar{x} = 137{,}000 + 292 = 137{,}292$; $s' = 204.59 = s$; MSD$'$ = 33485.2 = MSD and MAD$'$ = 163.2 = MAD

33. It is also common practice to code grouped data in order to simplify numerical computations. The grouped data of Problem 17 are duplicated below.

Midpoints:	115	145	175	205	235
Frequencies:	7	4	6	5	2

Code the data by the transformation

$$x' = \frac{x - 175}{30}$$

thereby centering the new data at 0 and taking into account the class interval width. Calculate $\bar{x}'$ and s'. Then use the fact that $x = 30\,x' + 175$ and Problem 29 to rediscover $\bar{x} = 163.75$, $s = 40.36$, and MSD = 1,560.94.

Ans: Use ST–03 (grouped) with new midpoints and frequencies

Midpoints:	−2	−1	0	1	2
Frequencies:	7	4	6	5	2

to verify $\bar{x}' = -0.375$, $s' = 1.345$, MSD$' = 1.734$.

Then $\bar{x} = 30\,\bar{x}' + 175 = 163.75$, $s = 30\,s' = 40.36$, MSD $= 900$ MSD$' = 1560.94$.

34. Whenever the data represent scores achieved by testing, the transformation $x' = (x - \bar{x})/s$ (see Problem 27) is often called a *z*-score, reminiscent of (but not the same as) standardizing a variable. Transform the scores of Problem 14 to *z*-scores using $\bar{x} = 79.3$ and $s = 9.69$ and calculate $\bar{x}'$ and s'. Keep the transformed data in storage for Problem 35.

Ans: $\bar{x}' = .0022$; $s' = .9998$ (the answers should always be close to 0 and 1, respectively)

35. In order to place scores on several tests on a common, easily managed scale, *z*-scores are often converted to what are called *T*-scores by the transformation $T = 10z + 50$. This puts all scores on a scale with a mean of about 50 and a standard deviation of about 10, making it easy to visualize scores on the typical scale of 0 to 100. Convert the *z*-scores of Problem 34 to *T*-scores and calculate the new values of $\bar{x}$ and s.

Ans: $\bar{x}' = 50.022$, $s' = 9.9979$ (the answers should always be close to 50 and 10, respectively).

Section 3.4: The Central Limit Theorem

We have seen that many of the statistics we have been using entail summing quantities involving the sample data values. For example, $\bar{x}$ involves summing terms of the form x_i/n, and s^2 involves summing terms of the form $(x_i - \bar{x})^2/(n - 1)$. It is useful to contemplate the values we might obtain for such statistics in advance of actually obtaining the sample and doing the computation.

While a study of the relationship between probability distributions of various statistics and those of the underlying population is generally rather complicated and theoretical, there is a simple result, called the *Central Limit Theorem,* which allows us to approximate the distributions of averages. The Central Limit Theorem asserts that, if random variables $X_1, X_2, \ldots, X_n$ constitute a random sample from some population having mean μ and variance σ^2 (regardless of the distribution of that population), the quantity

$$\bar{X} = \sum_{i=1}^{n} \frac{X_i}{n}$$

is approximately normal, with mean μ and variance σ^2/n, that is, standard deviation $\sigma/\sqrt{n}$. What is meant by "approximately normal" is that, when n is large enough, you can use the normal distribution (program ST–19 in your statistics module, for example) to calculate the probabilities of events involving $\bar{X}$. For practical purposes, the Central Limit Theorem gives useful approximations when n is larger than 40 (or even 30, for that matter).

EXAMPLE 3.5

A random sample of size 50 is to be taken from an exponential population having mean $\mu = 10$. Calculate $\Pr(\bar{X} > 8)$.

Solution The actual probability distribution of $\bar{X}$ for this case is complicated. But we can appeal here to the central limit theorem, since $n = 50$ is so large. Recalling that $\mu = \sigma = 10$ for this population, $\bar{X}$ has mean 10 and standard deviation $\sigma/\sqrt{n} = 10/\sqrt{50} = 1.414$. Standardizing the value of $x = 8$ to $(8 - 10)/1.414 = -1.414$, $\Pr(\bar{X} > 8) \doteq Q(-1.414) = .92$, using program ST–19.

With discrete populations, it is sometimes helpful to apply what is called a *continuity correction.* Reminiscent of how discrete values were treated against a continuous underlying scale when grouping data, this correction imagines discrete numbers as representing an interval of values extending from half-a-unit below to half-a-unit above the measured value. A common example of this application is the use of the normal distribution to approximate the binomial distribution. The latter arises in connection with sums ΣX_i of independent Bernoulli trials, where $E(X_i) = p$ and $V(X_i) = pq$ for each i. Of course $\bar{X}$ is just this sum divided by n, and so is related to the binomial distribution.

Since the Bernoulli population has mean p and standard deviation $\sqrt{p(1 - p)}$, that normal distribution which best approximates the binomial distribution of ΣX_i is the one having mean np and standard deviation $\sqrt{np(1 - p)}$. The actual distribution of $\bar{X}$ is discrete, with positive masses at

$$0, \frac{1}{n}, \frac{2}{n}, \ldots, \frac{n}{n}$$

These masses are approximated by *areas* under the normal curve. These areas may be interpreted as areas of rectangles with heights given by the normal density and bases given by a $\pm.5/n$ tolerance on either side of

$$0, \frac{1}{n}, \frac{2}{n}, \ldots$$

This situation is depicted in Figure 3.2, where we have taken $n = 10$ and multiplied $\bar{X}$ by n so that

$$n\bar{X} = \sum_{i=1}^{n} X_i$$

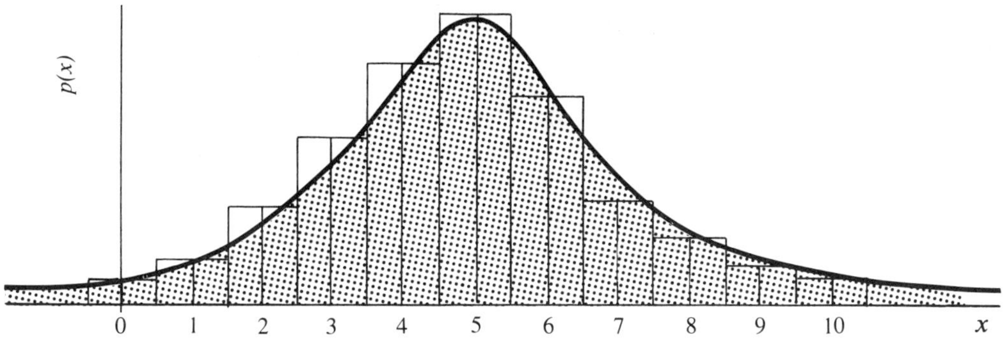

Figure 3.2. Normal approximation to a binomial (10,p) distribution, using a normal [$10p$, $10p(1-p)$] distribution.

is a binomial random variable. The actual binomial probability mass values are shown in Figure 3.2 as vertical bars at 0, 1, 2, . . . , 10, which are the same as the areas of rectangles with base 1 and height given by the mass function. The approximating areas are shown as crosshatched regions under the normal density having $\mu = 10p$ and $\sigma^2 = 10p(1 - p)$. The approximation involves taking the crosshatched area above a given interval, such as $4 \pm .5$, to represent the rectangle area above that interval.

The continuity correction involves the use of the half integer (.5 in our example) to form the rectangles and regions we are considering. If it is desired to approximate the probability of some event involving any discrete random variable W with positive probability masses at the integer points 0, 1, 2,..., use the normal distribution indicated by the Central Limit Theorem, and add (or subtract) half-interval widths (.5), so that the area under the normal curve represents the rectangle areas defined by the event in question. Thus, if $E(W) = \mu$ and $V(W) = \sigma^2$, and Y is a normal random variable, with mean μ and variance σ^2, then for $k = 0$, 1, 2,...,

$$P(k) \text{ is approximated by } \Pr(Y \leq k + .5)$$
$$Q(k) = \Pr(W > k) \text{ is approximated by } \Pr(Y > k + .5)$$
$$\Pr(k_1 \leq W \leq k_2) \text{ is approximated by } \Pr(k_1 - .5 \leq W \leq k_2 + .5)$$

and similarly for other events involving W. The approximate probabilities can be found using ST–19. Alternatively, you may use program ZS–2, if you will remember to input $\sigma/\sqrt{n}$ for the standard deviation.

EXAMPLE 3.6

In a reliability testing situation, 35 electronic components are to be selected and tested. What is the probability of obtaining at least 30 successes, if the probability, p, for success is .95? Compare the Central Limit Theorem approximation with the more exact answer.

Solution Using ST–20 with $n = 35$ and $p = .95$, $Q(29) = .9927$ is an exact (to 4 decimal places) answer. To apply the Central Limit Theorem, $\bar{X}$ is approximately normal with mean $p = .95$ and variance $\sigma^2 = pq/n = .00136$ or $\sigma = 0.04$. Obtaining at least 30 successes is equivalent to an average (continuous) value of at least $30/35 = .86$. Standardize this to $z = (.86 - .95)/.04 = -2.25$ and use ST–19 to obtain $Q(-2.25) = 0.9878$. The continuity correction requires a total of 30 to be viewed as extending as low as 29.5 for an average value of $29.5/35 = .84$, $z = (.84 - .95)/.04 = -2.75$ and $Q(-2.75) = .9970$ is in very close agreement with the more exact answer.

Problems

36. Suppose X has a binomial distribution with parameters $n = 40$ and p as shown in each part below. Use the normal approximation, with continuity correction, to calculate $\Pr(16 < X \le 30)$.

a. p = .5
b. p = .8
c. p = .9

Ans: With the continuity correction, evaluate $\Pr(16.5 \le X \le 30.5) = \Pr(.4125 \le \bar{X} \le .7625)$.

a. .865
b. .277
c. .0019

37. Work Problem 36 without the continuity correction.
Ans: Evaluate $\Pr(17 \le X \le 30)$.

a. .828
b. .215
c. .0008

38. Rework Problem 36, using the binomial distribution.
Ans: Use ST–20 and compute $P(30) - P(16)$.

a. .866
b. .268
c. .005

Note: The continuity correction in Problem 36 gives closer approximations than without in Problem 37. Note also that the approximations are best for p around .5.

39. An aircraft holds 80 passengers. Assume the weights of passengers have mean 125 pounds and standard deviation 35 pounds. Find the probability that a fully loaded aircraft has a total passenger weight between 9,500 and 10,500 pounds.

Ans: The total weight T is approximately normal with mean 10,000 and standard deviation 313, so that $\bar{X} = T/80$ is approximately normal with mean 125 and standard deviation $35/\sqrt{80} = 3.91$. Using ST–19 or ZS–2

$$\Pr(9{,}500 \leq T \leq 10{,}500) = \Pr(118.75 \leq \bar{X} \leq 131.25) = .89.$$

40. It is unsafe for the aircraft of Problem 39 to fly with average passenger weight in excess of 130 pounds. What is the probability that a random loading of the aircraft will have to be lightened by removing baggage and fuel?

Ans: $\Pr(\bar{X} > 130) = .10$

41. If, in Problem 40, it is desired to reduce the probability of an overload weight of passengers (10,400 pounds) to at most .01 by reducing the number of seats filled, how many passengers should be accommodated?

Ans: Use trial and error with n (and the corresponding mean $\mu = 125$ and standard deviation $\sigma = 35$) in ZS–2 to evaluate $\Pr(\bar{X} > 10{,}400/n)$ and discover $n = 77$.

42. A chain has links with mean length .4 inches and variance .001 inches.[2] It is required that a chain with 75 links have length between 29.1 and 30.1 inches. What proportion of the chains manufactured with these links will not meet the total *length* requirement?

Ans: $1 - \Pr(29.1/75 \leq \bar{X} \leq 30.1/75) \doteq 1 - .64 = .36$

43. Would the manufacturer of the chains in Problem 42 be better off to make chains with 74 links?

Ans: Yes, since only 6.6% of such chains would then fail to meet the specification.

44. How small would the individual link variance have to be, in order that 98% of the chains described in Problem 43 would meet the total length requirement? Hint: Use trial and error with values of σ.

Ans: $(.025)^2 = .000625$

45. A person claiming to have ESP powers was asked to identify (while blindfolded) in which hand different objects were held by 30 different people in the audience. Assuming he has no such powers, what is the probability of the person identifying the correct hand 20 or more times?

Ans: Using ST–20 with $n = 30$, $p = .5$, $Q(19) = .05$. Normal approximation yields $\Pr(\bar{X} \geq 19.5/30) = .05$.

46. Suppose Y is binomial with $n = 100$ and $p = .4$. For what value of k is $\Pr(Y > k) = .05$, approximately?

Ans: $\Pr(Y > k) = \Pr(\bar{X} \geq (k - .5)/100) = .05$ needs to be solved for k. Using ZS–2 with trial and error (start with $k = 50$), $k = 48$ or 49.

47. In a recent election, a polling service claimed that 63% of the population favored a particular candidate. If this is true, how likely would it be that fewer than 620 respondents out of a sample of 1,000 would favor the candidate?

Ans: .246 = $\Pr(\bar{X} \leq .6195)$

48. In the manufacture of drain pipe, the diameter of the pipe is a critical factor. The drawing machine is set to produce 3-inch pipe on the average, with a standard deviation of .1 inch. How likely is it that the average of the next 1,000 pipes will be between specification limits 2.99 and 3.01?

Ans: .9984

49. A particular battery is supposed to have a mean life of $\mu = 36$ months, with a standard deviation of 6 months. What is the probability that the average life of 100 batteries sold will exceed 37 months?

Ans: $Q(37) = .05$

50. It is claimed that 70% of the income tax returns from a particular district have one or more arithmetic errors. Out of a sample of 100 returns, what are the chances of finding fewer than 20 without errors?

Ans: Binomial (p = 0.3) answer $P(19) = .0089$
Approximation is given by $\Pr(\bar{X} \leq 19.5/100) = .0110$.

ZS–3: User Instructions

STEP	PROCEDURE	ENTER	PRESS	DISPLAY
1	Calculate the range of a sample when raw data have been entered using ST–03.		[C]	R
2	Calculate the range of a sample when data are grouped and have been entered using ST–03 (w = cell width).	w	[C']	R
3	Compute MAD when ungrouped data have been entered using ST–03.		[B]	MAD
4	Compute MAD when grouped data have been entered using ST–03.		[B']	MAD
5	Transform data by the transformation $x' = f(x)$:			
	a. Initialize.		[Pgm]03[E]	1
	b. Repartition, if necessary ($n \leq 9$).	n	[Op] [17]	$xxx.yy$
	c. Exit ST–03.		[RST]	
	d. Enter learn mode.		[GTO] [A'] [LRN]	$171yy$
	e. Enter $f(x)$, using parentheses where necessary, and always end with [INV] [SBR]. Exit learn mode.		⋮ [LRN]	
	f. (1) Keyboard entry (repeat for each i)	x_i	[A]	i
	(2) Original data stored by ST–03 (n = sample size)	n	[E]	n
	Note: Steps 1 and 3 apply following Step 5f.			
6	Recall transformed data.		[RST]	
			[D]	x'_1
	Note: May be repeated at any time.		[D]	x'_2
			[D]	x'_3
			⋮	⋮

Data Processing

1. For the ungrouped data below, calculate $\bar{x}$, s, MSD, MAD, and R. Then transform the data by $x' = 1/x$ and calculate the same statistics for the transformed data.

   5, 10, 6, 4, 3, 8, 12

   Recall the actual values of the first 3 data points.

ZS STEP	ENTER	PRESS	DISPLAY	COMMENTS
		[Pgm]03[E]	1.	Select ST–03 and initialize.
	5	[A]	1.	Enter data.
	10	[A]	2.	
	6	[A]	3.	
	4	[A]	4.	
	3	[A]	5.	
	8	[A]	6.	
	12	[A]	7.	Data Entry complete.
		[RST]	7.	Exit ST–03.
		[$\bar{x}$]	0.	Ignore output.
		[$x \rightleftarrows t$]	6.86	The value of the sample mean.
		[INV][$\bar{x}$]	0.	Ignore output.
		[$x \rightleftarrows t$]	3.29	The value of the sample standard deviation.
		[Op]11	0.	Ignore output.
		[$x \rightleftarrows t$]	9.27	Value of MSD
3		[B]	2.69	MAD calculated and displayed
1		[C]	9.	The range $R=9$
5a		[Pgm]03[E]	1.	Initialize ST–03 for data storage.
5c		[RST]	1.	Exit ST–03.
5d		[GTO][A′]		
		[LRN]	171*xx*	Prepare for transformation.
5e		[$1/x$]	172*xx*	Simple algorithim
5e		[INV][SBR]	173*xx*	Necessary return instruction
5e		[LRN]	1.	Exit learn mode for ZS–3 operation.
5f(2)	7	[E]	7.	Data automatically transferred and stored in R_{31}, R_{32}, ...
		[$\bar{x}$]	0.	Ignore.
		[$x \rightleftarrows t$]	.18	Value of $\bar{x}'$ rounded.
		[INV][$\bar{x}$]	2.52	Ignore output.
		[$x \rightleftarrows t$]	.089	Rounded value of s'
		[Op]11	5.43	Ignore output.
		[$x \rightleftarrows t$]	.007	Rounded value of MSD′
3		[B]	.0697	Rounded value of MAD′
1		[C]	0.25	Value of R', the new range
6		[D]	0.2	Recall value of $x'_1 = 1/x_1$.
		[D]	0.1	Recall value of $x'_2 = 1/x_2$.
		[D]	0.17	Recall value of $x'_3 = 1/x_3$.

2. For the grouped data below, calculate $\bar{x}$, s, MSD, MAD, and the range.

Frequency:	3	4	9	4	5
Class Interval:	0–10	10–20	20–30	30–40	40–50

Solution

		[Pgm] 03 [E′]	1.	Initialize ST–03 for grouped data entry.
	3	[B]	3.	Enter first frequency.
	5	[A]	1.	Enter first midpoint (running count displayed).
	4	[B]	4.	
	15	[A]	2.	Repeat for each pair.
	9	[B]	9.	
	25	[A]	3.	
	4	[B]	4.	
	35	[A]	4.	
	5	[B]	5.	
	45	[A]	5.	Data entry concluded
		[RST]	5.	Exit ST–03.
		[$\bar{x}$]	0.	Ignore output.
		[$x \rightleftarrows t$]	26.6	Grouped mean value $\bar{x}$
		[INV][$\bar{x}$]	0.	Ignore output.
		[$x \rightleftarrows t$]	12.81	Rounded value of s
		[Op]11	0.	Ignore output.
		[$x \rightleftarrows t$]	157.44	Value of MSD.
4		[B′]	10.048	Value of MAD.
2	10	[C′]	50.	Value of grouped range, R, based on a class width of 10

4
Estimation

Section 4.1: Point Estimators

In the last chapter, you saw how data may be processed to produce various estimates of parameters of interest. Such *point* estimates, as they are called, are invaluable in a problem where we have no idea what the parameter value is. You see, when you make an assumption about a population random variable, such as X is normal, or X is exponential, you thereby reduce a large part of the uncertainty in the experiment. Such an assumption specifies the manner in which the values of X are distributed and reflects your belief, however justified, in how you expect those values to behave. But such a specification is incomplete, for all of those distributions have one or more parameters, and these are typically unknown to the experimenter. So not all of the uncertainty is removed, and what is more natural than to perform the experiment involved and use the observations to guess or estimate the values of the unknown parameters?

Experiments are not free, however. At the very least, they involve an experimenter's time and effort, and at the most, they may involve destroying very expensive equipment to observe such phenomena as time to failure or the meeting of contract specifications. Because of such cost considerations and other limitations, the experimenter will never know exactly what the true values of parameters are, for he or she is limited to a small portion, a sample, of such a population. Not only that, but the values of estimates, being themselves values of a random variable, will vary or fluctuate from one set of trials to another. The experimenter thus faces the dilemma that the estimate computed from the sample is rarely, if ever, equal to the parameter being estimated and, worse, if the experiment were repeated, an even different value would be obtained.

Let us examine this last point more closely, for it is an important one. Say you wanted to sample from a normal distribution and, all considerations taken into account, you can only afford a sample of size $n = 5$. You are interested in estimating the mean of this distribution, so naturally you will be computing the value of the random variable $\bar{X}$, the sample mean. Unknown to you, the actual distribution from which you will be sampling has a population mean of $\mu = 50$ and a standard deviation of $\sigma = 10$. Of course, you have no idea what five numbers "nature" will yield to you, but if you agree to draw a random sample, the random number generator will simulate how things will turn out.

To illustrate, use a seed of 110 in program ST–02 to see that the five sample values you could reasonably expect to see (rounded) are 48.5, 43.4, 48.7, 33.6, and 51.3, yielding an average value (using the [$\bar{x}$] key) of $\bar{x} = 45.1$. Your best estimate of μ is of course this average value, 45.1, and, as it turns out, you are not all that close to the true value of $\mu = 50$. You saw this happening before in the Problems for Chapter 2. Moreover, if you were to repeat this experiment (or think of someone else doing the same experiment with a different seed), you would not expect to see the same five values or the same value of $\bar{X}$. This is the random nature of things, which you may observe by continuing with ST–02 to generate a second set of five values independent of the previous results. (Remember, ST–02 produces its own seed each time.) You may verify that the next 5 values are 28.1, 38.0, 54.3, 61.8 and 50.4, producing an average of $\bar{x} = 46.5$. Not surprisingly, this value of $\bar{X}$ differs from the last one, and neither equals 50. A third sequence of trials would yield 55.2, 48.2, 28.1, 54.2, 37.2, with $\bar{x} = 44.6$.

The lesson to be learned here is that $\bar{X}$ is truly a random variable. The population from which you sampled did not change, so the parameter $\mu = 50$ is exactly the same each time, and, of course, the sample size $n = 5$ is fixed and, therefore, cannot account for any fluctuations. Yet the values of $\bar{X}$ do indeed fluctuate. How? As you learned in the last chapter, the values of $\bar{X}$ behave randomly according to a probability distribution having the same mean μ as the population, but with a smaller variance, namely, σ^2/n. In our example, the mean of $\bar{X}$ is 50 and the variance of $\bar{X}$ is $100/5 = 20$. Moreover, theory tells us that, since we are sampling from a normal distribution, the distribution of $\bar{X}$ is itself normal. If n were large, the Central Limit Theorem (CLT) assures us that the distribution is approximately normal, no matter what population we sampled from. Figure 4.1 depicts in a rough way how these two distributions compare. Actually, we can gain even more insight into this behavior using the simulator.

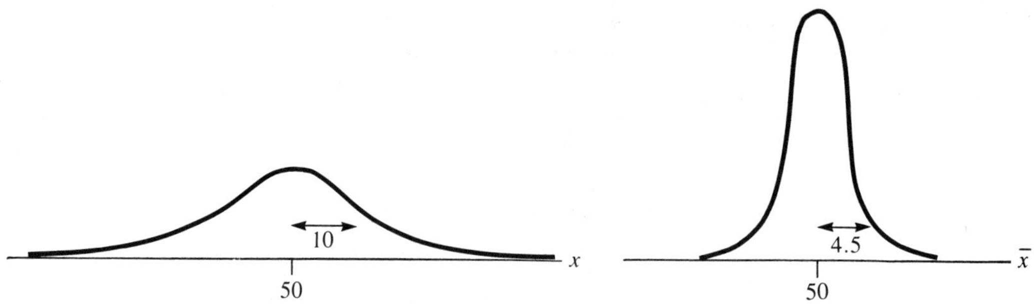

Figure 4.1. Distributions of X and $\overline{X}$ compared

You have seen three independent values of $\overline{X}$. Suppose you were to repeat that process until you had 100 values of $\overline{X}$. Certainly no one would claim that 100 such values constitute the entire distribution of $\bar{x}$-values. But if those values were conveniently grouped and placed into a histogram, surely the result ought to resemble the second graph in Figure 4.1. By selecting an arbitrary seed, you might want to verify* that 100 trials will yield a histogram like the one shown in Figure 4.2. There, a cell width of 3 was selected on the basis of the smallest observed $\bar{x}$-value of 41.7, along with a maximum of 58.3, to yield 7 cells. These 100 values all averaged out to 49.92, with a spread of 3.90 standard deviation units. As expected, these values came close to the theoretical values of $\mu = 50$ and $\sigma = 10/\sqrt{5} = 4.47$ for this case. Naturally, they do not coincide exactly—that would be a surprise—and s did not fare as well as $\bar{x}$. Never lose sight of the fact, however, that in most real-world situations, you do not have the privilege of repeating the experiment 100 times. You only get one chance, typically at some expense, and so the reality is that you only see one of these $\bar{x}$-values—any one of them might be reasonably expected. Moreover, you have no (or very little) idea what μ is; if you did, why would you be sampling in order to esti-

Figure 4.2. 100 $\bar{x}$-values

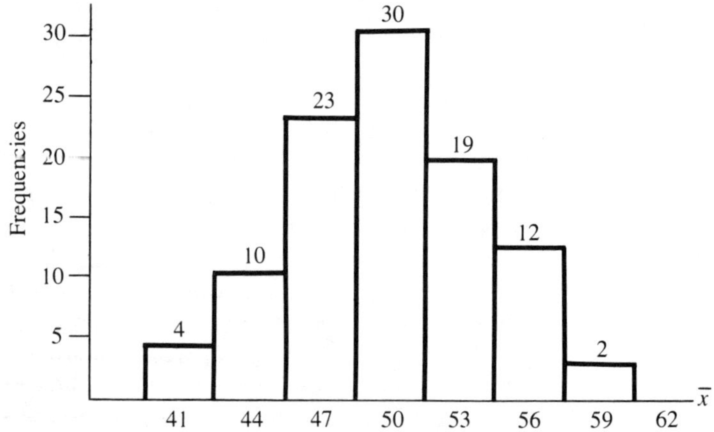

* If you do, keep a record of s^2, along with $\bar{x}$, for a later discussion.

mate μ? So, you do not know how close your one value is. But you have acted in good faith and evaluated a formula which provides the best way (by almost any standard) of estimating μ. Until more information is available, you live with your estimate as a best guess.

In summary then, the estimate of a parameter seldom, if ever, agrees exactly with the parameter being estimated. Indeed, such estimates are themselves observed values of random variables, and, as such, fluctuate according to some probability distribution. It is up to the theorists to determine what exactly (or approximately) those probability distributions are for various estimators, but simulation, along with some data processing tools, may be used to advantage in order to gain some insight into how those distributions behave. In what follows in this book, we will use a somewhat standard notation whenever we wish to discuss parameters in general. We will reserve the Greek letter θ to denote an arbitrary parameter. Thus, in one setting, θ might stand for the mean, μ, of a population like the normal or perhaps the Bernoulli parameter, p. In another setting, θ could stand for a variance or some other unknown distribution characteristic like a percentile. An *estimate* of θ is often denoted $\hat{\theta}$ by placing a carat or "hat" above θ. (We even read this symbol as "θ-hat.") It would be quite appropriate to write $\hat{\theta}$ as a function $\hat{\theta}\,(x_1, x_2, \ldots, x_n)$ of the data, for such is how we compute estimates. The corresponding *estimator* would be denoted $\hat{\Theta}$, utilizing capital letters in the usual way. As a random variable, $\hat{\Theta}$ has a theoretical probability distribution of its own, typically one that involves the unknown parameter θ itself. It may or may not be easy to determine what that distribution is, but we can usually simulate it using a large number of replications and examine a histogram of the result.

To illustrate the last point, we kept track of s^2, along with $\bar{x}$, in generating the 100 samples of size 5 from the normal distribution, with mean $\mu = 50$ and $\sigma = 10$. You were told in Chapter 3 that the random variable S^2 is *unbiased.* By this we mean that $E(S^2) = \sigma^2$ (100 in this example). And that explains why

$$\Sigma\,(X_i - \bar{X})^2$$

is divided by $n - 1$, instead of n. But that does not buy you very much, of and by itself. To see that point, we grouped the 100 s^2-values we obtained into a histogram using 10 cells of width 40, and the results are depicted in Figure 4.3. This was based on a range of values running from 8.64 to 376. That range in itself is striking by comparison with the $\bar{x}$-case, when you stop and consider again that you typically only see 1 value at a given cost to you. The fact that the single value you will obtain could range over such a wide interval should make you uneasy, for it is that 1 value you will be using to guess what σ^2 is.

Another striking difference in Figure 4.3 is the lack of symmetry, and you should notice that, while S^2 is unbiased, its mean is $\sigma^2 = 100$ only because of the fact that such large values in the 200's and 300's

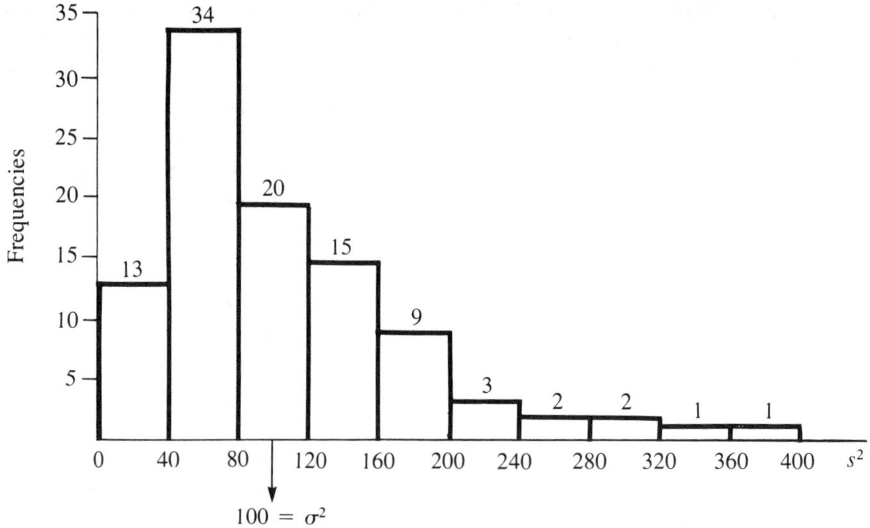

Figure 4.3. 100 s^2-values

may appear.* Actually, there is a bigger chance of getting a value less than 100 than there is of getting 1 larger, so if you were to bet on what you are going to see with a single sample, you should stay on the low side of 100. Noticing that most values were between 40 and 80, it is hard to take a lot of comfort in unbiasedness. In fact, these 100 s^2-values averaged out to 104.26, so on the average (if you could repeat the experiment so many times), you would have to come out on the high side. And if the 1 value you obtained was $s^2 = 237.45$, you might wonder why that should be more valuable to you than MSD, which would have been $99/100\ s^2 = 235.08$. Certainly the difference here of 2.38 square units is hardly worth making a fuss over. Both estimate the true value rather badly by about the same amount.

The main difference, then, between the results of Figures 4.2 and 4.3 is that, while both estimators are unbiased, S^2 has more variability than does $\overline{X}$, so that unbiasedness by itself is not a strong requirement for an estimator. In general, we say that $\hat{\Theta}$ is an unbiased estimator for θ, if $E(\hat{\Theta}) = \theta$, that is, if the mean of the distribution of $\hat{\theta}$ values is exactly θ. Since the variance of $\hat{\Theta}$ also plays a role in how "close" a $\hat{\theta}$ value is to θ, you will sometimes find *mean squared error* (MSE) used as a measure of usefulness of an estimator. This measure simply combines biasedness and variance and is given by

$$\text{MSE}(\hat{\Theta}) = V(\hat{\Theta}) + [E(\hat{\Theta}) - \theta]^2$$

As you can see, if $\hat{\Theta}$ is unbiased, its MSE is simply its variance. It turns out, for example, that MSD then has a smaller MSE than does S^2, so that if MSE is used as a criterion for selecting an estimator, then MSD

* Incidentally, while S^2 is unbiased, the sample standard deviation $S = \sqrt{S^2}$ is *not*.

should be used in favor of S^2. You will find this criterion singled out by some writers on the subject. You will even find other criteria used to select point estimates in various situations, but these matters are of little concern to us here. In most cases that arise in introductory courses, the formulas yielding point estimates are already hard-wired in the TI–59 or else are simple keyboard functions of those available. With that much said, we turn our attention in the next section to the more interesting problem of estimating a parameter by an interval of values.

Section 4.2: Interval Estimation

Regardless of how we decide to process data for a point estimate of θ, the one thing we can be sure of is that even in the best of circumstances, it is not very likely that the single value, $\hat{\theta}$, is exactly equal to θ. And, we made a case in the last section for the fact that $\hat{\theta}$ might very well be a great deal different from θ. An appealing alternative to being satisfied with a single, untrustworthy value is to report an interval of values within which we can be fairly confident that the elusive θ will be found. How valuable such a report is will, of course, depend on at least two things: the degree of confidence we attach to such an interval and the width of the interval, that is, the extent of the values contained therein. Surely, stating that I am confident that mean time to failure is between 10 and 1000 hours is not nearly as informative as stating that it is between 900 and 1000 hours, regardless of how I measure confidence.

In many cases, the modus operandi for computing a confidence interval is to add and subtract an "error" term, e, to and from $\hat{\theta}$, thereby determining an interval extending from a lower limit $l = \hat{\theta} - e$ to an upper limit $u = \hat{\theta} + e$. Often this is displayed simply as $\hat{\theta} \pm e$, reminiscent of measurement error you may encounter in some applied sciences. The error term e is chosen to reflect the degree of confidence to be attached in reporting that θ is in the interval (l, u) or, in other words, $l < \theta < u$. This is a matter of theory and can itself be a substantial problem.

To introduce confidence intervals, almost all introductory textbooks begin with an artificial case, namely, sampling from a normal population in which μ is unknown—hence to be estimated from data—but σ is known. In that case, $\bar{x}$ is used as a point estimate of μ and e turns out to be $z_{\alpha/2}\sigma/\sqrt{n}$ which, of course, could not be evaluated if σ were unknown. Here, $1 - \alpha$ is taken to reflect the degree of *confidence* we wish to inject into the problem and $z_{\alpha/2}$ is the standard normal percentile point. For example, a traditional choice for confidence—expressed as a percent—is 95%. As a probability, 95% = .95 represents odds of 19 to 1 and will reflect the degree to which we believe the interval we compute will in fact contain μ. Of course, 5% = .05 must then be a measure of "error," in stating that μ is in the interval in question, and this is what we select for α in computing the *error term e*. Thus, $z_{\alpha/2} = z_{.025} = 1.96$

for the traditional choice—a choice you should always default to when no other is given in a problem. The theory that yields the formula for e is quite straightforward and follows.

Before we ever do any evaluating, we know in these circumstances that $\bar{X}$ is normally distributed with mean μ and standard deviation $\sigma/\sqrt{n}$. (This is approximately true, even if X is not normal, when n is sufficiently large by CLT.) Then $\sqrt{n}(\bar{X} - \mu)/\sigma$ is, of course, a standard normal random variable, Z, and you know that $.95 = \Pr(-1.96 < Z < 1.96)$. In terms of $\bar{X}$,

$$.95 = \Pr\left(-1.96 < \frac{\sqrt{n}(\bar{X} - \mu)}{\sigma} < 196\right) \tag{4.1}$$

or, multiplying each inequality by *positive* $\sigma/\sqrt{n}$,

$$.95 = \Pr\left(\frac{-1.96\sigma}{\sqrt{n}} < \bar{X} - \mu < \frac{1.96\sigma}{\sqrt{n}}\right) = \Pr\left(|\bar{X} - \mu| < \frac{1.96\sigma}{\sqrt{n}}\right) \tag{4.2}$$

In this form, we see that the odds are 19 to 1 that the computed value of $\bar{X}$ will be within 1.96 $\sigma/\sqrt{n}$ units of μ. Sometimes you will find $\sigma/\sqrt{n}$ called a *standard error* (SE), and then this result states that you are very likely (19 to 1 odds) to find the value of $\bar{X}$ within 2 standard errors of μ. Or, if you like, you are *not* likely (probability .05) to find that difference exceeding 2 SE's.

Going one step further, you may subtract $\bar{X}$ throughout the inequality to write

$$.95 = \Pr\left(\frac{-1.96\sigma}{\sqrt{n}} - \bar{X} < -\mu < \frac{1.96\sigma}{\sqrt{n}} - \bar{X}\right)$$

and, finally, multiplying through by -1 (remember to reverse the inequalities)

$$\begin{aligned} .95 &= \Pr\left(\bar{X} + \frac{1.96\sigma}{\sqrt{n}} > \mu > \bar{X} - \frac{1.96\sigma}{\sqrt{n}}\right) \\ .95 &= \Pr\left(\bar{X} - \frac{1.96\sigma}{\sqrt{n}} < \mu < \bar{X} + \frac{1.96\sigma}{\sqrt{n}}\right) \end{aligned} \tag{4.3}$$

It is important to note that, in going from Equation 4.1 to Equation 4.3, only the *form* of the probability statement has been changed and $\bar{X}$ is the *only* random variable involved. To emphasize this, you may define

$$L = \bar{X} - \frac{1.96\sigma}{\sqrt{n}} \qquad \text{and} \qquad U = \bar{X} + \frac{1.96\sigma}{\sqrt{n}}$$

both simple functions of $\bar{X}$ (hence random variables themselves), and then you may write Equation 4.3 further as

$$.95 = \Pr(L < \mu < U) \tag{4.4}$$

Finally, if you remember that an equality string, such as $a < b < c$, in mathematics literally stands for two simultaneous inequalities, $a < b$ *and* $b < c$, then Equation 4.4 is simply the expression $\Pr(A \cap B) = .95$, where A is the event $(L < \mu)$ and B the event $(\mu < U)$. In this form, A expresses a *lower* bound or limit to the plausible values of μ, and B expresses an *upper* bound or limit to the plausible values of μ. In general, then, for any choice of α, a 100 $(1 - \alpha)$% *confidence interval* (CI) for μ is given by evaluating the random limits

$$\begin{aligned} L &= \bar{X} - \frac{z_{\alpha/2}\sigma}{\sqrt{n}} \\ U &= \bar{X} + \frac{z_{\alpha/2}\sigma}{\sqrt{n}} \end{aligned} \tag{4.5}$$

That is, the confidence interval is determined *after* data have been processed to produce l and u, and we often express the interval in the form (l,u). Also, you will often find the formulas for l and u written more compactly, as,

$$\bar{x} \pm \frac{z_{\alpha/2}\sigma}{\sqrt{n}} \tag{4.6}$$

If two random samples are drawn, each with the same size n, then you expect to see different values of $\bar{x}$ and accordingly two different pairs of values l and u. That is to say, you would not expect to see the same confidence interval each time, even though the *confidence* $1 - \alpha$ [or 100 $(1 - \alpha)$%] remains the same. It is important to be aware of the random nature of the limits L and U if you are to appreciate fully what is and what is not being stated here. We illustrate.

Imagine that the population is normal with $\sigma = 10$ and $n = 30$. For example, we might be modeling the random score, $\bar{X}$, on an achievement test for a given class. Then 1.96 SE units amounts to about 3.6 score units. If the test scores happen to average out to $\bar{x} = 85$, then a 95% CI for μ in this case would compute to (81.4, 88.6). Hence you are 95% confident that $81.4 < \mu < 88.6$; it is not likely (5%) that either $\mu \geq 88.6$ or $\mu \leq 81.4$. Certainly that sort of report tells you a lot about the class performance—more, incidentally, than just reporting the class average as 85. It is a mistake, however, to report $\Pr(81.4 < \mu < 88.6) = .95$, as though μ were a random variable having a probability distribution in which you discovered that 95% of the area is between 81.4 and 88.6—it is not, and you have not. Again, you report that you believe $81.4 < \mu < 88.6$. The

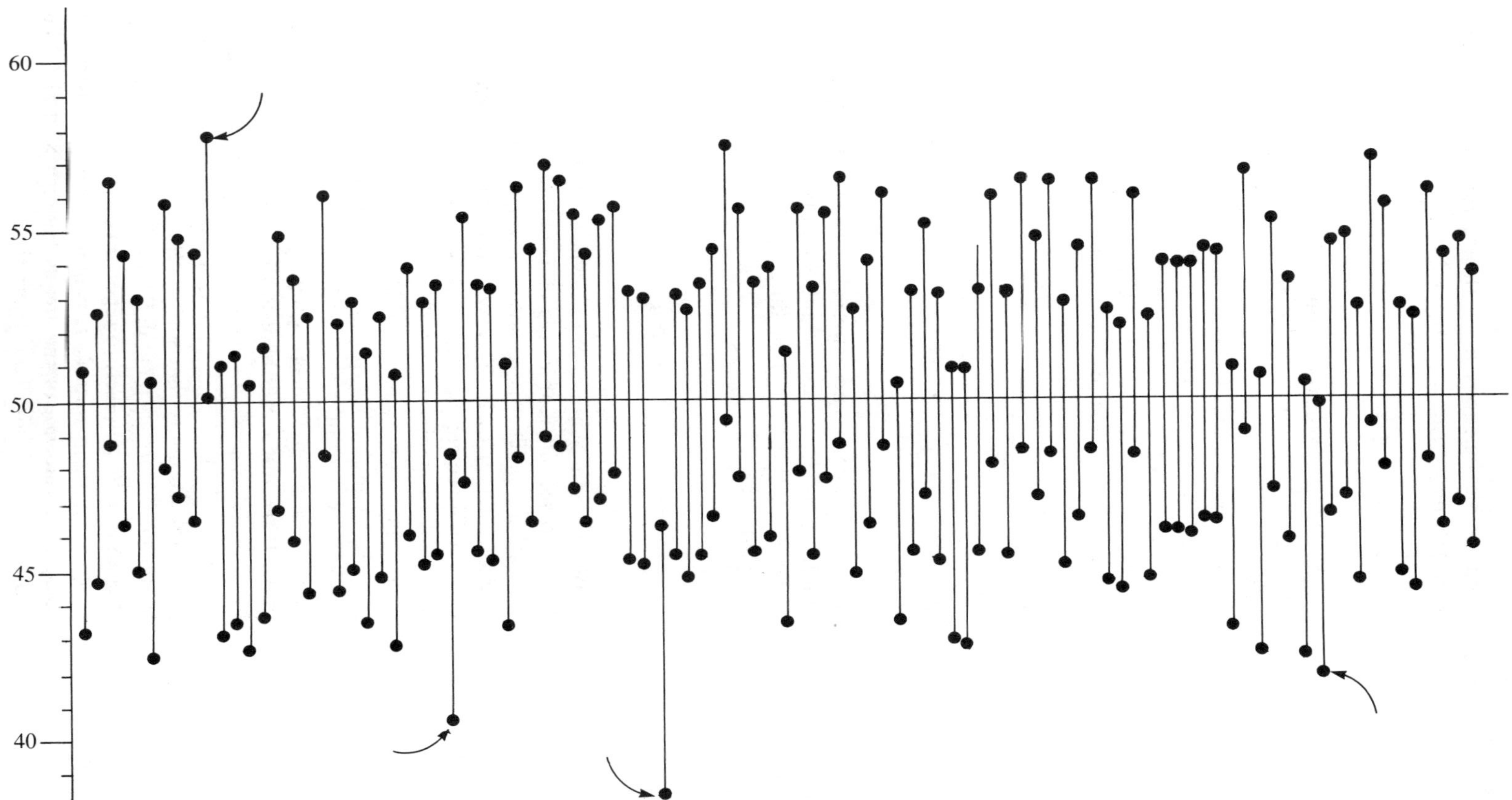

Figure 4.4 100 confidence intervals

chances are 95% that you are correct, and there is a 5% *risk* that you are wrong when you make such a report. In this sense, the interval of values from 81.4 to 88.6 constitutes a set of *plausible* values of μ, those that are compatible with the data $n = 30$, $\sigma = 10$, and $\bar{x} = 85$. It is not very likely (5%) that either values of $\mu \leq 81.4$ or $\mu \geq 88.6$ would produce $\bar{x} = 85$ in a sample of size 30, though it is possible.

Another way to look at it is that if you were to repeat this experiment a large number of times, with α, n, and σ fixed, reporting the interval (l, u) each time as covering the value of μ, you would find (if you really knew the value of μ) that about 95% of them do in fact include μ and about 5% do not. As a test of this point we used ST–02 to generate 100 random samples of size 25 each from a normal distribution having $\mu = 50$ (selected for convenience) and $\sigma = 10$. For each sample we computed the 95% confidence interval and graphed the resulting 100 intervals in Figure 4.4. As you can verify in that figure, 96 of the intervals cover the known mean value of $\mu = 50$ and 4 do not, giving a 96% coverage. With only 100 intervals, that is reasonably close to the expected 95%. You may repeat this experiment with your own seed, if you like, and discover that you will always be close to 95% coverage. A good point to make, however, is that you usually only have one experiment, and for all you know, nature may just deliver the 27th sample so that you would be in error in reporting $41.0 < \mu < 48.8$.

The reason for dividing the risk, 5%, into two parts is to guard against both the contingency that $\mu \geq u$ and also that $\mu \leq l$. Here, the risk is divided evenly between those two contingencies. There is no mandate that we so divide the risk, but it is a tradition you will find followed by almost all textbooks on the subject. In some circumstances, you may want to divide the risk a different way. Indeed, at the extremes, circumstances may call for not guarding against one of the contingencies $\mu \geq u$ or $\mu \leq l$ at all! In our achievement test example we could conceivably not care how *high* μ might be—the higher the better. In that case, we would only be interested in a lower bound on μ, and so there is no need to "waste" half the risk on an upper bound. This means we ought to be able to push the computed lower bound of 81.4 up a bit to make a more definitive statement about μ. Computationally, this amounts to using $z_{.05} = 1.645$ in place of $z_{.025} = 1.96$ in the evaluation of l and, of course, ignoring an upper limit u altogether. This gives us a new value of $l = 82.0$, predictably higher, but the confidence coefficient of 95% is still the same. So, in reporting this time that $\mu > 82$, the odds are 19 to 1 in our favor that this report is true, but of course there is a 5% chance that such a report is false, because in fact (unknown to us) $\mu \leq 82$.

This is an example of a one-sided confidence interval, more specifically, a *lower* one-sided interval, sometimes reported as (l, ∞). An *upper* one-sided interval is in general obtained by replacing $z_{\alpha/2}$ by z_{α} in the computation of u and ignoring l, thus reporting that $\mu < u$ [sometimes reported as $(-\infty, u)$], with confidence $100(1 - \alpha)\%$. In general, such an upper limit will be less than the corresponding upper limit of a two-sided interval, but is a more definitive statement to make about μ, when the contingency $\mu < l$ is either not viable or not of interest.

As for computing confidence intervals for this artificial case, we have programmed the necessary subroutines under code $N(\mu|\sigma)$ in the general program ZS–4. As you can see from the user instructions, the program allows you to enter raw data via ST–03 (ungrouped)* or, as is so often the case, a summarized form of the data; in that case, you need the sample size n and the sample average $\bar{x}$. You also need to enter the *known* standard deviation σ and store it in R_{48} before proceeding. If you cannot find the value of σ (not s or RMSD, mind you), then you should NOT be using this subroutine (see the next section). Finally, you need only enter half the risk, $\alpha/2$, and press [B] to discover the lower limit l of the CI for μ. The value of u is located in R_T and may be recovered from a press of [$x \leftrightarrows t$]. Try this with the foregoing example, where $\bar{x} = 85$, $n = 30$, and $\sigma = 10$, to verify the values presented. You may also verify the value of the lower one-sided interval by entering the full risk $\alpha = .05$ in place of $\alpha/2 = .025$ and ignoring R_T. In case you want an upper one-sided interval, you enter α, but ignore R_D, choosing the value u in R_T. For the example, $u = 88$, so we are 95% confident that $\mu < 88$, a bound that is slightly lower than the two-sided bound on u.

In Figure 4.4, you may have observed that the length of each confidence interval appears to be the same. In fact, for this case, that will always be true. Indeed, since l and u are determined from considering $\bar{x} \pm z_{\alpha/2}\sigma/\sqrt{n}$, you see that the length must be $2z_{\alpha/2}\sigma/\sqrt{n}$. Since α, σ, and n are fixed, this length is constant, and only the variable $\bar{x}$ values determine the variable confidence limits. This is an exceptional case (σ known), and you will soon see that most of the time this length is a random variable, varying from one set of trials to another just like l and u.

While we are on the subject of length of a confidence interval, one other matter needs to be discussed. Clearly the statement $84.9 < \mu < 85.1$ tells you a lot more about μ than the statement $71.2 < \mu < 99.6$. For this reason, many authors say the former statement or interval is more *precise* than the latter. That means, all other things being equal, we would prefer a confidence interval to be as short as possible. In our current case, with a length of $2z_{\alpha/2}\sigma/\sqrt{n}$ there are several ways of shortening a length. One obvious way is to increase the sample size. But samples are not free in the real world, as we have repeatedly pointed out. There is a cost associated with each piece of data, and that in itself may be a real constraint on the sample size you are allowed. Another way to shorten the length is to decrease σ, but you usually have no control over that parameter—indeed, you rarely even know its value at all.

A third way to shorten an interval would be to lessen the value of $z_{\alpha/2}$, and that can be accomplished by *increasing* the value of α. But α is

* You may not (and need not) repartition for full entry. Reset pointer with label [C] if necessary.

the risk of being *wrong* in reporting $l < \mu < u$! Clearly, there is a limit to how much you can afford to increase that value. In our example, with $\bar{x} = 85$, $n = 30$, and $\sigma = 10$, the 95% CI was (81.4, 88.6) with a length of 7.2 score units. If we increase α to 10%, thereby *lowering* the confidence to 90%, the interval is (82.0, 88.0), with a shorter length now of 6 units—not too different. Say you change α to 40%. The 60% CI is then (83.5, 86.5) and has length half as much—only 3 score units. The statement $83.5 < \mu < 86.5$ is more precise, but you only have 60% confidence in its validity. To go to extremes, if you let $\alpha = 100\%$, then the interval collapses to the single value $\bar{x} = 85$ (stating then that $\mu = 85$). Another way of putting this is that you can be 100% sure that $\mu = 85$ is a false statement, which is something we have been saying all along. At the other extreme, if you let $\alpha = 0$, then you can be 100% certain that $-\infty < \mu < \infty$, which is about as uninformative as can be. In this way you see that there is a natural trade-off between how much confidence (or how little risk) you impose on a problem and the resulting precision of the interval.

With that much said, you are invited to verify the answers to the problems that follow. These should provide enough practice to fix the general ideas. In succeeding sections, we turn our attention to the situations that are more frequently encountered (σ unknown, for example). The general theme will essentially be the same, however; namely, a point estimate, $\hat{\theta}$, is used as a focal point about which a confidence interval is constructed, using formulas found in most standard textbooks on the subject. Several isolated quantities of interest in a given problem may be found by consulting the register contents of ZS–4. You might observe that if ST–03 is used for raw data, the usual registers are used for data storage and IDB. We are not interested in permanent storage here, however, so we have listed only those contents of various data registers having to do with CIs. Any raw data that may have been there from Step DE will be written over.

Problems

(In all cases, assume the underlying population is normal when n is not large.)

1. In a study of income tax returns, a random sample of 500 returns was examined and revealed an average return to the taxpayer of \$53.62. Find a 95% CI for μ based on a reported σ-value from past history of \$8.92.

   Ans: $\$52.84 < \mu < \$54.40$

2. A random sample of 200 welfare recipients was processed to find an average annual income of \$2,510. Past records show that $\sigma = \$160$ for this population.

   a. Find a 95% CI for μ.

b. Compare the length in (a) with lengths determined for risk values of $\alpha = .5, .20, .10, .01$.

c. What is the most reasonable (99% confidence) upper one-sided limit for μ?

Ans: a. (2488, 2532)

b.

α =	.50	.20	.10	.05	.01
Length =	\$15	\$29	\$37	\$44	\$58

c. $\mu < 2536$

3. In checking blood pressure for 30 employees of a small industrial plant, an average systolic pressure of 119 was recorded. If $\sigma = 11$, what is the standard error for $\hat{\theta} = \bar{x}$? Calculate a 99% CI for mean systolic pressure.

   Ans: The 99% CI runs from 113.8 to 124.2. Register 32 contains a standard error value of SE = 2.008.

4. Fifteen gallons of paint covered an average of 382 square feet of wall. With a σ value of 50 given, within what limits would the true mean value lie with confidence 95%?

   Ans: $357 < \mu < 407$

5. The average score on an aptitude test for 100 inductees was 55. The test designers report a standard deviation of 25 score units for this test, with a population mean of 60. It appears that this group is below the "norm" on this test, and that is a trend perceived by many to be the case. If you were asked to place an upper bound on mean achievement based on these data, what would you report, if you wish to protect yourself by assuming only a small risk of 1% of being in error in your report?

   Ans: Report an upper one-sided 99% limit of 61, that is, the evidence does not support a real drop in aptitude, since $\mu = 60$ is in the region of plausibility.

6. Five specimens of coke were tested for porosity and showed weight gains (after soaking in water for a specified period) in pounds of 2.16, 2.19, 2.31, 2.30, and 2.21. The production process standard deviation for this type of coke is believed to be $\sigma = .25$. Find a 90% CI for the mean weight gain.

   Ans: (2.05, 2.42) $\bar{x} = 2.234$ (in R_{40})

7. The range of a certain type of rocket has a standard deviation of $\sigma = 50$ yards. A sample of 100 rockets was fired and yielded an average of 2,208 yards. Find a 99% CI for the true average range and report its length.

   Ans: The interval extends from 2,195 yards to 2,221 yards for a length of about 26 yards.

8. In problem 7, how many rockets would have to be fired to make the length of the interval 20 yards?

   Ans: For $\alpha = .01$, the key sequence .005, [SBR], [sin] displays

$z_{\alpha/2} = 2.58$. With length $= 2z_{\alpha/2}\sigma/\sqrt{n} = 257.6/\sqrt{n}$ specified as 20, $\sqrt{n} = 257.6/20 = 12.88$, so that n should be about 166.

9. The experiment in Problem 6 was repeated the next day, and the specimen weights were reported as 2.01, 2.05, 2.35, 2.51, and 2.42. Find the 90% CI for that day. Which one is more precise?

Ans: The new 90% CI is (2.08, 2.45) $(\bar{x} = 2.268)$ and its length is .37 pounds, which has to be the same length, since σ is unchanged. Notice that today 2.06 is not a plausible value, even though it was yesterday. Tomorrow. . .?

10. An instructor firmly believes that scores on a standard exam he administers are normally distributed with a mean of 48 and a standard deviation of 5, based on past records of the test administered to many students. He just administered the test to a new group and wonders if the nominal mean of 48 is still in effect. Using his test scores below, what would you report?

32	43	45	37	42	45	49	44	46	49
48	44	36	45	47	46	47	50	34	35
38	42	47	46	43	43	40	47	45	48
45	34	38	41	47	48	31	46	39	43

Ans: If the standard deviation of 5 is assumed to hold fast, a 95% CI for μ, based on the data, would be (41.3, 44.4). Since 48 is not in the interval, there is evidence that either μ is less than he guessed or perhaps this new class is not typical. In any case, you would have to say that a mean of 48 is not compatible with these data. This should be evaluated in the light of the risk being assumed. However, even the 99% interval (40.8, 44.9) doesn't make it, and that's a lot of confidence.

11. A sample of 50 families has been selected from a large metropolitan area to determine the average number of children in a family. The sample mean, $\bar{x}$, is found to be 3, and it is known that the standard deviation for the number of children in a family is 1 for the metro area. Determine a 99% CI estimate for the true population mean, μ.

Ans: (2.64, 3.36)

12. A distributor for a large tire manufacturer is concerned about the quality of tires he is receiving from the manufacturer. It is known that the standard deviation for the tread life of all tires produced by the manufacturer is 2,800 miles. If the distributor samples 30 of his tires and finds a mean trend life of 48,212 miles, what is a 95% CI estimate for the mean tread life of all tires produced by the manufacturer?

Ans: (47,209.8, 49,214.2)

13. Prior reading tests administered to third grade students in a certain state have been normally distributed with a mean of 84 and a variance of 32. If 30 of this year's students are tested and have an average score of 88, can this year's students be considered typical? Use $\alpha = .05$.

Ans: (86, 90) is the 95% CI, and 84 is not a plausible value.

14. Fifteen metal rods were measured for corrosion resistance. After 10 weeks, the average weight loss in grams was reported to be .2 g. Past records for similar measurements revealed a standard deviation of $\sigma = .005$ g. Find a 95% one-sided CI, both upper and lower, and compare the limits with the two-sided interval.

Ans: $\mu > .198 \qquad \mu < .202 \qquad .197 < \mu < .203$

15. A class of 35 scored an average of 75 on a test following an experimental method of teaching. Past experience with this standard test makes a choice of $\sigma = 13$ quite reasonable. The instructor is preparing a report and decides to report a 99% CI for the true mean score of such classes. He wishes to establish that the true mean is not likely to be less than the recorded lower limit of such an interval, thus making a case for his new method. Show that he would be slightly better off computing a one-sided interval, rather than a two-sided one.

Ans: The two-sided interval is (69, 81) so he may say that it is unlikely that μ is less than 69 score points. But the one-sided limit is $l = 70$, so he can report with the same risk that μ less than 70 is unlikely; he gains a score point by placing all the risk for error on one side.

Section 4.3: Normal Mean—Unknown Variance

When the variance or standard deviation, σ, is unknown in sampling from a normal population, the formulas developed for confidence intervals in the last section are of no use, since there is no value to substitute in $e = z_{\alpha/2}\sigma/\sqrt{n}$ for σ. And it is more often the case that we do not know σ any more than the mean μ we are attempting to estimate. You might suggest using the data to compute s or RMSD and use that numerical value for σ. Particularly if the sample size is large, you know that value should be close to right. But if the data are good enough to use s as though that were σ, then using $\bar{x}$ for μ is an even better idea, for you will recall from Section 4.1 that it is even a better bet (due to less variability) that $\bar{x}$ will be close to μ than that s will be close to σ. Still, if forced to give an interval of values for μ, using the formula $\bar{x} \pm z_{\alpha/2}s/\sqrt{n}$ is not a bad idea and gives good results if n is large (at least 30).

When n is not large, the procedure suggested is not a good idea, partly because of the large variability in s, but more particularly because the random quantity $\sqrt{n}(\bar{X} - \mu)/S$ from which the confidence limits L and U are derived (recall Equation 4.1 and those that follow) departs from the standard normal distribution severely enough that the required probability $1 - \alpha$ is not satisfied by $\Pr(-z_{\alpha/2} < \sqrt{n}(\bar{X} - \mu)/S < z_{\alpha/2})$.

For some time there was no alternative, until an English statistician by the name of W. S. Gossett discovered the probability distribution of this random variable around the turn of the century. He published the results under the pen name "Student," and to this day the result is called the Student *t*-distribution.

Actually, there is a whole family of *t*-distributions, one for every positive integer, and they arise in a variety of applications, of which the current is only one. This is no place to derive or even discuss the *theory* underlying the *t*-family. That is done in a variety of standard textbooks at different levels of mathematical maturity. At any level, like the normal, the CDF cannot be expressed in closed form, and so we have to be satisfied with the approximations given by numerical integration. We need to be able to find "tail" areas and the corresponding percentile points, in order to process the formulas for such matters as confidence intervals. And, since the TI–59 statistics module incorporates the *t*-family into program ST–21, it will suffice for our purposes to summarize the matters presented in ASM for you.

The basic discussion for the *t*-family is given on page 6–14 of ASM. The formula for the density function is given by $f(t)$ on that page, and you see that it is not a particularly pleasant one and, of course, is of no direct use to us. We should single out the presence of the parameter ν (Greek nu) in that formula, for this is the parameter that generates the family. You may think of ν as an abbreviation for *name*, so that members of this family are distinguished by a single name ν, taken to be one of the *positive integers*. For traditional reasons (there are some others), this name is almost universally called *degrees of freedom*, and you will find various other symbols, such as d.f. or even °F, in the literature. We will stay with ASM, since it is their program we will be using. Each choice of ν determines a similar, but distinct, density curve, and a few of them are pictured in Figure 6.7 of ASM. As you can see, using the standard normal density (marked $\nu = \infty$) as a guide, all *t*-densities resemble a normal, but are flatter at the top and hence have necessarily more area in the tails. They are all symmetric about 0 and, as ν increases from the first member of the family ($\nu = 1$), the curves begin to resemble more and more the standard normal. In fact, the resemblance is so strong from $\nu = 30$ onward that we rarely make a distinction in members after that point (they all look the same to us) and simply treat them all as standard normal.

Numerically, we are interested in capturing the usual quantities CDF, CCDF, and percentiles for various members of the *t*-family. You see from the user instructions on page 6–15 of ASM that the matter of finding the value of the CDF, $P(t)$, at an arbitrary point t, is quite simple. After sending the pointer to ST–21 at Step 1, you must call for a member of the family and input ν at Step 2, then press [A]. You may ignore the output and view this as a required initialization procedure.* At

* You may not let ν exceed 139, a point of minor concern, since we will rarely consider ν beyond 30 in the applications. Also, this step may be replaced by simply storing ν in R_{15}, for there is no interest in the output or the computation of $f(t)$.

Step 4, you may find $P(t)$ by simply entering t and pressing [E]. That step may be repeated for various values of t, as long as ν remains fixed, that is, you do *not* have to return to Step 2 unless the problem has changed to a different member of the family. You might observe that the program utilizes registers R_{15} through R_{23}, and that must be taken into account for any programming where this program might be used.

Naturally, if we have $P(t)$, we may find the CCDF by subtraction, since $Q(t) = 1 - P(t)$. If T represents the random variable having the t- distribution with ν degrees of freedom (d.f.), then, of course, $Q(t) = \Pr(T > t)$ or $\Pr(T \geq t)$, for that matter.* You know that 1.645 is the 95th percentile for the standard normal density. Using ν as a subscript on the CDF for various t-densities, you should verify that

$$P_2(1.645) = .88 \qquad P_5(1.645) = .92 \qquad P_{10}(1.645) = .93$$
$$P_{20}(1.645) = .94$$

while $P_{30}(1.645) = .945$, just about the standard normal value.

Unfortunately, there is no simple way to calculate a required percentile point t_α under any of the t-densities using ST–21, except perhaps trial and error. For example, you just verified that $P(1.645) = .88$ when $\nu = 2$, that is, $1.645 = t_{.12}$ for that case. Hence $t_{.05}$ must be a *larger* value. If we just guess 2 and check, we find $P(2) = .91$, while $P(2.5) = .94$. Since $P(3) = .952$, 2.9 is close enough to $t_{.05}$. This is not very satisfactory, and while there are routines similar to the one we used for z_α in ZS–4 (subroutine sin), we have not been able to find one manageable enough to leave sufficient programming steps for our purposes. That makes it necessary to read standard tables, and that in turn restricts us to certain selected, but standard, choices for α. (Those in turn can help with trial and error to find others, if the occasion arises, but you will rarely find α values other than .10, .05, .01 used in the textbooks currently available.) We have supplied such a table in the appendix, as well as near the programs where they are used. One such location follows the user instructions for ZS–4. If you will turn there, you will see the standard percentage choices published for values of ν from 1 through 30. In a row labeled $\nu > 30$, you will find the standard z-values published.

To return to the application at hand, whenever a random sample of size n is drawn from a normal population having mean μ and standard deviation σ, the random variable $\sqrt{n}(\bar{X} - \mu)/S$ (the quotient of two random variables you might observe) has a t-distribution with $\nu = n - 1$ degrees of freedom. Here, S is the random variable version of the sample standard deviation, namely

* While we are not directly interested in $A(t)$ as discussed on page 6–14 of ASM, you should make the same correction as with ST–19, namely $A(t) = \Pr(|T| \leq |t|) = 2P(|t|) - 1$, for it is intended that this be the area from $-|t|$ to $|t|$. (If $t < 0$, $2P(t) - 1 < 0$, so that in itself will be an error.)

$$S^2 = \sum_{i=1}^{n} (X_i - \bar{X})^2/(n - 1) \text{ and } S = \sqrt{S^2}$$

This allows us immediately to answer questions like those following.

EXAMPLE 4.1.

A random sample of size 18 is to be drawn from a normal population. What is the probability that the sample mean $\bar{X}$ will differ numerically from the population mean μ by as much as (less than or equal to) 1/3 of the sample standard deviation?

Solution Suppose we let T_{17} stand for a random variable having a *t*-density with $\nu = 17$ d.f.,* so that

$$T_{17} = \sqrt{n}(\bar{X} - \mu)/S.$$

Then,

$$\Pr(|\bar{X} - \mu| \leq 1/3\, S) = \Pr[|\sqrt{17}(\bar{X} - \mu)/S| \leq \sqrt{17}/3]$$
$$= \Pr(|T_{17}| \leq 1.37)$$

Using ST–21 with $\nu = 17$, this value is given by $A(1.37) = 2\,P(1.37) - 1 = .81$. More important than such applications as this, we can now derive confidence interval formulas for this case quite simply. With S replacing σ in the formulation leading from Equation 4.1 to Equation 4.5 and finally to Equation 4.6, the *t*-version of the latter would be

$$\bar{x} \pm t_{\alpha/2}\, s/\sqrt{n} \qquad (\nu = n - 1) \tag{4.7}$$

So a 100 $(1 - \alpha)$% CI for μ is obtained by adding and subtracting an error term of $e = t_{\alpha/2}\, s/\sqrt{n}$ to the estimate $\hat{\theta} = \bar{x}$ to produce an interval (l, u) quite like the σ-known case, but with important differences. Unless n is large, $t_{\alpha/2}$ will not be the same—indeed it is always larger, as you see from the *t*-table, than $z_{\alpha/2}$. All other things being equal, this produces a wider interval than the normal one and so is less precise, but reflects the fact that σ is unknown, and hence you should not expect to be able to be as precise in your statement about μ. (Incidentally, the standard error SE $= \sigma/\sqrt{n}$ is unknown now and can only be estimated. Yet you will find that many authors call the estimate $s/\sqrt{n}$ standard error, nevertheless, and we defer to them.)

But all other things, in fact, are *not* equal. That is, quite apart from the fact that $t_{\alpha/2} > z_{\alpha/2}$, s is the value of a random variable and it-

* Usually we will designate such a random variable by T_ν, when we need to pinpoint the degrees of freedom in a problem.

self subject to random fluctuations. Among other things, this makes the length of the confidence interval $2t_{\alpha/2}\, s/\sqrt{n}$, the value of a random variable, and even the length, then, will vary from one set of trials to another. Do you remember in Section 4.1 you learned that, even though S^2 is unbiased, S is not, and that in any case, you are more likely to see values of s smaller than σ than larger? Well, it just could happen that s is smaller enough than σ to offset the fact that $t_{\alpha/2} > z_{\alpha/2}$, so as to produce a confidence interval that is *shorter* than the normal interval (if we could actually compute the latter). In fact, if we were to compute a large number of such intervals, say all with $\alpha = 5\%$ and n fixed, we would still discover about 95% of the intervals would cover μ and about 5% would not. But the intervals would vary in length, no two of them being the same (except by sheer coincidence—usually explainable through rounding). Some will be shorter and some longer than the corresponding normal interval with σ known. (You are invited to verify this through simulation, since you have that capability by selecting an arbitrary μ and σ to use in the simulation.) We can say, however, that the *expected value* of that random length is greater than $2z_{\alpha/2}\sigma/\sqrt{n}$ for any fixed σ, so that *in the long run*, *t*-confidence intervals will be longer than their normal versions.

EXAMPLE 4.2

In Problems 6 and 9 ($n = 5$), it was later discovered that $\sigma = .25$ was just a wild guess, and that in fact σ is unknown. What effect does that have on the CI's?

Solution You may easily verify that, for Problem 6, $s = .067$, while $s = .225$ for Problem 9. Both are smaller than the alleged $\sigma = .25$, although the latter is close. With $\nu = 4$, $t_{.05} = 2.132$, so for Problem 6, $e = .064$, giving a 90% CI for μ of (2.17, 2.30), having length .13. For Problem 9, $e = .215$, which produces the interval (2.05, 2.48), having length .43. You see that the length in Problem 6 is less than the constant length .37 for both Problems 6 and 9, while in Problem 9, the interval is wider, hence less precise, than any of the others. All four have confidence 90%!

The confidence interval formulas have been programmed for you under code $N(\mu)$ in the ZS–4 program. You will see from the user instructions that data entry is quite like $N(\mu|\sigma)$, except of course there is no σ to enter. You start by pressing [A], whereupon $\nu = n - 1$ is displayed for you. Now you need to look up $t_{\alpha/2}$ from the *t*-table, and then simply press [R/S]. The usual worked-out example illustrating the calculator procedures will be found in the ZS–4 examples following the user instructions. The problems that follow should provide you with ample practice for a variety of situations that are typically encountered in basic textbooks.

Assume normality and σ unknown, unless explicitly given as σ.

16. A random sample of 3 is to be taken from a normal population having mean 0. What is the probability that $\bar{X}$ will be within 2 SE units from the mean?

Ans: Using the estimate $S/\sqrt{3}$ as the SE, $\Pr(|\bar{X}| < 2S/\sqrt{3}) = \Pr(|\sqrt{3}\ \bar{X}/S| < 2) = \Pr(|T_2| < 2) = 2P(2) - 1 = .82$ from ST–21.

17. A random sample of 6 12-ounce soft drink cans was taken and found to have an average of 12.01 ounces, with a standard deviation of .13 ounces. Find a 95% confidence interval for the mean fill.

Ans: (11.87, 12.15)

18. Five samples of 3-inch pipe were measured to the nearest thousandth of an inch yielding measurements

2.992 3.011 2.994 2.995 2.998

Find a 99% confidence interval for the mean diameter.

Ans: (2.982, 3.014)

19. A sensitive electronic component of a moon spacecraft must be able to withstand high temperatures in the process of flight. Because of the expense involved, only 3 components were tested to determine the temperature withstood before burnout. The recorded temperatures in degrees Centigrade were 2050, 2065, and 1995. Determine a 90% CI for μ, the mean temperature.

Ans: (1975, 2099)

20. A class of 28 students made the following scores on an achievement test given in a statistics course.

60	51	67	42	35	56	51
47	62	65	43	31	50	50
48	78	49	40	56	56	63
53	72	58	55	54	50	68

Find a confidence interval for the population mean of which these are a sample.

Ans: Using a default value of 95%, $50 < \mu < 58$.

21. A manufacturer is concerned about the 48-month guarantee he offers on his batteries. A sample of 20 such batteries yielded the data $\bar{x} = 47$ months, with a sample standard deviation of .9 months. Find a 99% upper one-sided confidence interval for the mean life of his batteries. Do you think on the basis of the results that he has cause for alarm?

Ans: Maybe. Confidence is 99% that $\mu < 47.5$, so he should be a little alarmed.

22. A random sample of 10 applicants for unemployment insurance were surveyed to determine the average number of days' delay before receipt of their first check. The data yielded

$$\sum_{n=1}^{10} x_i = 60 \text{ and } s^2 = 28.$$

Find a 95% confidence interval for the mean delay.

Ans: (2, 10)

23. Compute a 99% confidence interval for the data of the preceding problem and compare the lengths.

Ans: (1, 11); a length of 10 days compared to 8 in Problem 22

24. In order to test a delicate balance, 5 analysts independently took readings on a 5-gram weight and reported 5.001, 4.999, 5.002, 5.001, and 5.002. Find a 99% confidence interval for the mean reading. Based on this information, does it appear that the balance is measuring correctly?

Ans: (4.998, 5.004); appears to be all right.

25. Thirty prospective telephone operators were given a timed test of motor ability, yielding results

$$\bar{x} = 15 \text{ minutes and } \sum_{i=1}^{30} (x_i - \bar{x})^2 = 3500.$$

Find a 95% confidence interval for the mean of this test based on this much information.

Ans: (10.9, 19.1)

26. In a human factors laboratory, a test was given to 20 randomly selected subjects to see how long it takes on the average to carry out a sequence of checks before operating a computer. A sample average of 25 seconds was reported, along with a sample variance of 145. Find a 90% CI for the mean time.

Ans: (20.3, 29.7)

27. The athletic director for a large metropolitan high school decided to take a sample from the school records of the heights of male students. If he found an average height of 72 inches in a sample of 20 records, and it is known that the variance in height of all of the school's male students is 14 inches^2, what is a 95% confidence interval for the mean height of all male students at the school?

Ans: (70.36, 73.64)

28. A recent study on beer consumption included a random sample of 20 members of college fraternities. The sample resulted in an average consumption per week per individual of 120 ounces. If the standard deviation for the per week per individual beer consumption is 11.72 ounces, what is a 99% confidence interval for the mean beer consumption for members of college fraternities?

Ans: (112.5, 127.5)

29. Ten joggers from a large runners' association were found to jog at an average speed of 6.1 MPH, with a standard deviation of .9 MPH. Construct a 95% confidence interval for the mean speed of all members of the association.

    Ans: (5.46, 6.74)

30. The grade point averages of 10 senior students who were randomly selected from a large university were as follows: 3.42, 3.67, 3.19, 2.98, 3.0, 2.7, 3.85, 3.77, 3.62, 3.5. Calculate a 99% confidence interval for the mean grade point average of all seniors at the university.

    Ans: (2.97, 3.77)

31. The arm length of 30 randomly selected boxers averaged 70.1 inches, with a standard deviation of 1.9 inches. Construct a 99% confidence interval for the mean arm length of all boxers.

    Ans: (69.14, 71.06)

32. The National Blue Mountain skeet shoot competition draws more than 5,000 competitors annually. The 30 scores listed below were randomly selected from this year's competition. Construct a 90% confidence interval for the average score for this year's contestants.

40	52	61	43	55	69
39	51	63	40	58	49
58	67	48	52	61	50
49	61	47	52	45	39
59	47	60	50	48	70

    Ans: (50.04, 55.49)

33. Twenty-seven student nurses were randomly selected from a large class of student nurses. The average grade point average for these 27 nurses was 3.88 with a standard deviation of .09. Find a 95% confidence interval for the mean grade point average of all student nurses in the class.

    Ans: (3.84, 3.92)

34. A service department manager for a large automobile dealer wants to improve the efficiency of his operations. He decides to take a random sample of his customers to determine the average waiting time before a service writer prepares an estimate for the automobile repairs. From a sample of 41 customers, he finds the mean waiting time to be 28 minutes, with a variance of 54 minutes.[2] Determine a 95% confidence interval estimate for the actual mean waiting time.

    Ans: (25.75, 30.25)

35. A student's ability in a course is measured by scores on a test assumed to be normally distributed. His scores for the term were 86, 88, 91, and 94. What is a plausible range of values for a mean score μ of all such students?

    Ans: Defaulting to 95%, $84 < \mu < 95$.

In case σ (or σ^2) of a normal population is unknown, as in the last section, it makes sense to address the question of a confidence interval as an alternative to reporting a single point estimate, just as for μ. Fortunately, the theory for that case has been completed, and formulas for confidence limits for σ^2 are available. The formulas do entail another special family of density functions like the *t*—indeed programmed in ST–21, along with the *t*-family, and we examine that family first.

The distribution that applies here is called *Chi-square.* Like the *t*-family, there is a Chi-square density function for each positive integer, so that the index that distinguishes one member of this family from another is also denoted ν and called degrees of freedom. But there the resemblance to the *t*-family stops. This family is presented and briefly discussed on pages 6–11 through 6–13 of ASM. On page 6–11 you will see a picture of 3 Chi-square densities in Figure 6.6. The one labeled $\nu = 2$ meets the vertical axis at .5, and the curve for $\nu = 1$ looks about like this and never touches the vertical axis anywhere. None of the curves ever touch the horizontal axis on the right, as appears to be shown, and the two curves labeled $\nu = 6$ and $\nu = 10$ do present typical pictures of what one imagines when the phrase "Chi-square density" is mentioned. The formula for the density function appears at the bottom of page 6–11 and certainly looks formidable and hard to manage mathematically. It is for most cases. However, for $\nu = 2$, the formula may be reduced to*

$$f(\chi^2) = {}^1/_2 \exp\left[\frac{-\chi^2}{2}\right] \qquad \text{or} \qquad f(x) = .5e^{-.5x}$$

But you should recognize this as an exponential density having mean $\mu = 2$ or $\lambda = .5$ (see example 2.3). However, that is about the only easily managed case.

One of the of the first things to strike you about the Chi-square family is its lack of symmetry. This means among other things that when we wish to find 2 percentile points like $\chi^2_{.05}$ and $\chi^2_{.95}$, we will be forced to consider two distinct values, for one is not just the negative of the other, as in the case of both the normal and the *t*-families. In fact, you see from the graphs that negative values for a Chi-square variable are not admitted, that is, the range is from 0 to ∞. As for calculations, the CDF is intractible after $\nu = 2$, so once more we find extensive tabulation through numerical integration for this family. Program ST–21 is arranged, just as with the *t*-family, to output a good approximate value for the CDF at any value by pressing [C] instead of [E]. (As with *t*, ν is entered first with *A*—and is essential—so it is the same initialization. Step 3 is of no interest to us.) Thus, for example, if $\nu = 4$, the CDF at 2.16

* $\Gamma(1) = 1$ and χ^2 is the symbol for lower-case Greek Chi-square. The symbol for capital Greek Chi-square cannot be distinguished usually from Roman X^2.

would be given as $P(2.16) = .29$. Hence, 2.16 is the 71st percentile, $\chi^2_{.29}$, of that distribution.

As ASM notes, the program is slow running for large degrees of freedom. In that case, we may exploit the fact that even this nonsymmetric family of densities starts to look "normal" after awhile. Actually, our main concern, as with the t, is finding percentile values, and they cannot be found except by trial and error with ST–21 anyway. So we must resort once again to published tables and will be satisfied with the typical values of α. We have provided such a table at its point of use following the user instructions and again in the appendix for ease of reference. For selected values of a proportion P, we provide percentiles for all values of ν through 40. For $\nu > 40$, you see from the instructions in the table that ZS–4 may be used to calculate the approximate values from the normal approximation. After storing ν in R_{15}, you input any proportion P you desire, and press [D] to calculate and display χ^2_P.

As for the application at hand, the theoretical result we are interested in is the fact that $(n - 1)S^2/\sigma^2$, as a random variable, has a Chi-square distribution with $\nu = n - 1$ degrees of freedom. This is true when sampling from a normal population, and you see that the parameter μ (also unknown) does not enter into the picture. By judiciously selecting percentile points from the appropriate Chi-square distribution, we can effect bounds on σ^2 quite easily from the fact that

$$1 - \alpha = \Pr\left(\chi^2_{1-\alpha/2} < \frac{(n-1)S^2}{\sigma^2} < \chi^2_{\alpha/2}\right)$$

We can take reciprocals (remembering, then, to reverse inequalities) to obtain

$$1 - \alpha = \Pr\left(\frac{(n-1)S^2}{\chi^2_{\alpha/2}} < \sigma^2 < \frac{(n-1)S^2}{\chi^2_{1-\alpha/2}}\right)$$

This will give us the required formulas for the random limits L and U, namely

$$L = \frac{(n-1)S^2}{\chi^2_{\alpha/2}} \qquad U = \frac{(n-1)S^2}{\chi^2_{1-\alpha/2}}$$

from which the nonsymmetric confidence interval limits l and u may be calculated once S^2 has been replaced through data by s^2.

Example 4.3

Suppose 200 incoming freshmen at a certain college are checked at the health center for pulse rates. The sample variance turns out to be $s^2 = 100$. Find a 95% confidence interval for σ^2.

Solution First, since $\nu = 199$ is quite large, we will find percentile points by approximation. Storing 199 in R_{15}, an input of .975 yields $\chi^2_{.975} = 162$ (rounded) when [D] is pressed. An input of .025 yields $\chi^2_{.025} = 240$. Since $(n - 1)s^2 = 19{,}900$, $l = 19900/240 = 83$, and $u = 19900/162 = 123$, so we are 95% confident that $83 < \sigma^2 < 123$, or if you prefer, that $9 < \sigma < 11$.

In the example, we translated a confidence interval for σ^2 to one for σ (which may be more meaningful, since it is in the same units as the measurements themselves), by taking square roots throughout the inequalities. This is quite justifiable since all quantities are positive, $\sqrt{x}$ is a monotone increasing function, hence preserves the sense of the inequalities. You notice that the *upper* percentile point $\chi^2_{\alpha/2}$ becomes part of the computation of the *lower* confidence limit and vice versa. This is because the function $1/x$ is monotone *decreasing* and reverses the sense of inequalities. So if we want a one-sided lower confidence limit for σ^2, we should simply calculate

$$l = \frac{(n - 1)s^2}{\chi^2_\alpha}$$

By contrast, the upper one-sided bound on σ^2 would be given by

$$u = \frac{(n - 1)s^2}{\chi^2_{1-\alpha}}$$

Upper one-sided intervals are often called for in certain applications where σ^2 is interpreted as a measure of precision of an instrument, such as a delicate balance. In quality control or calibration matters, the main concern is that σ^2 not become too large, with little or no concern as to how small it might be.

EXAMPLE 4.5

Ten independent readings in pounds on a bathroom scale were taken and computations gave a value of $s^2 = .0084$. With $\nu = 9$, $\chi^2_{.95} = 3.33$, and hence

$$u = \frac{(n - 1)s^2}{\chi^2_{.95}} = .0227$$

We are reasonably confident that $\sigma^2 < .0227$ or $\sigma < .15$. It is not likely that σ is as large or larger than .15. Based on normality most readings should be within no more than $3\sigma < .45$ pounds of the true reading, which strikes us as quite satisfactory.

All of these simple matters have been programmed for you under $N(\sigma^2)$ in the ZS–4 program with instructions quite similar to those you already are familiar with. Naturally, it is a bother to have to enter 2 separate percentile points, but it is absolutely essential, because of the lack of symmetry of the distribution, and there is no simple relationship between the 2 points. Incidentally, dividing the risk in this case evenly between the tails is not the best way to proceed so far as minimizing the expected length of the interval. But the optimal strategy poses a tough mathematical problem, and the solution is not worth the effort in most cases. In any case, we provide the formulas you will find in almost all books that treat the subject. The following problems should provide you with adequate practice in implementing these standard formulas.*

Problems

36. Find the values of the following percentile points with degrees of freedom shown in parentheses.

$\chi^2_{.90}(30)$ $\chi^2_{.95}(22)$ $\chi^2_{.99}(18)$ $\chi^2_{.005}(36)$ $\chi^2_{.05}(60)$ $\chi^2_{.025}(130)$ $\chi^2_{.10}(500)$

Ans: 20.6, 12.3, 7.01, 61.6, 79.1, 163.5, 540.9

37. A sample of size 20 is to be taken from a normal population whose variance is 1. What is the probability that the sample variance will be greater than $1/2$?

Ans: $\Pr(S^2 > 1/2) = \Pr(19S^2 > 9.5) = Q(9.5) = 1 - P(9.5) =$ .9642 from ST–21

38. A nail machine is supposed to manufacture nails with small variability. A random sample of 30 such nails was found to have a standard deviation of $s = .03$ inches. Find an upper bound on σ, having confidence 99%.

Ans: $\sigma < .043$

39. A soup can filling machine is supposed to fill each can with 10 ounces of clear broth. A random sample of 20 cans yielded a sample variance of $s^2 = .04$. Find a range of likely values for σ^2, that is, compute a two-sided 95% confidence interval.

Ans: $.02 < \sigma^2 < .09$

40. In order to see if the variability in heights of graduating classes is changing, the heights of 100 randomly selected graduates were taken. The experimenter reported the value of the unbiased estimator, $s^2 = .95$. Find a 95% confidence interval for the population variance σ^2.

Ans: $.73 < \sigma^2 < 1.28$

41. Variability in weather is always a problem at resort areas. In one such area, a hotel offers a refund if the "variation" is more than ten

* When $\nu > 40$ you need not store ν in R_{15} before pressing [D] since [C] will automatically do this for you.

degrees. Last year's records for 365 days showed a sample standard deviation of $s = 14.14$ degrees. With confidence 99%, what is an upper bound on the true standard deviation, if we assume that the variance is really constant from year to year?

Ans: $\sigma < 15.47$

42. Four samples of 3-inch pipe were measured to the nearest thousandth of an inch yielding measurements

2.992 3.011 2.994 2.995

Based on these data, what is a lower bound on the standard deviation of pipe diameter using $\alpha = .01$?

Ans: $\sigma > .0045$

43. Using a 2-gram standard, repeated readings were taken on a sensitive balance and yielded a record of 2.01, 2.02, 1.98, 1.99, and 2.00. Find a 90% confidence interval for σ^2.

Ans: $.0001 < \sigma^2 < .0014$

44. Acme Corporation has been receiving customer complaints about one of their new food processors. Apparently a small screw works loose while the machine is operating, causing a malfunction. A sample of 19 of the screws from the production supply bin yielded a variance of .08 in the diameter of the screw. Construct a 95% confidence interval for the variance in the diameters of all of the screws in the large production bin.

Ans: (.0457, .1750)

45. The following are the test scores for 19 students who took the last Graduate Management Aptitude Test:

420 580 601 420 403 519 500 520 480 475
470 510 530 415 420 516 540 425 430

Determine a 95% confidence interval for the variance in the scores of all students who took the test.

Ans: (1997.1, 7643.8)

46. If a sample of 31 test scores from a given population has a variance of 100 ($s^2 = 100$), construct a 95% confidence interval for the population variance, σ^2.

Ans: (63.83, 178.57)

47. The following scores represent a random sample for a test given to a large statistics class: 80, 85, 88, 62, 91, 74, 56, 100, 91, 73, 78, 82, 88, 90, 80. Calculate a 95% confidence interval for the variance in the scores of the whole class.

Ans: (71.51, 331.51)

48. In a recent election in a large city, 21 voters were polled as to how long they had to wait in line to cast their ballots. It was observed that the sample variance in waiting time was 10 minutes. Find a

95% confidence interval estimate for the variance in waiting time for all voters in the city.

Ans: (5.8, 20.9)

49. A cardiovascular fitness test was recently administered to 10 athletes who followed a very popular training program. The test results were: 85, 80, 95, 90, 55, 60, 75, 90, 80, 75. Construct a 95% confidence interval for the variance in cardiovascular fitness of all athletes who use the same training program.

Ans: (79.08, 556.48)

50. The management of a large company is considering the use of a new, less expensive production process for their standard product. Since a large variability in length of the product is not acceptable, a sample of 21 has been conducted to measure the product length using the new production process. If the sample standard deviation is .3 of an inch, what is a 95% confidence interval for the variance in length of the company's product using the new process?

Ans: (.053, .188)

Section 4.5: Exponential Mean

Sample variance is not the only setting in which the Chi-square family arises, by any means. Curiously enough, it also arises in finding the distribution for the sample mean $\bar{X}$ when sampling from an exponential population. You were introduced to the exponential family of density functions in Example 2.3 of Chapter 2. You learned that it is often used as a model for random time to failure (TTF), and, in that context, λ, the single parameter of the family, is known as the failure rate. You also were told that $\mu = 1/\lambda$ in every case, and in TTF models, μ is known as the mean time to failure (MTTF).

So important is μ and associated concepts in these types of applications that you will often find the density written as

$$f(x) = \frac{1}{\mu} e^{-x/\mu} \qquad x > 0$$

instead of

$$f(x) = \lambda e^{-\lambda x} \qquad x > 0$$

The form that is used is simply a matter of taste, since they are equivalent.

The exponential family has a particularly simple expression for the CCDF, namely,

$$Q(x) = e^{-\lambda x} \equiv exp[-\lambda x]$$

Not only that, but since $Q(x) = \Pr(X > x)$, it is especially convenient to interpret the CDDF in TTF models as the probability that the item in question will survive x units of time. For that reason, $Q(t)$ is often called the *reliability* at time t and, to emphasize that interpretation, is often written $R(t)$. Naturally, you cannot know the value of $R(t)$ for a fixed t unless you are told what to use for λ or, equivalently, μ. And that is where statistics enters the picture.

If you draw a random sample from an exponential population, then a natural estimate of μ is $\bar{x}$. In that case, you would use $1/\bar{x}$ as an estimate of λ. Since reliability may be written

$$R(t) = exp\left[\frac{-t}{\mu}\right]$$

it would follow that

$$\hat{R}(t) = exp\left[\frac{-t}{\bar{x}}\right]$$

would yield a point estimate of the realiability at time t.

As usual, we do not expect point estimates to coincide with the parameters they estimate and so we seek a way of expanding the estimate to an interval of plausible values, and that is where the Chi-square distribution comes in. The theory that allows us to find confidence intervals is based on the fact that

$$2n\lambda\bar{X} = 2\lambda \sum_{i=1}^{n} X_i$$

has a Chi-square distribution, with $\upsilon = 2n$ degrees of freedom when the sample mean $\bar{X}$ is computed from a random sample. With that in mind, the probability statement

$$1 - \alpha = \Pr(\chi^2_{1-\alpha/2} < 2n\lambda\bar{X} < \chi^2_{\alpha/2})$$

can be quickly converted to random limits that cover λ (or μ) with that probability. The resulting inequalities (in observed data form) would be

$$\frac{\chi^2_{1-\alpha/2}}{2n\,\bar{x}} < \lambda < \frac{\chi^2_{\alpha/2}}{2n\,\bar{x}}$$

$$\frac{2n\,\bar{x}}{\chi^2_{\alpha/2}} < \mu < \frac{2n\,\bar{x}}{\chi^2_{1-\alpha/2}} \qquad (4.8)$$

and these provide very simple formulas, indeed, for the 100(1 − α)% CI for either λ or μ. If you have one, you have the other, so we need not generate formulas for both, and we have selected the one for μ for programming in the ZS–4 program under the Code Exp(μ).

As you can see from the user instructions, data entry is very much like the other cases in this program and all you have to do is remember to press [E]. This will yield $v = 2n$, whereupon you enter percentiles, just as in the $N(\sigma^2)$ case for a CI on μ. If you want the CI for λ, just take the reciprocals. Of course $1/l$ becomes the *upper* limit on λ and $1/u$ the *lower* limit. You can hardly make a mistake, because you know the lower number has to be *l*.

EXAMPLE 4.6

Five samples of times to failure for equipment that is supposed to have a failure rate of 1/1,000 gave values of 30, 253, 1392, 1873, and 639. Find an interval of plausible values for λ.

Solution Following the user instructions for *Exp*(μ), input the data through ST–03. (Be sure to press [RST] to exit ST–03 before pressing [D′].) Now press [E] to see $\nu = 10$. The Chi-square table yields $\chi^2_{.975} = 3.25$, which you should store in R_{41} and $\chi^2_{.025} = 20.5$, which you should store in R_{31}. Press [R/S] and you find that a 95% CI for μ runs from 408.5 to 2576.6. Taking reciprocals, $.0004 < \lambda < .0024$. This interval includes .001, so we feel comfortable about the assumption. Recalling R_{40} reveals that $\bar{x} = 837.4 = \hat{\mu}$, so .0012 is the point estimate of λ.

In this example, we could ask for the reliability at, say, 500 hours. That would be $R(500) = exp[-500\lambda]$. If the assumption $\lambda = .001$ is correct, then $R(500) = .61$, so there is a 61% chance the item will survive 500 hours. If the assumption is suspect for any reason, we may use our data to estimate $R(500)$ and, accordingly, $\hat{R}(500) = exp[-500/\bar{x}] = .55$, not too different from .61. We could even specify confidence bounds on $R(500)$, since we have bounds on μ or λ. That is, since we are 95% confident that $.0004 < \lambda < .0024$, we are 95% confident that $.2 < 500\lambda < 1.2$ or $-1.2 < -500\lambda < -.2$. Hence, we must have the same confidence that $.30 < R(500) < .82$, so you see the actual reliability at 500 hours could reasonably be anything from 30% to 82%. It is not likely that it is as high as 90% though, for example.

All of this can be generalized to produce a formula, namely

$$\text{if} \quad l < \mu < u \quad \text{then}$$
$$exp\left[\frac{-t}{l}\right] < R(t) < exp\left[\frac{-t}{u}\right] \tag{4.9}$$

and you may use this fact to compute a CI for reliability at time t. Among other things, you will be asked to find such confidence intervals in the problems that follow.

Problems

51. Six very expensive pieces of electronic equipment for a lunar spacecraft were tested in the development program and yielded the following times to failure (in hours): 1956, 1232, 1803, 1979, 1862, and 2050. Find a lower 95% CI on the mean time to failure.

Ans: $l = 1036$

52. Time to failure of a certain piece of support equipment for a space vehicle calls for a mean value of least 1,000 hours. A sample of 10 such items yielded a total failure time of 13,420 hours. Find a lower one-sided confidence bound on mean time to failure.

Ans: Defaulting to 95%, $\mu > 855$ hours

53. An air compressor offered for sale is advertised as being good for at least 10 years. If lifetime is assumed to be exponential, and 9 of these compressors had lifetimes of 4, 3, 10, 11, 2/3, 5, 1, 6, and 17 years, find an upper 95% confidence bound on $R(10)$. Interpret the result.

Ans: An upper 95% bound on μ is given by $u = 12.28$, hence $R(10) < exp\,(-10/12.28) = .44$. The odds are 19 to 1 that fewer than 44% of these compressors will survive longer than 10 years.

54. The time it takes for an order to arrive at a dock from the warehouse is assumed to be an exponential random variable. It is crucial that orders arrive within 48 hours, and a record taken yesterday shows varying arrival times of 35, 52, 41, 40, 44, 39, 19, and 8. On this basis, what percentage of the orders placed today would you think will arrive within 48 hours?

Ans: An upper 95% confidence bound on $R(48)$ is a measure of the percentage of orders exceeding 48 hours arrival. That number is $R(48) < .50$, that is, $\Pr(T > 48) < .50$. Hence, $\Pr(T < 48) > .50$. Thus, with confidence 95%, we infer that more than 50% of the items will arrive within 48 hours.

55. Times to failure of 50 electronic tubes in hours were recorded as,

166	28	252	76	300	84	37	4	84	221
34	9	15	143	16	36	151	30	77	97
5	241	394	230	17	17	127	4	54	7
299	18	42	102	20	322	22	117	82	99
117	62	120	8	131	48	84	105	20	33

Is 100 hours a plausible value for μ?

Ans: The 95% CI for μ extends from 74.2 hours to 129.5 and includes 100 as one of the plausible values.

ZS–4: User Instructions

STEP	PROCEDURE	ENTER	PRESS	DISPLAY
DE	Original Data Entry			
1	Enter data using ST–03. (Do not repartition)			
2	Exit ST–03.		[RST]	n
3	Process data for storage.		[D′]	0
N(μ)	*Normal Mean — σ Unknown*			
1	Enter data using DE *or*:			
	a. Enter sample size.	n	[STO]03	n
	b. Enter sample mean.	$\bar{x}$	[STO]37	$\bar{x}$
	c. Enter sample standard deviation.	s	[STO]38	s
2†	*Test H_0: $\mu = \mu_0$.*			
	a. Enter H_1-code*.	H_1-code	[A′]	H_1-code
	b. Enter μ_0, and compute *P*-value.	μ_0	[R/S]	P
3	*CI for μ*			
	a. Calculate degrees of freedom.		[A]	υ
	b. Enter $t_{\alpha/2}$ with d.f. $= \upsilon$,	$t_{\alpha/2}$	[R/S]	l
	and calculate limits.		[$x \rightleftharpoons t$]	u
	Note: For one-sided intervals, enter t_α at Step 3b and ignore l or u, as the case may be.			
N($\mu\|\sigma$)	*Normal Mean — σ Known*			
1	Enter data using DE *or*:			
	a. Enter sample size.	n	[STO]03	n
	b. Enter sample mean.	$\bar{x}$	[STO]37	$\bar{x}$
2	Enter population standard deviation.	σ	[STO]48	σ
3†	*Test H_0: $\mu = \mu_0$.*			
	a. Enter H_1-code.	H_1-code	[B′]	H_1-code
	b. Enter μ_0, and compute *P*-value.	μ_0	[R/S]	P
4	Calculate $100(1 - \alpha)\%$ *CI for μ*	$\alpha/2$	[B]	l
			[$x \rightleftharpoons t$]	u

$$* \ H_1\text{-code} = \begin{cases} 1 \text{ if } H_1{:}\theta > \theta_0 \\ 0 \text{ if } H_1{:}\theta \neq \theta_0 \\ -1 \text{ if } H_1{:}\theta < \theta_0 \end{cases}$$

† Test routines apply to procedures to be developed in Chapter 5.

STEP	PROCEDURE	ENTER	PRESS	DISPLAY
	Note: For one-sided intervals, enter α and ignore l or u, as the case may be.			
$N(\sigma^2)$	*Normal Variance*			
1	Enter data using DE *or*:			
	a. Enter sample size.	n	[STO]03	n
	b. Enter sample standard deviation.	s	[STO]38	s
2†	*Test* H_0: $\sigma^2 = \sigma_0^2$.			
	a. Enter H_1-code.	H_1-code	[C′]	H_1-code
	b. Enter σ_0^2, and compute P-value.	σ_0^2	[R/S]	P
	CI for σ^2			
3	a. Calculate degrees of freedom.		[C]	υ
	b. Enter Chi-square percentiles, ($\upsilon = n - 1$), and calculate limits.	$\chi^2_{1-\alpha/2}$ $\chi^2_{\alpha/2}$	[STO]41 [STO]31 [R/S] [$x \gtrless t$]	$\chi^2_{1-\alpha/2}$ $\chi^2_{\alpha/2}$ l u
	Note: For upper one-sided interval, enter $\chi^2_{1-\alpha}$ in R_{41} and R_{31}, then find u in R_D. For lower one-sided interval, enter χ^2_{α} in R_{41} and R_{31}, then find l in R_D.			
$Exp(\mu)$	*Exponential Mean*			
1	a. Enter sample size.	n	[STO]03	n
	b. Enter sample mean.	$\bar{x}$	[STO]37	$\bar{x}$
2†	*Test* H_0: $\mu = \mu_0$.			
	a. Enter H_1-code.	H_1-code	[E′]	H_1-code
	b. Enter μ_0, and compute P-value.	μ_0	[R/S]	P
3	*CI for* μ			
	a. Calculate degrees of freedom.		[E]	υ
	b. Enter Chi-square percentile ($\upsilon = 2n$), and calculate limits.	$\chi^2_{1-\alpha/2}$ $\chi^2_{\alpha/2}$	[STO]41 [STO]31 [R/S] [$x \gtrless t$]	$\chi^2_{1-\alpha/2}$ $\chi^2_{\alpha/2}$ l u
	Note: See Previous Note for One-sided Intervals.			

(*Note:* For any fraction P in R_D, [SBR] [sin] displays z_P.)

† Test routines apply to procedures to be developed in Chapter 5.

Register Contents

00 ↓	10 ↓	20	30 ts	40 $\hat{\theta}$
01	11	21	31 $t_{\alpha/2}$, $\chi^2_{\alpha/2}$	41 $\chi^2_{1-\alpha/2}$
02	12	22	32 SE	42
03 n	13	23	33 Used	43
04 Used	14 Used	24	34 θ_0	44
05	15 ν	25	35	45
06	16	26 Used	36 H_1-code	46
07	17	27	37 $\bar{x}$	47
08	18	28	38 s	48 σ
09 ↓	19 ↓	29 Used	39 $e(\hat{\theta} \pm e)$	49

TABLE T. STUDENT –t PERCENTILES

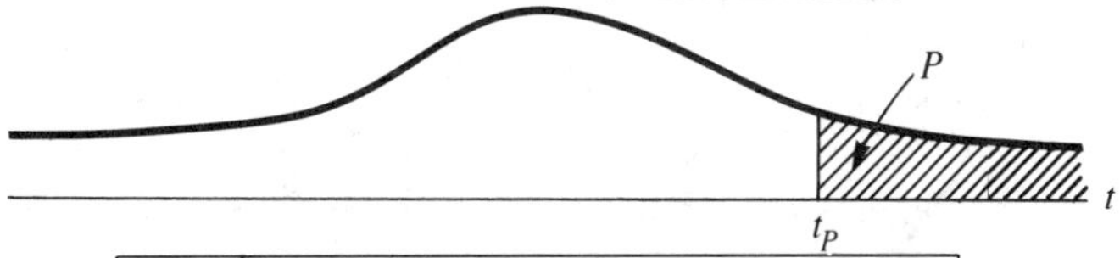

	P				
ν	.10	.05	.025	.01	.005
1	3.078	6.314	12.71	31.82	63.66
2	1.886	2.920	4.303	6.965	9.925
3	1.638	2.353	3.182	4.541	5.841
4	1.533	2.132	2.776	3.747	4.604
5	1.476	2.015	2.571	3.365	4.032
6	1.440	1.943	2.447	3.143	3.707
7	1.415	1.895	2.365	2.998	3.499
8	1.397	1.860	2.306	2.896	3.355
9	1.383	1.833	2.262	2.821	3.250
10	1.372	1.812	2.228	2.764	3.169
11	1.363	1.796	2.201	2.718	3.106
12	1.356	1.782	2.179	2.681	3.054
13	1.350	1.771	2.160	2.650	3.012
14	1.345	1.761	2.145	2.624	2.977
15	1.341	1.753	2.131	2.602	2.947
16	1.337	1.746	2.120	2.583	2.921
17	1.333	1.740	2.110	2.567	2.898
18	1.330	1.734	2.101	2.552	2.878
19	1.328	1.729	2.093	2.539	2.861
20	1.325	1.725	2.086	2.528	2.845
21	1.323	1.721	2.080	2.518	2.831
22	1.321	1.717	2.074	2.508	2.819
23	1.320	1.714	2.069	2.500	2.807
24	1.318	1.711	2.064	2.492	2.797
25	1.316	1.708	2.060	2.485	2.787
26	1.315	1.706	2.056	2.479	2.779
27	1.314	1.703	2.052	2.473	2.771
28	1.313	1.701	2.048	2.467	2.763
29	1.311	1.699	2.045	2.462	2.756
30	1.310	1.697	2.042	2.457	2.750
> 30	1.282	1.645	1.960	2.326	2.576

TABLE C. CHI-SQUARE PERCENTILES

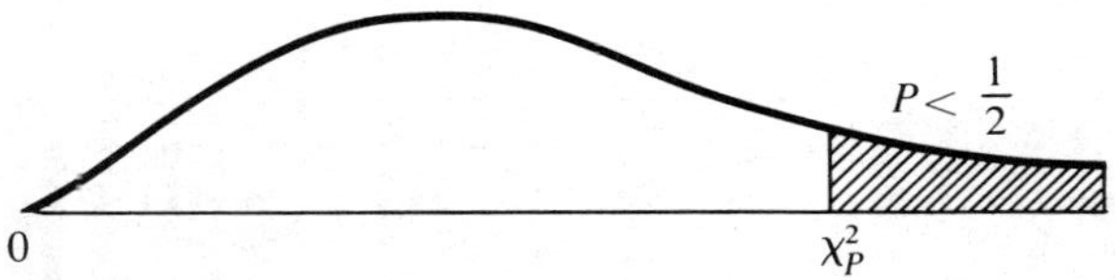

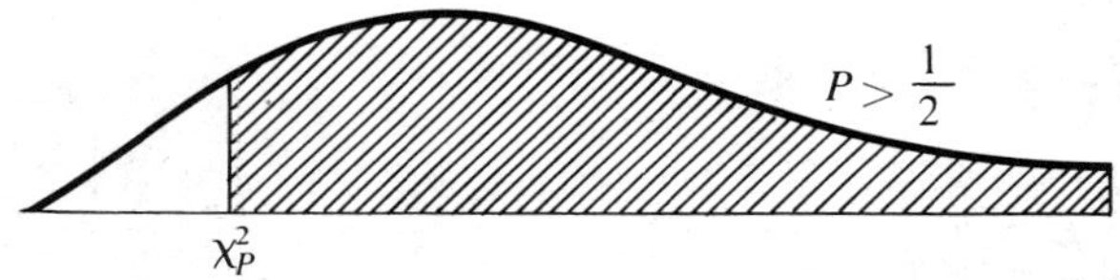

ν	P									
	.995	.99	.975	.95	.90	.10	.05	.025	.01	.005
1	.000	.000	.001	.004	.016	2.71	3.84	5.03	6.63	7.88
2	.010	.020	.051	.103	.211	4.61	5.99	7.38	9.21	10.6
3	.072	.115	.216	.352	.584	6.25	7.81	9.35	11.3	12.8
4	.207	.297	.484	.711	1.06	7.78	9.49	11.1	13.3	14.9
5	.412	.554	.831	1.15	1.61	9.24	11.1	12.8	15.1	16.7
6	.676	.872	1.24	1.64	2.20	10.6	12.6	14.4	16.8	18.5
7	.989	1.24	1.69	2.17	2.83	12.0	14.1	16.0	18.5	20.3
8	1.34	1.65	2.18	2.73	3.49	13.4	15.5	17.5	20.1	22.0
9	1.73	2.09	2.70	3.33	4.17	14.7	16.9	19.0	21.7	23.6
10	2.16	2.56	3.25	3.94	4.87	16.0	18.3	20.5	23.2	25.2
11	2.60	3.05	3.82	4.57	5.58	17.3	19.7	21.9	24.7	26.8
12	3.07	3.57	4.40	5.23	6.30	18.5	21.0	23.3	26.2	28.3
13	3.57	4.11	5.01	5.89	7.04	19.8	22.4	24.7	27.7	29.8
14	4.07	4.66	5.63	6.57	7.79	21.1	23.7	26.1	29.1	31.3
15	4.60	5.23	6.26	7.26	8.55	22.3	25.0	27.5	30.6	32.8
16	5.14	5.81	6.91	7.96	9.31	23.5	26.3	28.8	32.0	34.3
17	5.70	6.41	7.56	8.67	10.1	24.8	27.6	30.2	33.4	35.7
18	6.26	7.01	8.23	9.39	10.9	26.0	28.9	31.5	34.8	37.2
19	6.84	7.63	8.91	10.1	11.7	27.2	30.1	32.9	36.2	38.6
20	7.43	8.26	9.59	10.9	12.4	28.4	31.4	34.2	37.6	40.0

ν	P									
	.995	.99	.975	.95	.90	.10	.05	.025	.01	.005
21	8.03	8.90	10.3	11.6	13.2	29.6	32.7	35.5	38.9	41.4
22	8.64	9.54	11.0	12.3	14.0	30.8	33.9	36.8	40.3	42.8
23	9.26	10.2	11.7	13.1	14.8	32.0	35.2	38.1	41.6	44.2
24	9.89	10.9	12.4	13.8	15.7	33.2	36.4	39.4	43.0	45.6
25	10.5	11.5	13.1	14.6	16.5	34.4	37.7	40.6	44.3	46.9
26	11.2	12.2	13.8	15.4	17.3	35.6	38.9	41.9	45.6	48.3
27	11.8	12.9	14.6	16.2	18.1	36.7	40.1	43.2	47.0	49.6
28	12.5	13.6	15.3	16.9	18.9	37.9	41.3	44.5	48.3	51.0
29	13.1	14.3	16.0	17.7	19.8	39.1	42.6	45.7	49.6	52.3
30	13.8	15.0	16.8	18.5	20.6	40.3	43.8	47.0	50.9	53.7
31	14.5	15.7	17.5	19.3	21.4	41.4	45.0	48.2	52.2	55.0
32	15.1	16.4	18.3	20.1	22.3	42.6	46.2	49.5	53.5	56.3
33	15.8	17.1	19.0	20.9	23.1	43.7	47.4	50.7	54.8	57.6
34	16.5	17.8	19.8	21.7	24.0	44.9	48.6	52.0	56.1	59.0
35	17.2	18.5	20.6	22.5	24.8	46.1	49.8	53.2	57.3	60.3
36	17.9	19.2	21.3	23.3	25.6	47.2	51.0	54.4	58.6	61.6
37	18.6	20.0	22.1	24.1	26.5	48.4	52.2	55.7	59.9	62.9
38	19.3	20.7	22.9	24.9	27.3	49.5	53.4	56.9	61.2	64.2
39	20.0	21.4	23.7	25.7	28.2	50.7	54.6	58.1	62.4	65.5
40	20.7	22.2	24.4	26.5	29.0	51.8	55.8	59.3	63.7	66.8

$\nu > 40$ Store ν in R_{15}, enter desired P, and press [D] in ZS–4.

Estimation

1. To study the effects of a drug, 9 athletes were timed in a series of physical tests and yielded an average of $\bar{x} = 10.13$ minutes. It was assumed in the study that $\sigma = 1$ and that reaction times are normally distributed.
   a. Find a 90% CI for the mean reaction time μ.
   b. Determine a 99% lower one-sided interval for μ.
   c. Find a one-sided upper bound on μ having risk 15%.

Solution 1.

ZS STEP	ENTER	PRESS	DISPLAY	COMMENTS
N $(\mu\|\sigma)$				
1a	9	[STO]03	9.	Places the sample size in R_{03}
1b	10.13	[STO]37	10.13	Stores the sample average in R_{37}
2	1	[STO]48	1.	Stores known σ-value in R_{48}
4	.05	[B]	9.58	Enter $\alpha/2 = .10/2$; display l.
		$[x \leftrightarrows t]$	10.68	Exchange and display u: 90% CI for μ is (9.58, 10.68).
4	.01	[B]	9.35	Enter $\alpha = .01$ and find $l = 9.35$, so confidence is 99% that $\mu >$ 9.35, solving (b) (R_T is not examined).
4	.15	[B]	9.78	Using $\alpha = .15$, l is calculated, but ignored.
		$[x \leftrightarrows t]$	10.48	The T-register yields required upper limit on μ, solving (c).

2. Four specimens of an expensive cloth were subjected to strength tests, and the breaking strengths in psi were recorded as 181, 173, 176, 175. The standard deviation based on past experience is 5 psi. Assume normality.
   a. Find a 95% CI for μ, the mean breaking strength.
   b. What is a lower one-sided bound for μ with confidence 90%?

Solution 2.

ZS STEP	ENTER	PRESS	DISPLAY	COMMENTS
DE		[Pgm]03	*xxx*	
1		[E]	1.	Initialize ST–03 (ungrouped) for data entry.
	181	[A]	1.	First breaking strength entered
	173	[A]	2.	Second breaking strength entered
	176	[A]	3.	Third breaking strength entered
	175	[A]	4.	Fourth breaking strength entered
2		[RST]	4.	Data entry concluded; exit ST–03.
3		[D′]	0.	Data processed
N($\mu\|\sigma$)				
2	5	[STO]48	5.	Stores known σ-value in R_{48}
4	.025	[B]	171.3	Entering $\alpha/2$ for $\alpha = .05$; l is displayed.
		[$x \rightleftarrows t$]	181.2	T-register yields u. CI: (171.3, 181.2) is reported and (a) is resolved.
4	.10	[B]	173.	The 90% lower limit for (b) of 173. is found using $\alpha = .10$.

3. (Problem 6 revisited) Five specimens of coke tested for porosity showed weight gains of 2.16, 2.19, 2.31, 2.30, and 2.21, all in pounds. The variance of the process is unknown. Find a 90% CI for the mean weight gain. Find estimates of μ and σ and SE.

Solution

ZS STEP	ENTER	PRESS	DISPLAY	COMMENTS
DE		[Pgm]03	*xxx*	
1		[E]	1.	Initialize ST–03 (ungrouped) for data entry.
	2.16	[A]	1.	First weight entered
	2.19	[A]	2.	Successive weights entered
	2.31	[A]	3.	
	2.30	[A]	4.	
	2.21	[A]	5.	
2		[RST]	5.	Exit ST–03.
3		[D′]	0.	Process data.

ZS STEP	ENTER	PRESS	DISPLAY	COMMENTS
N(μ)				
3a		[A]	4.	Display $\nu = 4$.
3b	2.132	[R/S]	2.17	Lower confidence limit displayed
		[$x \rightleftarrows t$]	2.30	Upper limit retrieved from R_T
		[RCL]40	2.234	Retrieve $\hat{\mu} = \bar{x}$, the estimate of μ.
		[RCL]38	.067	Retrieve $\hat{\sigma}$, the estimate of σ.
		[RCL]32	.030	Retrieve $s/\sqrt{n}$, the estimate of SE.
	Report	$2.17 < \mu < 2.30$	or	90% CI for μ is (2.17, 2.30).

4. (Problem 9 revisited) The data of Problem 9 produce $\bar{x} = 2.268$, $n = 5$ and $s = .225$. Determine a 90% lower one-sided CI for μ and an upper 99% CI for μ.

Solution

ZS STEP	ENTER	PRESS	DISPLAY	COMMENTS
N(μ)				
1a	5	[STO]03	5.	Enters the sample size in R_{03}
1b	2.268	[STO]37	2.268	Enters the sample average in R_{40}
1c	.225	[STO]38	0.225	Enters the sample s.d. in R_{38}
3a		[A]	4.	Display $\nu = 4$.
3b	1.533	[R/S]	2.11	Lower limit is displayed;
3a		[A]	4.	R_T ignored.
3b	3.747	[R/S]	(1.89)	$t_{.01}$ entered and lower limit ignored;
		[$x \rightleftarrows t$]	2.65	R_T yields the upper one-sided limit.

5. (Example 4.3) Two hundred incoming freshman were checked for pulse rates and $s^2 = 100$ recorded. Find a 95% CI for σ^2 and σ.

Solution

ZS STEP	ENTER	PRESS	DISPLAY	COMMENTS
N(σ^2)				
1a	200	[STO]03	200.	Entering summary data
1b	100	[$\sqrt{x}$]	10.	
		[STO]38	10.	Given s stored in R_{38}
3a		[C]	199.	Since $\nu = 199 > 40$, D must be used to calculate $\chi^2_{.975}$
3b	0.975	[D]	161.81	Lower percentile point calculated
		[STO]41	161.81	and stored in R_{41}
3b	0.025	[D]	239.97	Upper percentile point calculated
		[STO]31	239.97	and stored in R_{31}
		[R/S]	82.93	l displayed
		[$x \rightleftarrows t$]	122.98	Value of u retrieved from R_D
		[$\sqrt{x}$]	11.09	Value of u for CI on σ.
		[$x \rightleftarrows t$]	82.93	Value of l for σ^2 recovered from R_T
		[$\sqrt{x}$]	9.11	Value of l for CI on σ.

ZS STEP	ENTER	PRESS	DISPLAY	COMMENTS

6. Five analysts took readings on a 5-gram weight using a delicate balance and reported 5.001, 4.999, 5.002, 5.001, and 5.002. Find a 95% upper one-sided confidence limit on σ^2 and σ.

Solution

ZS STEP	ENTER	PRESS	DISPLAY	COMMENTS
DE		[Pgm]03	*xxx*	
1		[E]	1.	Initialize ST–03 for ungrouped data entry.
	5.001	[A]	1.	First data value entered
	4.999	[A]	2.	Succeeding data points entered
	5.002	[A]	3.	
	5.001	[A]	4.	
	5.002	[A]	5.	
2		[RST]	5.	Data entry concluded; all flags lowered
3				
N(σ^2)				
3a		[D′]	0.	Data processed
3b		[C]	4.	Display $\nu = 4$.
	.711	[STO]41	0.711	$\chi^2_{.95}$ with $\nu = 4$ from Chi-square table
		[STO]31	0.711	entered in R_{41}, and in R_{31}
		[R/S]	.0000084	Upper one-sided limit for σ^2 in R_D
		[$\sqrt{x}$]	.0029	Upper one-sided limit for σ

7. Times to failure for 6 expensive pieces of electronic equipment were recorded in hours as 233.6, 1402.7, 3119.0, 612.9, 258.3, and 2211.2.
   a. Find a 95% CI for mean time to failure.
   b. Determine a point estimate and a lower one-sided 95% confidence limit on the reliability at 500 hours.

Solution

ZS STEP	ENTER	PRESS	DISPLAY	COMMENTS
DE		[Pgm]03	*xxx*	
1		[E]	1.	Initialize ST–03 for raw data entry.
	233.6	[A]	1.	First time to failure entered
	1402.7	[A]	2.	Succeeding times to failure
	3119	[A]	3.	entered and processed
	612.9	[A]	4.	
	258.3	[A]	5.	
	2211.2	[A]	6.	
2		[RST]	6.	Exit ST–03 and return to ZS-4 control.
3		[D′]	0.	Date processed

ZS STEP	ENTER	PRESS	DISPLAY	COMMENTS
Exp(μ)				
3a		[E]	12.	Display $\nu = 2n$.
3b	4.4	[STO]41	4.4	Storing lower percentile in R_{41}
	23.3	[STO]31	23.3	Storing upper percentile in R_{31}
		[R/S]	673.	Display l
		[$x \rightleftarrows t$]	3563.	Find u, so CI is (673,3563).
		[RCL]40	1306.	$\hat{\mu} = \bar{x} = 1306$
		[$1/x$]	.0008	$\hat{\lambda} = 1/\bar{x} = 0.0008$
		[$\times$]	.0008	Multiplying by
	500	[=]	.3828	500 to find $500\hat{\lambda}$
		[+/−]	−.3828	Change sign for exponentiation
		[INV] [*lnx*]	.68	to yield estimate of $R(500)$.
3a		[E]	12.	Display ν to start new problem.
3b	21.	[STO]41	21.	Store required $\chi^2_{.05}$ in R_{41}
		[STO]31	21.	and in R_{31} for one-sided limit.
		[R/S]	746	Display required lower limit on μ.
		[$1/x$]	.0013	Upper limit on λ
		[$\times$]	.0013	Multiplying by 500 to
	500	[=]	.6698	find upper limit on 500λ
		[+/−]	−.6698	Lower limit on -500λ
		[INV] [*lnx*]	.51	Lower bound on $R(500)$

5

Hypothesis Testing

Section 5.1: Basic Concepts

Many times the goal of statistical analysis is not so much to estimate the value of a parameter directly, but rather to decide whether or not some prescribed statement about that parameter is true or false. This leads to another area of statistical analysis known as testing hypotheses. For example, a plastics engineer may wish to know whether or not the product under development meets the specifications of having a mean (μ) breaking strength of 50 pounds per square inch (p.s.i.). Ten samples of the product yielded an average ($\bar{x}$) breaking strength of 47.6 p.s.i. with a standard deviation of 3.81. Here, the value $\mu = 50$ is prescribed, and the question is really one of whether or not the observed average value is different enough from $\mu = 50$ (some difference is to be expected) to conclude that $\mu \neq 50$. Here the prescribed value is called a hypothesis about μ and is denoted $H_0{:}\mu = 50$. A decision whether or not to reject H_0 is called a *test* of the hypothesis. In general, using our generic notation θ for a parameter, a hypothesis is a statement of the form $H_0{:}\theta = \theta_0$ where θ_0 is some specific value in the range of possibilities under study.

There are many *procedures* one might use to test a hypothesis H_0. One that suggests itself on the basis of the last chapter is the following. In the example, the engineer might proceed (assuming normality) to construct a 95% CI for μ. Using the data and program ZS–4, you may verify that the interval is (44.9, 50.3). To the extent that this interval represents the set of plausible values of μ and the fact that $\mu = 50$ is among those values, the hypothesis is an acceptable statement about the state of nature, μ. Had the computed interval been, say (42.7, 48.1), then the engineer would have evidence that $\mu \neq 50$ and would have a basis for rejecting H_0 since 50 is not among the plausible values.

The procedure just outlined is a standard one that may be used to test $H_0{:}\theta = \theta_0$ when a $100(1 - \alpha)\%$ CI formula is available for θ. It is often reexpressed in a form as follows. For the plastics example, you recall that the formula for the CI for μ is given by $\bar{x} \pm t_{\alpha/2}\, s/\sqrt{n}$. Accepting (or failing to reject) $H_0{:}\mu = 50$, if and only if 50 may be found in the interval, is equivalent to observing whether or not $\sqrt{n}(\bar{x} - 50)/s$ is between $-t_{\alpha/2}$ and $t_{\alpha/2}$. In this case the quantity $ts = \sqrt{n}(\bar{x} - 50)/s$ is called the value of a *test statistic*, *ts*, and testing $H_0{:}\mu = 50$ is one of observing whether or not *ts* lies between certain computed percentiles, namely, the percentiles of a *t*-distribution, the distribution of $TS = \sqrt{n}\,(\bar{X} - 50)/S$ (note the capital letters), when in fact $\mu = 50$. Then the confidence coefficient $1 - \alpha$ (95% in the example) is the probability of accepting H_0 when H_0 is true, and, of course, α is then the probability of rejecting H_0 when H_0 is true ($\mu = 50$). In this context α is called the *level* or *size* of the test.

Since, in fact, it will never be known whether or not H_0 is true, there is always a risk that the wrong action will be taken whenever a decision is based on sample data. There are two possible errors here, and they are given traditional names. The error of rejecting H_0 when it is true is called a *Type I error*, and α is the probability of committing such an error. The other error, called *Type II*, is one of failing to reject a false hypothesis, and the probability of committing such an error is denoted β or, more generally, $\beta(\theta)$, to emphasize that its size depends on which value, other than θ_0 (the hypothesized value), is leading to the error.

In most books on the subject, you will usually find the 4 contingencies the experimenter faces in a hypothesis testing situation conveniently summarized, as in Table 5.1. The risks of the 2 types of errors are inevitable, and they are related in a rather subtle way. All other things being equal, lowering α only increases β and vica versa. This is difficult to illustrate, because in general the computation of β is no simple task. Usually this is illustrated only with the artificial case when σ is known and μ is unknown. Since we will not be able to program more general formulas, we will not explore such details here. In that same general context, you will often find the probability of rejecting H_0, as a function of θ, called the *power* of the test. The power at $\theta = \theta_0$ is of course just α; otherwise, it is a measure of the ability of the test to "dis-

TABLE 5.1. FOUR POSSIBLE RESULTS OF A TEST OF H_0

Actual State \ Action	Reject H_0	Do not reject H_0
H_0 is true	*Type I error* Probability = α = level of the test	*No error* Probability = $1 - \alpha$ = confidence level
H_0 is false	*No error* Probability = $1 - \beta$	*Type II error* Probability = β

tinguish" values of θ different from θ_0 so as to lead to rejection. Similarly, the probability of accepting H_0, as a function of θ, is called the *operating characteristic* of the test, and, when $\theta \neq \theta_0$, produces the Type II error for that choice of θ. Again, these matters will not be considered here, but rather we will concentrate on processing formulas and procedures in the standard situations you will normally encounter.

The general procedure for testing $H_0{:}\theta = \theta_0$, then, is to select a test statistic *TS* (a random variable usually involving the hypothesized value θ_0) and a region, *C*, called a *critical region*, so that if the observed value, *ts*, of the test statistic falls in *C*, H_0 is rejected—and only then. The region *C* depends upon the probability distribution of *TS* when H_0 is true, and is dictated by the requirement α = Pr (rejecting $H_0 | H_0$ is true). The choice of *C* also depends upon another aspect of the problem, namely, what values other than $\theta = \theta_0$ are to be considered. The set of such values is called the *alternative* to the hypothesis and is denoted H_1. There are only three situations that are usually considered, and they are $H_1{:}\theta \neq \theta_0$, $H_1{:}\theta > \theta_0$, and $H_1{:}\theta < \theta_0$.

The last two alternatives are called, for obvious reasons, *one-sided*, and they deserve some comment. You saw one-sided confidence intervals in the last chapter, and those are related to one-sided alternatives. Take the plastics example once again. If the concern of the engineer is whether $\mu < 50$ rather than $\mu \neq 50$ (the higher the better), he would be better off computing the upper one-sided CI for μ and rejecting H_0, if and only if 50 falls above the upper limit. With $n = 10$, $\bar{x} = 47.6$, and $s = 3.81$, that interval is given by $(-\infty, 49.8)$. This suggests that H_0 should be rejected, since 50 (and any value larger) is not one of the plausible values; indeed, the plausible values are more compatible with $H_1{:}\mu < 50$. Had the interval extended past 50, then the data would not have been able to distinguish 50 from other plausible values of μ, and then H_0 would not be rejected.

If this procedure is followed, then rejecting H_0, if and only if $50 > u = \bar{x} + t_{.05}\, s/\sqrt{n}$, is equivalent to rejecting H_0 if $\bar{x} < 50 - t_{.05}\, s/\sqrt{n}$, that is, if $\bar{x}$ (the point estimate of μ) is "too small." Equivalently, H_0 would be rejected when $ts = \sqrt{n}(\bar{x} - 50)/s < -t_{.05}$ in terms of the chosen test statistic. The critical region, C, here is just the interval from $-\infty$ to $-t_{.05} = -1.833$. It should be clear that to test H_0 against $H_1{:}\mu > 50$, a *lower* one-sided interval (l, ∞) would be computed and H_0 rejected, if and only if $50 < l$. (In the example, $l = 45.4$, so H_0 would not be rejected.) This would be equivalent to rejecting H_0 precisely when $\bar{x}$ is "too large," or, equivalently, when the test statistic, ts, falls in the critical region, C, from $t_{.05} = +1.833$ to $+\infty$.

When confidence interval formulas are available for θ, the foregoing procedures may be safely carried out for testing $H_0{:}\theta = \theta_0$ against the three forms of H_1 and produces what some writers call classical tests or accept/reject tests. If the confidence level selected is $1 - \alpha$, then α is precisely the size or level of the test and measures the Type I error. Very often the corresonding test statistic is in the form of $ts = (\hat{\theta} - \theta_0)/\mathrm{SE}$, where SE is the (estimated) standard error (standard deviation) of $\hat{\theta}$ when H_0 is true. (In the plastics example, $\hat{\theta} = \bar{x}$, $\theta_0 = 50$, and $\mathrm{SE} = s/\sqrt{n}$.) Tests of H_0 are sometimes described as tests of whether $\hat{\theta}$ differs *significantly* from θ_0, the hypothesized value. In that context, when H_0 is rejected, the departure of $\hat{\theta}$ from θ_0 is said to be (statistically) significant. The reason for the parenthetical remark is to emphasize that the observed difference need not be of any practical or real significance. Thus, the discovery that 49 is a more likely value of μ than 50, because the former is in the CI and the latter is not, may mean very little to the engineer. By the same token, failing to detect such a difference may not be of any practical consequence either.

This leads to another form of testing H_0, called *significance tests* by many authors. The idea is to examine the data relative to the hypothesis and decide how "extreme" the data are. More specifically if H_1 is the one-sided version, $\theta < \theta_0$, H_0 would be rejected, if the observed test statistic ts is too small. So, the probability of observing a value that small or smaller, $\Pr(TS \le ts)$, is computed using the probability distribution (parameter θ_0), when H_0 is true. Such a value is called the *significance level* of the test, for it represents the smallest value of α for which H_0 would be rejected. This probability is also referred to as *probable value, prob-value,* or simply *P-value.* We defer to the latter. Thus, $P = P(ts)$ *when* H_1 is of the form $\theta < \theta_0$.

EXAMPLE 5.1 (PLASTICS EXAMPLE)

The engineer observed $\bar{x} = 47.6$, $s = 3.81$ in a sample of $n = 10$. To test $H_0{:}\mu = 50$ against $H_1{:}\mu < 50$, the test statistic $ts = \sqrt{n}(\bar{x} - \mu_0)/s = -1.99$ is to be used. When $\mu = 50$, TS has a t-distribution with 9 degrees of freedom. The P-value is just this t-CDF at -1.99, that is, $P = P(-1.99) = .04$. Hence, if the experimenter had selected $\alpha = .04$ or higher, H_0 would be rejected, but for any $\alpha < .04$, H_0 would not be rejected.

When H_1 is in the form $\theta > \theta_0$, significance level is defined to be $P = \Pr(TS > ts) = Q(ts)$ and evaluates the likelihood of observing a value as large or larger than ts, if H_0 is true. When H_1 is two-sided, some special problems are created. If the underlying probability distribution is symmetric (like the t and normal), significance level is usually defined to be P $= \Pr(|TS| \geq |ts|), = 2Q(|ts|)$ so that H_0 will be rejected when $Q(|ts|) < \alpha/2$. [Observe that when $ts > 0$, $P = 2Q(ts)$ while if $ts < 0$, $P = 2Q(-ts) = 2P(ts)$ for this case.] But when the distribution of TS is nonsymmetric (such as the Chi-square), there is no general agreement, as of this writing, as to how the significance level should even be defined.* Several alternatives have been considered, but none are universally accepted. To be definitive and consistent with the symmetric case, we will in this book mimic that case and simply double the one-sided P-value. Thus, if $\Pr(TS \leq ts) > .5$, $P = 2Q(ts)$, while if $\Pr(TS \leq ts) < .5$, $P = 2P(ts)$.

Reporting a significance level or P-value for a problem has its advantages and disadvantages. On the one hand, it is a nice way of summarizing the data in a problem. When H_0 is rejected just because ts is "large" or "small" in an accept/reject or classical test, little is revealed about how extreme the observed ts value is. On the other hand, merely reporting P as that value at which H_0 would just be rejected does not settle the issue of whether or not H_0 should be rejected. If P should turn out to be .000001, for example, almost anyone would agree that H_0 should be rejected, for the data then are too incompatible with H_0 to believe that H_0 is true, and that such extreme values would arise by chance —another possible explanation of the result. If P should be .999999, by the same token, almost anyone would accept H_0, or fail to reject it. Somewhere between these extremes, there should be a value below which we would tend to reject and above which we would tend not to reject H_0. But that is precisely where the size α enters the picture. While there might be many ways to select α, the traditional choices are $\alpha = .05$ (a default value), .01, and .10. You will find these choices in nearly every textbook you encounter, and we will defer to these also. In summary, then, we will combine the 2 approaches by computing a P-value for the test, and then comparing it to a selected α (typically 5%), rejecting H_0, if $P \leq \alpha$, and accepting or failing to reject H_0, if $P > \alpha$. This is the procedure we will follow in nearly all of our problem solutions.

EXAMPLE 5.2

A manufacturer of TV tubes is considering a new method of producing tubes. In the past, mean time to failure of the tubes has been established as 2,197 hours. He thinks this new method will increase MTTF, but is not positive, so he samples and finds that a sample of 30 tubes produces an average time to failure of 2,201 hours, with a standard deviation of

* A nice summary of the problem may be found in Jean D. Gibbons and John W. Pratt, "*P*-value: Interpretation and Methodology," *The American Statistician*, Vol. 29, No. 1 (1975).

24 hours. He wonders if such an increase can be ascribed to chance, or has μ really increased?

Solution Assuming normality, the manufacturer tests $H_0{:}\mu = 2197$ against $H_1{:}\mu > 2197$. The proper test statistic to use is $ts = \sqrt{n}(\bar{x} - 2197)/s = .91$ for his data. The significance level of the data, then, is given by $P = Q(.91) = .18$, using the t-distribution with 29 degrees of freedom. Such a value lends a little support to H_0. In any case, if $\alpha = .05$ is used, H_0 should *not* be rejected, and the resulting observed high average may be ascribed to chance.

Incidentally, this does not *prove* that H_0 is true. It merely establishes that the data could have reasonably arisen from a population having $\mu = 2197$, at least the event is not that rare ($P{=}.18$).

All of these matters can be automated for calculator solutions and will be amply illustrated in the ensuing material. Before turning to details, however, two matters need to be discussed. First, a classical test of the more complicated $H_0{:}\theta \leq \theta_0$ vs. $H_1{:}\theta > \theta_0$, or $H_0{:}\theta \geq \theta_0$ vs. $H_1{:}\theta < \theta_0$, is, for the cases we consider, precisely the same procedure as testing the simpler $H_0{:}\theta = \theta_0$ in each case. Only the value θ_0 enters the computation. The difference is that the prescribed α is an upper bound on the Type I error, Pr (rejecting $H_0 | H_0$ is true). So, you automatically obtain a more complicated test when you follow the procedure for the simpler one in these one-sided cases. It should be noted that, regardless of how the problem is posed, θ_0 is always located in H_0, *not* H_1.

A related question is the choice of H_1 in the formulation of the problem. It is the one aspect that we cannot automate, because it is not a mathematical consideration. For instance, in Example 5.2, suppose the manufacturer claims the new method increases MTTF, μ. To test this claim, you might make the hypothesis $H_0{:}\mu \geq 2197$ (remember, θ_0 always goes in H_0), thereby making the alternative $H_1{:}\mu < 2197$. In this form, rejecting H_0 is a rejection of the claim, and α is the protection, then, against rejecting the claim when it is true. That protection may be prescribed (at the expense of β, of course), and you see it protects the manufacturer against rejecting his claim at the expense, possibly, of β, the probability of accepting his claim when it is false. In many situations, it is common practice to do this, and, for obvious reasons, a prescribed value α is called the *producer's risk*, and β, the *consumer's risk*.

In the same context, however, you might have reasonably tested $H_0{:}\mu < 2197$ against $H_1{:}\mu > 2197$. This places the claim in H_1, rather than H_0, and now *rejecting* H_0 means *accepting* the claim, whereas failing to reject H_0 is equivalent to rejecting the claim. Now α is the protection against accepting a false claim, which means that the claim would be accepted (H_0 rejected) only if the data are extreme. This is the way Example 5.2 was carried out, and you saw there that a P-value of .18 was not judged small enough to reject H_0, hence support the claim. This puts the burden of proof on the data to support the claim and is more in line with the so-called scientific method. It rather resembles our legal standard of "innocent until proven guilty." It also protects against making hasty changes, for if H_0 is not rejected, typically the status quo is maintained. You will find that this formulation is used in the vast majority of cases where claims or statements of change are involved or simply where objectivity of evidence is to be emphasized. In this sense, H_0 is a disclaimer, a denial of the claim, and this is one source of the term *null* hypothesis.

But there are no hard and fast rules in the matter of selecting H_1, and you will find problems formulated either way, often with no justification one way or the other. Does it make a difference? Indeed it does, but it is a subtle difference; namely, since α is usually all we can compute easily, hence control, the choice of H_1 determines which of the 2 errors is protected against. Thus, H_1 should be selected in such a way that α guards against the more serious of the 2 consequences. That may be difficult to decide, or it may be a matter of indifference. In this book we will generally place *claims* in H_1, thereby evoking the scientific method and aligning with most authors. Occasional departures will be noted when they arise. In most problems, we will calculate and display the significance level, an appropriate confidence interval, and carry out a prescribed (usually 5%) classical test, thereby presenting you with all the options.

Section 5.2: Tests for Means

You may have noticed that program ZS–4 is already arranged for carrying out tests for each parameter considered there. Hence the same user instructions may be consulted, which is quite appropriate, since often you will want to examine confidence intervals along with tests. At the end of this chapter, you will find model problems implementing this part of ZS–4. In each case, an H_1-code is defined to identify for the calculator subroutine which alternative is being considered. The number +1 is used for $H_1{:}\theta > \theta_0$, −1 for $H_1{:}\theta < \theta_0$, and 0 for two-sided alternatives. Following that entry, you must enter θ_0 itself, which is the hypothesized value of θ for $H_0{:}\theta = \theta_0$. As previously remarked, the same procedure tests the more complicated cases, $H_0{:}\theta \leq \theta_0$ and $H_0{:}\mu \geq \theta_0$. No matter how the problem is formulated, remember that θ_0 goes in H_0 and is the value entered at this second step. A simple press of [R/S] then produces and displays the significance level or *P*-value. After execution of each program, you will find certain key quantities stored for you. For example, $\hat{\theta}$ will always be found in R_{40}, and the value of *ts*, the test statistic, will be found in R_{30}. These may be of interest to you.

EXAMPLE 5.3

Weights of filled cereal boxes are normally distributed with mean μ ounces and standard deviation $\sigma = 1$ ounce. The vendor claims that the mean weight is at least 10 ounces. A random sample of 11 boxes yields the following weights: 10.2, 11.1, 10.4, 10.0, 9.9, 10.3, 9.7, 10.9, 11.3, 9.9, 8.9. Test the vendor's claim.

Solution Following our guidelines, we would make H_0 the disclaimer $H_0{:}\mu \leq 10$ and test this against $H_1{:}\mu > 10$. (Notice that the boundary value of 10 went into H_0, even though the claim is, technically, $\mu \geq 10$.) This puts the burden of proof on the data to reject H_0 and thereby accept the claim.

Since σ is known, the $N(\mu|\sigma)$ procedures in ZS–4 should be used, with the data entered first using ST–03. At Step 2, the known value of $\sigma = 1$ is stored in R_{48}. At Step 3 enter an H_1-code of +1, and press [B′]. Follow this by an entry of $\mu_0 = 10$, and press [R/S] to see a P-value of $P = .22$, which is very little basis for rejecting H_0; the conclusion is that the vendor's claim is suspect. This is based upon a test statistic value of $ts = .78$ found in R_{30}, and a point estimate of $\bar{x} = 10.2$ found in R_{40}.

Using the default value of $\alpha = 5\%$, a lower one-sided 95% CI on μ may be calculated using ZS–4. The lower limit is $l = 9.7$, and since $\mu_0 = 10$ exceeds this limit, there is no basis for rejecting H_0 at the 5% level of significance. Indeed $P = .22$ indicates that 22% is the lowest level of significance at which H_0 would be rejected.

If you wish to test hypotheses about the mean of a normal population and the variance is *unknown*, then a "t-test" is used, following program $N(\mu)$ of ZS–4. In this case, the test statistic is $ts = (\bar{x} - \mu_0)/(s/\sqrt{n})$. When $\mu = \mu_0$, TS has a t-distribution with $n - 1$ degrees of freedom. For a significance test, the form of the alternate hypothesis is again selected by entering an H_1-code.

EXAMPLE 5.4

Seven observations of measured radiation intensity at a nuclear plant were 3.6, 4.2, 4.0, 4.1, 3.8, 3.9, 4.0. Conduct a one-tailed t-test of $H_0{:}\mu \geq 3.8$ vs. $H_1{:}\mu < 3.8$, with size $\alpha = .05$.

Solution Using the $N(\mu)$ program of ZS–4, you may verify that an upper 95% confidence limit on μ is 4.09. Since this limit does not exclude the boundary point 3.8, H_0 is not rejected. Entering an H_1-code of −1, press [A′]. Then enter 3.8, and press [R/S] to discover $P = .95$, hardly a basis for rejecting H_0, certainly not at $\alpha = .05$.

Significance tests or tests of hypothesis about the mean of an exponential population can be handled in an analogous manner. Use program Exp(μ) in ZS–4 to compute confidence limits or intervals for one-sided and two-sided tests, respectively. Or follow the user instructions to calculate the significance level. The test statistic is $ts = 2n\bar{x}/\mu_0$, where μ_0 is the hypothesized value. When $\mu = \mu_0$, *TS* has a Chi-square distribution with $2n$ degrees of freedom. Sometimes you will encounter problems involving the parameter λ of an exponential population, rather than its mean, μ. Since $\mu = 1/\lambda$, it is easy to convert tests about λ into equivalent tests about μ, so ZS–4 can still be used. For example, to test $H_0{:}\lambda \leq 2$ vs. $H_1{:}\lambda > 2$, rephrase this as a test of $H_0{:}\mu \geq .5$ vs. $H_1{:}\mu < .5$, and conduct the test with an upper confidence limit from ZS–4, or conduct a significance test using an H_1-code of -1.

EXAMPLE 5.5

Times required to complete a certain task on an assembly line were observed over a 15-day period. It is desired to test whether the mean time is greater than the desired upper limit of 1 minute. Conduct a significance test, with the following data: 2.3, .7, .6, .6, .8, 1.0, .4, .5, .6, .5, .9, 2.8, .1, .9, 1.5.

Solution Enter the data using DE in program ZS–4. Use $H_0{:}\mu \leq 1$, so enter H_1-code $= 1$, press [E′]. Then enter the boundary value (1) and press [R/S]. Following this procedure, you will find $P = .55$, which tends to support H_0, rather than H_1. A lower one-sided 95% CI on μ is $(.65, \infty)$, which includes the value $\mu_0 = 1$ and leads to the same conclusion.

The following problems will illustrate all of the matters discussed in this section. Remember, to test with confidence intervals, compute a *lower* one-sided interval for $H_1{:}\theta > \theta_0$, and an *upper* one-sided confidence interval for $H_1{:}\theta < \theta_0$. In either case, reject H_0, if and only if θ_0 is *not* in the interval.

Problems

1. Use the sample results given below to test the null hypothesis $\mu \geq 120$ against the alternative hypothesis $\mu < 120$. Conduct a significance test supposing $n = 81$, $s = 36$, $\bar{x} = 100$.

   Ans: $P = .0000003$

2. Conduct a test of size $\alpha = .01$ for the data of Problem 1.

   Ans: Reject H_0, since $P < \alpha$.

3. A sample of 30 welders yielded a mean age of 36 years, with a standard deviation of 8 years. If the industry ages for welders are normally distributed, with a mean of 39 years, and you wish to draw a conclusion about the welders represented by the sample, what hypothesis setup should you use?

   Ans: $H_0{:}\mu = 39$ vs. $H_1{:}\mu \neq 39$ (Do not let the data suggest the hypothesis value μ_0.)

4. Conduct a significance test for the data of Problem 3.

Ans: $P = .049$

5. Test the hypothesis in Problem 3, with a test of size .05.

Ans: Since $P < .05$, reject H_0.

6. The manufacturer of a battery-operated calculator has made modifications in its electronic display. One result of the modification seems to be an extended lifetime for the battery. From past data, it is known that the battery in the calculator had an average operating life of 120 hours. To investigate if the modification did improve battery life, a random sample of 100 calculators with the new display was obtained, yielding a mean operating lifetime of 122 hours and a standard deviation of 9.5 hours. What are the hypotheses under test?

Ans: $H_0{:}\mu \leq 120$ vs. $H_1{:}\mu > 120$ (120 is the status quo, and H_0 is the disclaimer.)

7. In Problem 6, can we infer that the battery lifetime was extended at the .05 level?

Ans: Since $P = .018 < .05$, reject H_0, and conclude that lifetime was extended.

8. A poultry grower claims that the average weight of a particular group of fryers is at least 30 ounces. Before agreeing to a purchase, a particular customer selected a sample of 25 fryers, which yielded a sample mean of 29.75 ounces and a standard deviation of 1.5 ounces. What are the customer's hypotheses?

Ans: $H_0{:}\mu \leq 30$ vs. $H_1{:}\mu > 30$ with H_0 being the disclaimer.

9. Should the claim in Problem 8 be rejected at the .05 level?

Ans: No; $P = .79$ certainly exceeds 5%.

10. A gasoline retailer has maintained records indicating that weekly sales have averaged 1,600 gallons. In an attempt to increase sales, the retailer issues a small prize with purchases of 10 gallons or more. If, after a 16-week period, the average weekly sales have increased to 1650 gallons, with a standard deviation of 150 gallons, is it wise to continue the prizes; are any special assumptions necessary?

Ans: With $H_0{:}\mu \leq 1600$ vs. $H_1{:}\mu > 1600$, $P = .10$. This is mild evidence in favor of continuing. We are assuming a random sample of size 16 from a normal population.

11. A machine used to fill oil cans has a setting of 32 ounces. The manufacturer wants to test whether or not the machine fills properly. A random sample of 59 cans produced a sample mean of 31.9 ounces and a standard deviation of .40 ounces. At the .05 level, what can you conclude about the machine?

Ans: Test $H_0{:}\mu = 32$ vs. $H_1{:}\mu < 32$ (in which case, $\alpha = .05$ is the producer's risk) The data produce $P = .027 < .05$, so H_0 should be rejected, and the conclusion is that the machine is not achieving its goal.

12. Suppose the situation in Problem 11 had included only 16 cans in the sample. a. Would a t-test be appropriate? b. If the t-test is appropriate, would your conclusion be different?

Ans:

a. Yes, assuming contents are normally distributed.

b. Yes, for now $P = .17$.

13. A lawn mower engine has been designed to have a trouble-free life of at least 2,000 hours. A random sample of 100 engines experienced malfunction at an average of 1,600 hours. If $\alpha = .10$, what can you conclude about the trouble-free life? (Use an exponential model.)

Ans: Test $H_0{:}\mu \geq 2{,}000$ vs. $H_1{:}\mu < 2{,}000$, letting α be the producer's risk; $P = .02 < \alpha$, so reject H_0, and conclude that the lawn mower does not meet specifications.

14. A report claimed that the amount of tar per cigarette for a particular brand was at least 13 grams. If a random sample of 100 cigarettes of this brand yielded a mean of 12.85 grams and a standard deviation of .75 grams, is there evidence the company should refute the claim?

Ans: Test $H_0{:}\mu \leq 13$ vs. $H_1{:}\mu > 13$; P = .98; (refute the claim at almost any reasonable level.)

15. A plumbing company maintains that the time required for a daytime service call does not exceed 25 minutes. If a random sample of 16 calls yields a sample mean of 26.25 minutes, with a standard deviation of 3 minutes, what can be concluded? (Use a test of size .05.)

Ans: Test $H_0{:}\mu \geq 25$ vs. $H_1{:}\mu < 25$, making H_0 the disclaimer; do not reject H_0, since $P = .94$.

16. In problem 15, if the sample size had been 36, would your conclusion change?

Ans: No, the value of $P = .99$ supports H_0 even more.

17. Sweet, Inc., has been manufacturing Sweet Fertilizer for years and has many distributors across the country. Recently, complaints that the 50-pound bag is light have been received from some dealers. To investigate these complaints, Sweet has selected a sample of 116 bags and obtained $\bar{x} = 49.5$ and $s = 1.6$ pounds. At the .05 level, how do you, an analyst for Sweet, react to the complaints?

Ans: Test $H_0{:}\mu \geq 50$ vs. $H_1{:}\mu < 50$; reject H_0 ($P = .0004$).

18. A process is set to produce at the rate of 8.9 tons per hour. The process, however, has produced at a higher and lower rate in the past. The most recent sample of 24 hours showed an average yield of 8.81 tons, with a standard deviation of .20 tons. Does it appear that the process average has changed?

Ans: Test $H_0{:}\mu = 8.9$ vs. $H_1{:}\mu \neq 8.9$, perhaps; with $P = .04$.

19. A certain developing solution is advertised to last at least 25 days under normal use. If a random sample of 36 customers yielded a sample mean of 23 days, is the advertising claim supported? Set $\alpha =$ .01, and assume the producer's point of view.

Ans: Test $H_0{:}\mu \geq 25$ vs. $H_1{:}\mu < 25$, using the exponential model; do not reject H_0 ($P = .33$), and conclude that the claim is supported.

20. Analyze the data of Problem 19 from the consumer's point of view, using a significance test.

Ans: Test $H_0{:}\mu \leq 25$ vs. $H_1{:}\mu > 25$; $P = .67$; do not reject H_0, and conclude the claim is not supported.

21. A teacher's union claims that teachers who are union members have average salaries of at least $30,000 per year. A random sample of 30 union teachers resulted in a mean salary of $28,500 with a standard deviation of $3,000. Test the null hypothesis at the 10% level of significance that the mean salary is $30,000, against the alternative hypothesis that the salary is more than $30,000 for union members.

Ans: $H_0{:}\mu = 30{,}000$; $H_1{:}\mu > 30{,}000$. A 90% lower confidence limit for μ is 27,782. Thus, H_0 is not rejected (the union claim is not compelling). The significance level is $P = .99$, lending a lot of support to H_0.

22. The mean score on an English proficiency test, given for many years to all incoming freshman at a large midwest university, is 72. The chairman of the English department wants to determine if this year's class is average, so he decides to take a random sample of the class scores. Given the following scores, determine at the .05 level if the year's class is average: 97, 72, 86, 46, 85, 84, 81, 98, 52, 96, 55, 78, 78.

Ans: $H_0{:}\mu = 72$; $H_1{:}\mu \neq 72$. 95% confidence interval for μ: (67.2, 87.9) Do not reject H_0; conclude that this year's class is not significantly different; $P = .27$ is the lowest value of α at which H_0 would be rejected.

23. A manufacturer of baseball bats claims that he can supply bats at the same cost as those currently sanctioned for Little Leagues throughout the United States, and that his will last longer than those presently being used. Little League records show that their bats have an average life of 2 years, with a standard deviation of 2 months. Little League officials tested thirty of the manufacturer's bats and found a sample average useful life of 25 months. Should the Little League officials accept the manufacturer's claim at the 5% level of significance?

Ans: $H_0{:}\mu \leq 24$; $H_1{:}\mu > 24$, using an exponential model. Lower 95% limit on μ is 19; $P = .39$. Do not reject H_0, and conclude the manufacturer's claim is not plausible.

24. A random sample of 80 accounts of a major oil company resulted in a mean balance of \$89, with a standard deviation of \$24. The credit manager has advised top management that the average balance of all of the company's accounts is less than \$90. Using a test size of .05 and the sample results, determine if the credit manager is justified in his claim.

Ans: $H_0{:}\mu \geq 90$; $H_1{:}\mu < 90$. Upper 95% limit on μ : 93.41; $P =$ .35. Do not reject H_0, and conclude that the credit manager's claim is not compelling.

25. A quality assurance inspector for the packaging process of Brand Cereal must insure that each box of cereal weighs at least 15 ounces. If he samples 25 boxes and finds an average weight of 14.5 ounces, with a standard deviation of 3 ounces, should he conclude that remedial action is necessary? Assume he has set $\alpha = .01$ for his test.

Ans: Use $H_0{:}\mu \leq 15$ vs. $H_1{:}\mu > 15$. (*Note*: The inspector is a "watchdog," representing the consumer's point of view.) Lower 99% limit on μ is 13.00; $P = .79$. Do not reject H_0; conclude that remedial action might be appropriate.

26. A certain expensive electronic component is supposed to have a mean life of 1,000 hours and a standard deviation of 50 hours. Five components were tested and found to expire at times: 1075, 1085, 1060, 998, and 995. Test the hypothesis at level 10% that the mean has not changed, assuming the standard deviation has not.

Ans: With $\sigma = 50$ known and $H_0{:}\mu = 1{,}000$ vs. $H_1{:}\mu \neq 1{,}000$, $N(\mu|\sigma)$ yields $P = .057$. At the 10% level, H_0 should be rejected.

27. In a study of 20 divorce cases, a marriage counselor found that the total length of time in months from marriage to divorce was 1763, with a standard deviation of 3 months. Test the claim that mean time to divorce exceeds 90 months.

Ans: With the claim $H_1{:}\mu > 90$, the significance level is 99%, lending very little credence to the claim.

28. Selective Service records of 1,000 draftees revealed an average height of 69.83 inches with a standard deviation of 2.67 inches. Could the population mean of draftees at that time have reasonably been 70 inches?

Ans: A 95% CI for μ extends from 69.665 inches to 69.995 inches and fails to include 70 as a plausible value. The *P*-value for $H_1{:}\mu \neq 70$ is $P = .044$, which also fails to support $H_0{:}\mu = 70$, at least at the 5% level.

29. It is claimed that a certain drug will lower man's temperature within 10 minutes of ingestion. Five subjects having normal temperatures were given the drug, and their temperatures taken 10 minutes later. The resulting sample summaries of $\Sigma x_i = 491$ and $\Sigma x_i^2 = 48{,}219$ was all that was recorded. Test the claim with these data.

Ans: Making $H_0{:}\mu \geq 98.6$ the disclaimer, $\bar{x} = 98.2$ and

$$s^2 = \frac{\Sigma x_i^2 - n\bar{x}^2}{n - 1} = .7 \qquad P = .17$$

suggesting the claim is false.

Section 5.3: Tests of Variance

Significance tests or tests of hypotheses about the variance of a normal population can be carried out also with program $N(\sigma^2)$ of ZS–4. Inferences about σ^2 are appropriate in connection with questions about the precision of a measuring device, the consistency of produced items, or the repeatability of examination scores.

The test statistic used in tests of σ^2 in a normal population is

$$ts = \frac{(n - 1)s^2}{\sigma_0^2}$$

where, as usual, σ_0^2 is the hypothesized value in the null hypothesis about σ^2. The statistic *TS* then has a Chi-square distribution with $n - 1$ degrees of freedom.

EXAMPLE 5.6

Suppose the radiation intensity data of Example 5.4 has been collected to examine the precision of a new radiation measurement technique. It is desired that the variance of the readings will be no more than .05. Conduct a test of significance. The data are 3.6, 4.2, 4.0, 4.1, 3.8, 3.9, and 4.0.

Solution Enter the data with DE in ZS–4. To put the burden of proof on the data, let $H_0{:}\sigma^2 \geq .05$ and $H_1{:}\sigma^2 < .05$. Then with H_1-code $= -1$ and $\sigma_0^2 = .05$, implement $N(\sigma^2)$ to find $P = .42$. Thus, H_0 is supported, and, therefore, the instrument ought to be examined possibly for calibration or adjustment to achieve a smaller σ^2.

Naturally, a test for the value of σ may be rephrased as a test for σ^2. Since the program is designed for σ^2 and not σ, this will have to be taken into account in the following problems. While most problems regarding σ^2 involve one-sided alternatives, occasionally they do not. Again, we would remind you that in such cases the reported P-value from the program should be accepted with some reservation for the underlying Chi-square distribution for this case is not symmetric. (See footnote, page 114.)

30. The manufacturer of a special instrument for measuring the fetal heart rate in hospital delivery rooms claims that the standard deviation in the accuracy of the instrument does not exceed .0002. An analyst hired to test the instrument for a hospital which is interested in purchasing it found a sample standard deviation of .0003 in testing 10 instruments. Using a 5% level of significance, would you reject the manufacturer's claim?

Ans: Use $H_0{:}\sigma^2 \geq (.0002)^2$ vs. $H_1{:}\sigma^2 < (.0002)^2$. Do not reject $H_0(P = .98)$. An upper one-sided 95% bound on σ is $u = .0005$. Either way, the manufacturer's claim should be rejected.

31. A large retailer of roofing nails claims that the lengths of his nails have a standard deviation which does not exceed .06 inches. You take a sample of 15 of his nails and find a standard deviation of .08 inches. Would you reject the retailer's claim at the 5% level of significance?

Ans: Use $H_0{:}\sigma^2 \geq (.06)^2$ vs. $H_1{:}\sigma^2 < (.06)^2$. Do not reject $H_0(P = .96)$. The upper 95% CI for σ is the interval (0, .12). Either way, the retailer's claim should be rejected.

32. If a sample of size 12 has a variance of 20, can the null hypothesis that $\sigma^2 = 9$ be rejected using a .05 level of significance?

Ans: Use $H_0{:}\sigma^2 = 9$ vs. $H_1{:}\sigma^2 \neq 9$. A 95% confidence interval for σ^2 is (10.0, 57.6), and P = .02 is the significance level. Thus H_0 is rejected.

33. A manufacturer of resistors for electronic ignition systems in automobiles guarantees that his resistors will last 4 years with a standard deviation of $1^1/_2$ years. If 8 of the resistors are found to last 3, 3.6, 4.8, 4.6, 4.2, 6.0, 2.5, 4.4 years, should the manufacturer still be convinced of a standard deviation of $1^1/_2$ years? What is the significance level? (Assume normality.)

Ans: Using α for producer's risk, test $H_0{:}\sigma^2 \leq (1.5)^2$ vs. $H_1{:}\sigma^2 > (1.5)^2$. With $P = .81$, H_0 should not be rejected.

34. A publisher of road and travel guides samples the cut size of 20 finished guides and finds a standard deviation of .062, following his quality assurance test. Should he reject the null hypothesis that σ^2 is equal to $(.07)^2$, in favor of an alternate hypothesis that $\sigma^2 > (.07)^2$?

Ans: No. ($P = .73$)

35. The copper content, in parts per million (ppm), is measured as part of a spectral analysis of used engine oil. In order to make decisions about the condition of the engine from which the oil was sampled, the spectrometer must have a precision (σ) of .5 ppm. The following measurements were made on a standard sample of oil: 7.3, 5.6, 7.0, 6.0, 6.4, 5.8, 6.9, 6.5, 5.8, 6.2, 5.9, 6.2. Does this constitute evidence that the spectrometer has adequate precision?

Ans: Testing $H_0:\sigma^2 \leq (.5)^2$ vs. $H_1:\sigma^2 > (.5)^2$, you find $P = .32$. The data do not provide strong evidence that the spectrometer is inadequate.

36. In handloading rifle bullets for match shooting, it is desirable to produce loads giving consistent muzzle velocity. A batch of loads was tested at a rifle range and the following muzzle velocities, in feet per second, were observed: 2,830, 2,845, 2,829, 2,838, 2,855, 2,854, 2,825, 2,839. What is the level of significance in testing $H_0:\sigma^2 \leq 64$ vs. $H_1:\sigma^2 > 64$?

Ans: $P = .052$.

37. It is feared a computer program, proposed for use as a random number generator, gives results which are correlated, and thus the variance of the output is too small. If the generator is working correctly, the variance should be $^1/_{12}$. A sample of size 100 from the generator had standard deviation .24. Does this constitute evidence the proposed generator may be giving correlated results?

Ans: Test $H_0:\sigma^2 = {}^1/_{12}$ vs. $H_1:\sigma^2 < {}^1/_{12}$; $P = .008$, so there is evidence the variance of the generator output is not as high as $^1/_{12}$.

38. A nail machine is supposed to manufacture 1-inch nails with a standard deviation of $\sigma = .025$ inch. A random sample of 30 nails yielded a standard deviation of .030 inch. Does this apparent increase warrant shutting down the machine? Use $\alpha = .05$.

Ans: Test $H_0:\sigma^2 = (.025)^2$ vs. $H_1:\sigma^2 > (.025)^2$, and find $P = .059$. At the 5% level H_0 would not be rejected, and the machine should not be tampered with.

39. In order to see if the variability of heights of graduating classes is really changing over a previously assumed value of 1 inch, the heights of 100 randomly selected graduates was recorded, and the experimenter reported $s = .95$ inch. Has variability changed?

Ans: With $H_1:\sigma^2 \neq 1$, $P = .51$ (with reservation). A 95% CI for σ^2 runs from .70 to 1.22 and includes 1 as a plausible value. Conclude that variability has not changed.

40. The precision of a certain delicate balance is measured by its standard deviation. The balance reads to the nearest milligram, and if $\sigma > .001$, the balance should be calibrated. Independently, 5 analysts took readings with a 5.000 gram standard and reported,

5.002   4.999   5.002   5.000   5.002

Should the balance be calibrated? Use $\alpha = .10$.

Ans: Test $H_0:\sigma^2 \leq (.001)^2$ vs. $H_1:\sigma^2 > (.001)^2$, thereby calling for calibration only if the data reject H_0. The significance level is $P = .09$, so that calibration should be called for.

Hypothesis Testing

1. (Example 5.6 revisited) Seven observations of measured radiation intensity at a nuclear plant were 3.6, 4.2, 4.0, 4.1, 3.8, 3.9, 4.0. Conduct a significance test of $H_0{:}\mu \le 3.8$ against $H_1{:}\mu > 3.8$.

Solution

ZS STEP	ENTER	PRESS	DISPLAY	COMMENTS
DE				
1		[Pgm]03[E]	1.	Select and initialize ST–03.
	3.6	[A]	1.	Enter data.
	4.2	[A]	2.	
	.			
	.			
	.			
	4.0	[A]	7.	
2		[RST]	7.	Exit ST–03.
3		[D′]	0.	Process data.
N(μ)				
2a	1	[A′]	1.	Enter H_1-code (+1).
2b	3.8	[R/S]	.053	Enter boundary value, and compute $P = .053$ from t-density.

2. A water meter has variance 14 (cu. ft.)^2. Twenty monthly readings indicate a sample mean of 1,284 cu. ft. per month.

   a. Test the hypothesis $H_0{:}\mu = 1286$ against $H_1{:}\mu \neq 1286$, using $\alpha = .05$.

   b. Calculate the significance level for the one-sided alternative H'_1: $\mu < 1286$.

Solution

ZS STEP	ENTER	PRESS	DISPLAY	COMMENTS
N($\mu \vert \sigma$)				
1a	20	[STO]03	20.	Enter sample size.
1b	1284	[STO]37	1284.	Enter sample average.
1c	14	[$\sqrt{x}$][STO]48	3.742	Enter known σ.
4	.025	[B]	1282.36	l
		[$x \rightleftharpoons t$]	1285.64	u [Since (l, u) does not contain μ_0, H_0 is rejected.]
3a	−1	[B′]	−1.	H_1-code for part (b).
3b	1286	[R/S]	.0084	P-value (data are inconsistent with H_0)

3. The standard deviation in GRE scores nationwide has been 40 points. The GRE scores for 86 Smith High School students this year has (sample) standard deviation 35.2. What is the significance of this result?

Solution

ZS STEP	ENTER	PRESS	DISPLAY	COMMENTS
$N(\sigma^2)$				
1a	86	[STO]03	86.	Enter sample size.
1b	35.2	[STO]38	35.2	Enter sample standard deviation.
2a	0	[C′]	0.	Enter H_1-code for $H_1{:}\sigma \neq 40$.
2b	40	[x^2][R/S]	.122	P-value (data are somewhat consistent with H_0: $\sigma = 40$)

4. Times to failure of a sample of 12 unused D-cells were (in weeks) 27, 41, 29, 33, 30, 33, 26, 37, 29, 11, 20, 29. The shelf life is claimed to be at least 35 weeks. Conduct a significance test of $H_0 : \mu \leq 35$ vs. $H_1{:}\mu > 35$.

Solution

ZS STEP	ENTER	PRESS	DISPLAY	COMMENTS
DE				
		[Pgm]03[E]	1.	Select and initialize ST–03.
	27	[A]	1.	Enter data.
	41	[A]	2.	
	.	.	.	
	.	.	.	
	.	.	.	
	29	[A]	12.	
2		[RST]	12.	Exit to main program.
3		[D′]	0.	Process data.
Exp(μ)				
2a	1	[E′]	1.	Enter H_1-code.
2b	35	[R/S]	.71	P-value (data are consistent with H_0)

6

Bivariate Populations

Section 6.1: General Considerations

A very common situation that arises in statistical analysis is that of comparing two populations. The two may in fact be physically separate populations, such as voters in two separate urban areas, represented by two different states or conditions on the same actual population, such as weights measured for several individuals before and after dieting. In either case, the question naturally under investigation is whether or not there is any difference in the two. Indeed, the denial of any difference—the assertion of *no* difference—seems to account for the expression "null hypothesis," although that term is now used, as you have seen, in a more general sense. How we analyze such comparisons depends, of course, on the assumptions we make, and that, in turn, affects and is affected by the manner in which the data are gathered.

Whatever the situation, we try, as always, to describe the 2 populations by a pair of random variables, say, X and Y. For example, if we were trying to examine the effect on achievement scores of a new method of teaching, we might divide students into a control group and an experimental group. We could then let X stand for the random score of the control group and Y the random score of the experimental group. Or again, we might be examining the effects of a diet so that X is the weight before dieting and Y is the weight after dieting on the same individual selected at random from a given group of "volunteers." Following the preceding considerations for single random variables, it is very convenient when matters can be reduced to comparing parameter values of the (assumed) distributions of X and Y.

If you studied the probability details of bivariate distributions, such as the exposition in Chapter 5 of *Probability by Calculator* in this series, you learned that a pair of random variables X and Y in general have a joint distribution, and, in addition to the means and variances (or standard deviations) of X and Y which remain of interest, a measure of the joint relationship between X and Y, covariance σ_{XY} (or, equivalently, correlation ρ), is a fifth parameter of interest.

It should come as no surprise that when X and Y are assumed to be normally distributed, then theoretical results which admit practical applications are readily available. And, fortunately, the normal assumptions are quite common in lots of experimental circumstances where a comparison of two populations is to be made. One bonus for making normal assumptions is the fact that X and Y are then independent, if and only if $\rho = 0$ (or $\sigma_{XY} = 0$), whereas, in general, $\rho = 0$ need not imply independence. This makes a test of independence—obviously one we might have a great deal of interest in—reducible to a single parameter. We will explore that matter further in Chapter 9. For now, we content ourselves with comparing means or variances in this chapter.

Section 6.2: Data Entry; Point Estimates

In order to exploit fully the TI–59 Statistics Module for the matters we wish to treat here, we will use Program ST–04 for entry of bivariate data. This program is discussed on page 3–7 of ASM. As you can see after the typical initialization with label E′, you are allowed to repartition the calculator in order to store up to 34 pairs of x-y values. But, for our applications, we have no need for permanent storage, and we do need the programming steps, so you should always IGNORE Step 3 in ST–04 for the problems in this book! The program then allows you to enter data either in pairs or as 2 separate sets of data, and we need to pause and discuss this momentarily.

How we consider the data and enter them in ST–04 depends upon the assumptions regarding the population random variables X and Y for which we are given observed values. But, there are only two situations we will be considering here. First, if X and Y are assumed to be independent, then the data should be viewed as two separate lists of observed values $x_1, x_2, x_3, \ldots, x_n$ of X and $y_1, y_2, y_3, \ldots, y_m$ of Y. In that case, the sample sizes n (also denoted n_x or n_1) and m (also denoted n_y or n_2) need not be the same. This would be exemplified by the situation where we wish to compare the mean time to failure of two entirely different satellite systems, and we have available just 5 times to failure for one, and 30 for the other. The other situation is exemplified in the study of weights before (X) and after (Y) dieting, measured for several individuals. In that case, X and Y cannot be reasonably assumed independent, and, for each x_i observation of X, there is a matching y_i observation for Y. In other words, the data are *paired* (x_i, y_i), and there will be exactly as many x's as y's—the sample sizes are equal.

To enter paired data in ST–04, after initialization, you see from the user instructions that you have only to enter each x_i and press [A]; then enter y_i and press [B], starting with $i = 1$ and continuing until you enter the nth, or last, pair. In standard partitioning, which we insist upon, this will only allow $n = 14$ pairs. An attempt to enter a 15th pair will result in a flashing display. Simply press [CE], and then [D], to reset the pointer. Then continue entering data starting with $i = 15$. Naturally, this will write over the first few pairs entered, and hence requires some care on your part in entering the data to begin with. We are only interested in the IDB, so you may enter as many pairs as you like—with care. (If you have a printer, the entire record of the data will be printed for you, so, in that case, nothing is lost.) Once the data have been entered, the IDB is sufficient to output the usual estimates of the separate means and variances and/or standard deviations using [$\bar{x}$] (with or without [INV]) and [Op] 11. You should be aware that the first press of any of these routines produces Y characteristics, with the corresponding X characteristics stored in R_T! We will illustrate shortly.

When data are paired, as in the case under discussion, the fifth parameter of interest, the correlation coefficient ρ, may be estimated by what almost all authors call the *sample* correlation coefficient and denote r or r_{xy}. It is evaluated in terms of data, as

$$\hat{\rho} = r = \frac{\sum_{i=1}^{n} (x_i - \bar{x})(y_i - \bar{y})}{\sqrt{\sum_{i=1}^{n} (x_i - \bar{x})^2 \quad \sum_{i=1}^{n} (y_i - \bar{y})^2}}$$

This rather unpleasant formula is the output of a special Op Code in the TI–59, and the value of r may be found by a simple press of [Op] 13, once the data are entered using ST–04 as paired data.

EXAMPLE 6.1

Five subjects were tested over mathematics ability (X) and motor skill (Y), with scores on the test reported in (x, y) pairs, as: (47,30), (61,29), (91,37), (74,23), (92,32). Find point estimates of μ_x, μ_y, σ_x^2, σ_y^2, and ρ.

Solution Initialize Program ST–04, and enter the 5 pairs of data as per instructions. At the conclusion, you should see $n = 5$ in the display register. Now press [$\bar{x}$] to verify that $\bar{y} = 30.2$; follow this with [$x \rightleftarrows t$] to see

$\bar{x} = 73$. Now press [INV][$\bar{x}$] and discover $s_y = 5.07$, while $s_x = 19.4$ may be retrieved from R_T by pressing [$x \rightleftarrows t$]. You may use [Op] 11 in the same way to verify $MSD_y = 20.56$ while $MSD_x = 301.2$. Now press [Op] 13 to discover $r = .43$ to 2 decimal places.

When data are not paired, but are independent, program ST–04 may still be used with slight modification for data input and compilation, although it is not advertised as such in ASM (until a later use in ST–13). You initialize the program in exactly the same way, but you enter each x with successive presses of [A], until all n_x data points are entered, *and then* you enter each y with successive presses of [B], until all n_y data points are entered. In standard partitioning, this allows a total input of $n_x + n_y = 28$ data points. But, again, if $n_x + n_y > 28$, just press [D] to reset the pointer when the storage registers become full and continue entering data.

There is, however, one other small problem created *when* $n_x \neq n_y$. In that case, as you can see from the IDB listing on page 3–1 of ASM, n_x is stored in R_{15}, while n_y is stored in R_{03}, where we are used to finding sample sizes. This means that [$\bar{x}$], [INV][$\bar{x}$], and [Op] 11 will produce, respectively, $\bar{y}$, s_y, and MSD_y, as in Example 6.1.* Following these with [$x \rightleftarrows t$] will *not*, however, produce the corresponding X characteristics as before, since n_x is not in R_{03}. The solution is easy enough, however. Just temporarily store n_x in R_{03}, and proceed as in Example 6.1 to find the moments for X. This may create a flashing display, or even negative displays, when you press [INV][$\bar{x}$] or [Op] 11, but if that happens, just ignore the results (stop any flashing by pressing [CE]), and press [$x \rightleftarrows t$]. The results will be valid. Of course, this could all be avoided by just treating the x and y data separately, using ST–03, but we would lose some results in the IDB for ST–04 that we need in our applications. You will probably want to replace n_y in R_{03} at the conclusion, just to be sure you are back where you started.

Since we are assuming that X and Y are independent in this case, the computation of r using [Op] 13 is meaningless for $\rho = 0$. But the calculator has no way of knowing this, and so [Op] 13 will calculate a number using the register contents, just as though paired data had been entered. The output should be ignored, of course.

EXAMPLE 6.2

Two small classes were (independently) tested on a standardized test in the hopes of establishing no difference between the two groups. The test results were:

Group 1: $x =$ 82, 87, 91, 54, 97, 76, 64, 98, 92, 57, 80, 53, 64
Group 2: $y =$ 91, 62, 94, 92, 87, 79, 86, 75, 90, 73, 83, 93, 65, 89, 68, 52

Find the usual summary statistics for the 2 groups.

* Clear and ignore any flashing display with [CE].

Solution Initialize Program ST–04 as before. Then enter the x-data with successive presses of [A]. At the conclusion, you should see $n_x = 13$ in the display. (You may also verify that n_x is stored in R_{15}.) Now continue entering data, starting with $y = 91$, but press [B] instead of [A]. Notice that the data-count in R_D starts all over at 1 again. When you arrive at 52, you will encounter a flashing display. That is because a total of 28 data points have already been entered. So, clear the flashing, and press [D], whereupon you should see 32 in R_D, indicating that the pointer has been reset. Now reenter 52 and press [B] for the last time, observing a y-count of $n_y = 16$ at the conclusion (also to be located in R_{03}). If you now press [$\bar{x}$], [INV][$\bar{x}$], and [Op] 11, you may correctly read out the respective values of $\bar{y} = 79.94$, $s_y = 12.798$, and $MSD_y = 153.56$. Now recall R_{15} to see again $n_x = 13$, and press [STO] 03. Then press [$\bar{x}$][$x \rightleftarrows t$] to discover $\bar{x} = 76.54$. When you press [INV][$\bar{x}$], you will obtain a flashing result, which you should ignore, and clear with [CE]. Now [$x \rightleftarrows t$] will produce the correct value of $s_x = 16.425$. Pressing [Op] 11 shows an absurd negative result. Ignore it, and press [$x \rightleftarrows t$] to find $MSD_x = 249.02$. Finally, restore 16 in R_{03}, in case you need to use the program again; otherwise, proceed with another problem.

That is all there is to Program ST–04 that is of any direct use to us. Following is a small sample of problems to allow you some practice with the points we have discussed. You will have an opportunity for a great deal more practice when we turn to the more important applications in succeeding sections, where data entry will be required.

Problems

1. The following pairs of data are observed IQ's (x) and scores on a mathematics exam (y) for 20 individuals. Calculate the 5 parameter estimates $\bar{x}$, $\bar{y}$, s_x, s_y, and r_{xy}.

   (130,78) (115,71) (104,62) (150,89) (153,92)
   (164,97) (141,85) (119,73) (129,78) (101,58)
   (118,71) (104,62) (125,74) (116,71) (144,87)
   (110,67) (109,65) (118,72) (126,75) (139,82)

   Ans: $\bar{y} = 75.45$ s_y — 10.59 $MSD_y = 106.45$ $\bar{x} = 125.75$ $s_x = 17.81$ $MSD_x = 301.39$ $r = .994$

2. To study whether or not you gain weight if you quit smoking, 10 subjects were weighed before (x) and after (y) they quit. Process the data for $\bar{x}$, $\bar{y}$, s_x, s_y, and r_{xy}.

   (100,166) (191,197) (138,136) (182,190) (159,160)
   (205,192) (201,212) (180,187) (160,155) (140,140)

   Ans: $\bar{y} = 173.5$ $s_y = 25.69$ $MSD_y = 594.05$ $\bar{x} = 165.6$ $s_x = 32.87$ $MSD_x = 972.24$ $r_{xy} = .754$

3. Twenty taxicabs were divided into 2 separate groups of 10 each and mileages compared over a period of time. Calculate summary statistics for the 2 groups.

x:	13.4	14.1	13.5	13.4	15.0	12.2	13.0	13.1	14.1	14.0
y:	14.1	14.4	13.6	13.4	14.9	13.0	13.1	13.3	14.0	14.5

Ans: $\bar{y} = 13.83$ $s_y = .646$ $\text{MSD}_y = .3761$ $\bar{x} = 13.58$ $s_x = .766$ $\text{MSD}_x = .5276$
Even though $n_x = n_y = 10$, the data are not paired, so r is meaningless.

4. A test measuring color perception was given to a control group (x) and an experimental group (y). The test results were as follows. Find $\bar{x}$, $\bar{y}$, s_x, and s_y, MSD_x and MSD_y

x:	16.3	14.7	12.3	13.5	16.0	17.1	17.3
y:	14.0	16.5	17.7	15.9	18.0	16.3	

Ans: $\bar{y} = 16.4$ $s_y = 1.434$ $\bar{x} = 15.3$ $s_x = 1.885$ $\text{MSD}_x = 3.0469$ $\text{MSD}_y = 1.7133$

5. Times to failure for two different electronic components of similar type were recorded as follows. Find the point estimates of the failure rates for each.

x:	166	28	252	76
y:	5	241	394	

Ans: $\bar{y} = 213.33$ and $\bar{x} = 130.5$ so $\hat{\lambda}_y = 1/\bar{y} = .005$ and $\hat{\lambda}_x = 1/\bar{x} = .008$

Section 6.3: Paired Data

So much for data entry and processing. Let us turn to the more interesting problem of statistical inference for the populations under study. The first case we will examine assumes the (x, y) data are paired observations from normal random variables. Naturally, we imagine that X and Y are then dependent random variables, with some correlation other than 0. But even so, we turn our attention to merely comparing the 2 means μ_x and μ_y. In particular, we examine the question, are they the same or not? Lots of problems fall into precisely this category.

The theory for resolving this particular question is based simply on the fact that $D = X - Y$ is normally distributed with mean $\mu_x - \mu_y$. So, if we imagine the data as arising from n independent pairs, then the successive differences $d_1, d_2, \ldots, d_n$ may be viewed as a random sample from D. Accordingly, inference concerning the mean of D, $\mu_x - \mu_y$, may be effected after that simple step by the methods of Chapters 4 and 5, related to the mean of any normal random variable having an unknown variance. For the record, although it is not essential for the computation of either confidence intervals or tests, the variance of D is given by

$$\sigma_d^2 = \sigma_x^2 + \sigma_y^2 - 2\rho\,\sigma_x\sigma_y$$

and is estimated by

$$s_d^2 = s_x^2 + s_y^2 - 2r\,s_x s_y = \sum_{i=1}^{n} \frac{(d_i - \bar{d})^2}{(n-1)}$$

and thereby incorporates the sample correlation coefficient r.

In summary, then, confidence intervals will be based on the t-distribution having $\nu = n - 1$ degrees of freedom, just as before. The procedure is outlined in the program ZS–6 provided for you under the coding PN (for paired normal), and has user instructions quite like those you are already familiar with. The program allows for entering the original paired data with ST–04, or else summary data in the form of $\bar{d} = \bar{x} - \bar{y}$, and the sample standard deviation of the differences s_d. You may enter $\bar{x}$ and $\bar{y}$ separately or else as a difference $\bar{d}$. You must have s_d given, if no data are available. As you have just seen, it is not enough to know s_x and s_y to compute s_d.

As for tests, we may use normal theory to test $H_o{:}\mu_x - \mu_Y = \theta_0$, where θ_0 is any specified difference of interest, and, in particular, when this hypothesized value is selected to be 0, we have a test equivalent to the hypothesis that the 2 population means are the same, $H_o{:}\mu_x = \mu_y$. The 3 usual alternative situations we may examine are then $H_1{:}\mu_x - \mu_y < \theta_0$, $H_1{:}\mu_x - \mu_y > \theta_o$, and $H_1{:}\mu_x - \mu_y \neq \theta_o$. The test statistic takes the form $TS = \sqrt{n}\,(\bar{D} - \theta_0)/s_D$ and has a t-distribution with $\nu = n - 1$ degrees of freedom when H_o is true.

When the data are entered in pairs through ST–04, Program ST–13, discussed on pages 4–12 through 4–15 of ASM, compiles the necessary data for calculating *ts*. In that manual, Δ is used for D, except at Step 8, where it has a different meaning (to be discussed in the next section). Steps 4–6 in the user instructions apply to the current situation, and may be used for carrying out tests based on comparing *ts* (their t) with the appropriate t-percentile. No provision is made directly for calculating an observed significance level, however, to carry out a significance test. So we have arranged all of this for you in ZS–6, making use of Program ST–21 for calculating P-values. An example should help clarify the instructions.

EXAMPLE 6.3

In a study of the effects of dieting, 10 women were selected, weighed (x), and placed on a certain diet for 2 weeks. At the end of that time they were weighed again (y), with the following paired results.

x:	119	122	136	130	129	136	134	133	119	115
y:	114	119	134	126	119	137	124	127	119	107

Analyze the data for inference concerning $\mu_x - \mu_y$.

Solution First, enter the data with ST–04 as previously outlined (see also ZS examples). Be sure to press [RST] at the conclusion to exit the module and enter the card program ZS–6. Now press [D'] to compile the data and store the necessary quantities for future use. If you press [B], you will eventually have $\nu = n - 1 = 9$ degress of freedom displayed for you for this case. A 95% CI for $\mu_x - \mu_y$ may then be found by entering $t_{.025} = 2.262$, and pressing [R/S]. You should discover that the limits run from 1.94 to 7.46. It is not likely that $\mu_x - \mu_y = 0$, for 0 is not in the interval. The weight loss more reasonably varies from about 2 to about 7.5 pounds.

We can also complete a significance test. We know from the confidence interval that we would likely reject equality of means (no weight loss) at the usual levels. Suppose we examine the claim that the weight loss exceeds 2 pounds. Making $H_o{:}\mu_x - \mu_y \leq 2$ the disclaimer, the alternative becomes $H_1{:}\mu_x - \mu_y > 2$. With an H_1-code of +1, we can conduct a significance test for $\theta_o = 2$ using the PN routine. Following the instructions, you should be able to verify a *P*-value of .027. We would reject H_o in favor of the claim, if $\alpha = .05$.

You may practice the PN routine with the problems that follow. In all cases, you should assume that the difference between the random variables is normal.

Problems

6. A group of 10 randomly selected students were tested before (x) and after (y) participation in a rigorous physical conditioning program. Their strength measures on a standard test before and after the conditioning program follow:

x:	1.1	1.2	0.9	1.7	2.8	1.3	1.4	1.2	1.1	3.1
y:	1.2	1.7	1.1	1.8	3.1	1.8	1.7	1.5	1.2	3.4

Calculate a 95% CI for the strength change.

Ans: (−.38, −.16)

7. Eight people selected at random had their pulse rates measured before (x) and after (y) a 3-minute step exercise with the following results:

x:	68	75	72	60	85	70	72	75
y:	115	125	130	110	140	135	130	128

Calculate a 99% confidence interval for the difference in pulse rate after the step tests.

Ans: (−61.7, −47.3)

8. A group of 10 shoppers was randomly selected to determine if there was a difference in selected spending habits before and after enact-

ment of new federal monetary regulations. The sample results follow:

Shopper:	1	2	3	4	5	6	7	8	9	10
Before:	1800	2150	1700	950	2450	1620	1840	1930	1900	1870
After:	1620	1800	1480	950	2100	1185	1620	1700	1625	1530

Calculate a 90% confidence interval for the difference in spending habits for the population represented by the 10 shoppers after enactment of the monetary regulations.

Ans: (190,330)

9. Ten automobiles were randomly selected as they departed the manufacturer's assembly line. Each car was driven for 100 miles and tested for its miles per gallon rating. Each car was then driven 1,000 additional miles with a MPG rating conducted on the last 100. Results are listed below.

Car:	1	2	3	4	5	6	7	8	9	10
MPH—1st 100 miles:	18.7	17.9	20.0	18.0	17.6	18.9	20.1	20.1	17.9	18.5
MPH—2nd 100 miles:	19.4	18.1	21.7	19.1	19.1	18.9	19.9	21.8	20.3	19.1

Calculate a 95% confidence interval for the change in MPG resulting from the additional break-in mileage.

Ans: (−1.58, −.36)

10. A group of 10 school children were randomly selected to participate in a rigorous English spelling program. The 10 children were given a 1,000 word spelling test before and after the program. Test results are listed below.

Student:	1	2	3	4	5	6	7	8	9	10
Before program: number of words missed:	346	708	659	801	209	310	418	521	123	604
After program: number of words missed:	210	81	37	41	101	102	73	41	48	107

Calculate a 95% interval for the change in the number of words missed resulting from the spelling program that the children underwent.

Ans: (209.83, 561.77)

11. Eight pairs of observations (X, Y) resulted in the data below. Show that even though $\bar{x} - \bar{y}$ is 44, nevertheless the hypothesis of equality of means should be accepted at the 5% level of significance.

(60,200) (140,20) (180,20) (80,180) (20,150) (210,40)
(200,10) (225,145)

Ans: 95% CI $(-75.6, 163.1)$ includes 0 For $H_1: \mu_x \neq \mu_y$ $P = .41$ $ts = .87$

12. The following paired observations are known to come from normal distributions. Test the hypothesis at level 1% that the means are identical.

(19,63) (31,9) (3,23) (12,21) (12,30) (9,21) (15,91) (33,12) (22,5) (31,2)

Ans: For $H_1: \mu_x \neq \mu_y$, $P = .41$, $ts = -.86$. Do not reject: 99% CI: $(-43, 25)$

13. It is alleged that if you quit smoking, you will surely gain weight. But not every quitter gains weight. Here is a sample of 10 men who quit smoking, showing their weights before they quit (x) and their weights after they quit (y). Is the change significant at the 5% level?

x:	100	191	138	182	159	205	201	180	160	140
y:	166	197	136	190	160	192	212	187	155	140

Ans: $H_1: \mu_x - \mu_y < 0$ (claim) $P = .14 > .05$; Do not reject. 95% CI: $(-\infty, 4.6)$ includes 0.

14. Does the color of paper make a difference in typing speed? Nine typists were given material to type, first on white paper (x), and then on yellow paper (y), measured in words per minutes. The results show a decrease in typing speed. Is it safe to assume that the difference is real? Try $\alpha = .01$.

x:	62.6	61.8	62.4	63.8	64.0	62.1	61.2	60.9	62.7
y:	61.6	60.0	62.0	63.1	63.2	60.4	59.8	60.5	62.2

Ans: $H_1: \mu_x > \mu_y$ (apparent decrease), $P = .0004$. Reject H_o. 99% CI: $(.44, \infty)$

15. A certain gasoline additive is supposed to *increase* mileage. We are dubious. Using 8 different cars over a racetrack, each car is checked for mileage in miles per gallon with the same amount of gasoline first without additives (x), and then with additives (y). The results are these:

x:	16.44	16.88	16.80	16.95	16.18	16.75	16.09	16.21
y:	16.35	16.97	16.92	16.82	16.07	16.87	16.05	16.39

Does the additive increase mileage? Use $\alpha = .01$.

Ans: For $H_1: \mu_x < \mu_y$, $P = .35$, hence do not reject $H_o: \mu_x \geq \mu_y$.

16. It is claimed that wheel balancing will add at least 2,000 miles to the wear of an automobile tire. To support this claim, 8 automobiles were used in which one front wheel (X) was balanced every 5,000 miles, and the other (Y) was not. At the end of the experiment, the total wear in miles as measured by tread depth was reported as follows.

X:	22,500	29,800	20,510	22,660	26,400	27,500	21,800	24,300
Y:	19,700	28,300	20,100	21,700	22,300	25,000	19,200	23,400

Can the claim be justified?

Ans: For $H_1: \mu_x - \mu_y > 2{,}000$, $P = .53$. Reject claim in favor of H_o.

17. Correlated data representing scores on 2 traits were reported as

(11,75) (4,29) (2,74) (5,62) (18,83) (11,53) (22,33)

Estimate the correlation coefficient, and find a 99% confidence interval for the difference of means. Would the null hypothesis be accepted?

Ans: $r = -.07$, CI:$(-80, -16)$, $P = .001$ with $ts = -5.6$; No.

18. In a study of the effects of dieting, it was reported that a mean weight loss of 2.5 pounds resulted from an experiment involving 70 patients. A standard deviation for differences in weight was also recorded as 15 pounds. The report claimed that because of the large sample size, the evidence for the diet was substantial. Is such a report justified?

Ans: Letting X denote before weight and Y the weight after dieting, the claim is that $\mu_x - \mu_y \geq 0$. Making $H_o: \mu_x - \mu_y \leq 0$ the disclaimer, a significance test against $H_1: \mu_x - \mu_y > 0$ gives a P-value of $P = .08$. Hence H_o (no weight loss) would not be rejected at the typical 5% level. Report not justified.

19. To test the effectiveness of a new incentive program tried at a large retail outlet, a pilot study was performed using 80 employees. Before the program began, the average number of hours reported on sick leave was $\bar{x} = 59.7$, and that figure dropped to $\bar{y} = 42.3$ after the program. Is the reduction real, or just due to chance, given a value of 1,800 for s_d^2?

Ans: If X is the status before the program, a test of $H_o: \mu_x - \mu_y \leq 0$ vs. $H_1: \mu_x - \mu_y > 0$ yields a P-value of .0001, too small to attribute the effects to chance alone. Reject H_o in favor of H_1, and conclude that the program appears to be producing real results.

Section 6.4: Independent Data

When data represent independent samples, several difficulties arise in contrast to the paired case. Indeed, the difficulties are so insurmountable, that, at this writing, there exists no complete solution to the general problem. You will often find the dilemma discussed under the heading of the Behrens-Fisher problem. There are partial solutions, however, some exact and some approximate, each of which levies its own toll in assumptions. In every case, we suppose that the population random variables X and Y are normal and independent. In that case, they are uncorrelated, and you would think that would make life easier with the elimination of a covariance term to estimate. Not so, and the presence of the two separate variances σ_x^2 and σ_y^2 is the source of the problem.

Of course, if the two variances are *known*, then there is no problem at all, but we are rarely in that set of circumstances. After all, $\overline{X}$ and $\overline{Y}$, based on two possibly different sample sizes n_x and n_y, are themselves independent normal random variables, and the quantity

$$\frac{\overline{X}-\overline{Y}-(\mu_x-\mu_y)}{\sqrt{\dfrac{\sigma_x^2}{n_x}+\dfrac{\sigma_y^2}{n_y}}} \tag{6.1}$$

then, has a standard normal distribution. Such a random variable may be used for deriving confidence interval formulas and serves as a basis for testing hypotheses about $\mu_x - \mu_y$, as well. This would be true regardless of the sizes of n_x and n_y. If the latter were both large and σ_x^2, σ_y^2 unknown, you might justify replacing them by their sample estimates s_x^2 and s_y^2 and treating the result

$$Z = \frac{\overline{X}-\overline{Y}-(\mu_x-\mu_y)}{\sqrt{\dfrac{s_x^2}{n_x}+\dfrac{s_y^2}{n_y}}} \tag{6.2}$$

as though it were Equation (6.1), with $\sigma_x^2 = s_x^2$ and $\sigma_y^2 = s_y^2$, and proceed with standard normal theory. We did this in the one sample case, and it is common practice here as well. We voice no strong objection, but do add once again the comment that the theory permitting σ_x^2 and σ_y^2 to be replaced by s_x^2 and s_y^2 allows μ_x and μ_y to be replaced by $\bar{x}$ and $\bar{y}$ with even less error.

That aside, we have programmed this case for you in the ZS–6 program under the code LSN for "large sample normal." As you can see from the user instructions, the program allows only for summary data entry, because it would rarely be used with small samples. In any case, you store n_y in R_{03} and n_x in R_{13}, where we will be storing n_x in the data processing routine. The next step is to enter the two sample means, and we have reserved registers R_{37} and R_{47} for them. Finally, you enter σ_x and σ_y in R_{48} and R_{38}, respectively, if you know their values; otherwise, enter s_x and s_y there, *provided* n_x and n_y are both large. In the absence of either of these circumstances, you should NOT be using this program. You may now test the hypothesis $H_0{:}\mu_x - \mu_y = \theta_0$ for any value θ_0 (Δ in ASM) for the population mean difference, using label A' in much the same way that you are used to, except you must set flag 5 at Step 2b, in order to signal the calculator that this is a known-variance case. For confidence intervals, you just press [A] and ignore the output, which would normally be degrees of freedom for t, and input $z_{\alpha/2}$ (or z_α for one-sided cases) to compute a confidence interval. An example should suffice to illustrate the mechanics.

EXAMPLE 6.4

A paint manufacturer is experimenting with a new, less expensive paint. He finds that in a sample of 144 gallons, the average coverage, $\bar{x}$, was 480 square feet, while his old paint covered an average of 500 square feet, using 138 gallons. He believes the true standard deviations in either case is 30 square feet. How likely is it that the new paint does not cover as well? He is anxious to implement the new paint, for it is cheaper.

Solution With $n_x = 144$, $\bar{x} = 480$, $\sigma_x = 30$, $n_y = 138$, $\bar{y} = 500$, $\sigma_y = 30$, we could test $H_o{:}\mu_x - \mu_y \geq 0$ against the alternative $H_1\mu_x - \mu_y < 0$, putting the burden of proof on the data to reject H_o and support the situation $\mu_x < \mu_y$ that the old paint is better. Following the instructions for LSN, 144 is stored in R_{13}, while 138 is stored in R_{03}. Now place $\bar{x}$ = 480 in R_{47} and $\bar{y}$ = 500 in R_{37}. Finally, enter 30 in both R_{48} and R_{38}. That concludes data entry. Using an H_1-code of -1, press [A′], and then [St flg] 5. Since $\theta_0 = 0$, enter that, and press [R/S] to find a significance level of $P = 1.1 \times 10^{-8}$. For any reasonable level α, H_o should be rejected. Indeed, a 99% upper CI for $\mu_x - \mu_y$ may be found by pressing [A], entering 2.326, then [R/S] to find an upper limit of -11.7. It certainly appears that $\mu_x < \mu_y$, and he should not switch to the new paint.

If the variances are unknown (as is typical), but somewhere it can be justified that they are equal (as in the last example), then there is an exact procedure based on a *t*-distribution available for inference about $\mu_x - \mu_y$. This procedure will work for any sample size and is based on the fact that if $\sigma_x^2 = \sigma_y^2 = \sigma^2$, say, then the entire data base can be used to estimate this *common* variance by what is usually called a *pooled* estimate, s_p^2, given by

$$s_p^2 = \frac{\sum_{i=1}^{n_x} (x_i - \bar{x})^2 + \sum_{j=1}^{n_y} (y_j - \bar{y})^2}{n_x + n_y - 2} \equiv \frac{(n_x - 1)s_x^2 + (n_y - 1)s_y^2}{n_x + n_y - 2} \tag{6.2}$$

The denominator stems from the fact that s_x^2 is associated with $n_x - 1$ d.f., while s_y^2 corresponds to $n_y - 1$, so the combined d.f. is $n_x - 1 + n_y - 1$. The numerator, of course, is just the *total* sum of squared deviations available in the data, or, viewed another way, combines the two estimates s_x^2 and s_y^2—both estimates of σ^2.

Accordingly, then, theory yields that

$$\frac{\bar{X} - \bar{Y} - (\mu_x - \mu_y)}{\sqrt{\frac{1}{n_x} + \frac{1}{n_y}}\; S_p} \tag{6.3}$$

has a t-distribution with $n_x + n_y - 2$ degrees of freedom. This allows us to generate confidence intervals using the formulas

$$\bar{x} - \bar{y} \pm t_{\alpha/2} s_p \sqrt{\frac{1}{n_x} + \frac{1}{n_y}} \tag{6.4}$$

or to test hypotheses of the basic form $H_o{:}\mu_x - \mu_y = \theta_o$, using as a test statistic

$$ts = \frac{\bar{x} - \bar{y} - \theta_o}{\sqrt{\frac{1}{n_x} + \frac{1}{n_y}}\, s_p} \tag{6.5}$$

Once those facts are established, the procedures are quite like any t-test or confidence interval, and you will find the TI–59 procedures outlined in ZS–6 under the Code INA for independent normal, part A. After data entry, tests are controlled by [C′] and confidence intervals by [C] with the usual instructions. We illustrate.

EXAMPLE 6.5

A test of color perception was administered to a control group (x) and an experimental group (y) with measured (normal) results.

x: 16.3 14.7 12.3 13.5 16.0 17.1 17.3
y: 14.0 16.5 17.7 15.9 18.0 16.3

Following the DE procedure, the data should be entered as independent data using ST–04 and processed with [D′]. Now, *assuming* $\sigma_x = \sigma_y$, press [C] to have $\nu = 11$ displayed. Entering $t_{.025} = 2.201$, a 95% CI for $\mu_x - \mu_y$ extends from -3.16 to .99, and includes 0. It is safe to conclude $\mu_x = \mu_y$, or, perhaps better, it is not safe to reject $\mu_x = \mu_y$. If you prefer, you can test $H_o{:}\mu_x - \mu_y = 0$ against $H_1{:}\mu_x - \mu_y \neq 0$ to find $P = .27$, little basis for rejection.

After any of these routines, you will find the pooled standard deviation s_p in R_{33}. As you can see from the user instructions, this may also be entered, along with other summary data, if available. If you only know s_x and s_y, the program allows for that entry also, and s_p^2 will be computed using Equation 6.2 for you, provided you ensure that 0 is in R_{33}, which you may do by pressing [CMs] first.

EXAMPLE 6.6

A random sample of 50 students scored an average of 58 on a test with standard deviation 4.5. Another section of the class having 35 members scored 62, with a standard deviation of 5.5. How do the two classes compare?

Solution Entering $n_x = 50$, $\bar{x} = 58$, $n_y = 35$, and $\bar{y} = 62$ as usual in INA Steps 1a and 1b, then step 1c requires that you enter $s_x = 4.5$ in R_{48} and $s_y = 5.5$ in R_{38}. Store 0 in R_{33}, since s_p must be computed. Now press [C] to find $\nu = 83$, large enough to enter $z_{.025} = 1.96$ to compute a 95% CI for $\mu_x - u_y$ as $(-6.1, -1.9)$. Conclude $\mu_x - \mu_y \neq 0$. Alternatively, enter 0, press [C′], then [R/S], to find $P = .0002$ for a test of $H_0{:}\mu_x - \mu_y = 0$ vs. $H_1{:}\mu_x - \mu_y \neq 0$. H_0 should be rejected. In R_{33} you will now find $s_p = 4.93$, which you may verify directly, if you like, using Equation 6.2.

Of course, you may find it hard to justify the assumption $\sigma_x = \sigma_y$, but the foregoing procedure requires that this assumption be made. Put another way, when you use INA, you automatically imply that assumption, whether or not you care to be explicit about it. And, in practice, you will find most small sample problems solved this way. If you absolutely cannot justify the assumption, there are not a lot of options open to you. You might use the PN routine, if it should happen that $n_x = n_y$ (differences are still normal), but you cannot count on that. Besides, you lose in degrees of freedom, for $n_x - 1$ will be smaller than $n_x + n_y - 2$, and that is generally viewed as a loss of information, because you do not just have n_x pairs in this case, but rather $n_x + n_y$ independent pieces of data. So, this option is not to be recommended.

There are several, somewhat similar, approximate procedures available. One of these, the so-called Welch-Aspin test, is based on using Equation 6.2 and backing it with a theory that shows the result has *approximately* a *t*-distribution. The degrees of freedom, however, is a random variable dependent on the sample results, and you have no control over that. In the absence of an exact test procedure, it may be used if you recognize that the results are, accordingly, only approximate. The details are programmed under INB, with directions quite like those for INA.

EXAMPLE 6.7

Suppose in Example 6.5, we do not assume $\sigma_x = \sigma_y$. With unequal sample sizes, we may proceed to test H_o using the Welch test. Reenter that data and 0 again as an H_1-code. Press [A′], then [R/S], to find a significance level of $P = .27$ agreeing with the previous case to 2 decimal places. If you will recall R_{15}, you will find that $\nu = 10$, so only one degree of freedom was lost in applying the approximate test. If you press [A], you will also see ν displayed, whereupon you may enter $t_{.025} = 2.228$ to recompute a 95% CI of $(-3.14, .97)$, a result not too unlike the exact result.

In this example, it did not make a lot of difference which procedure we used. But you will not always be that fortunate, since the result depends on the sample outcomes. In any case, you should opt for the exact case, assuming $\sigma_x = \sigma_y$ whenever there is any justification, however remote. Choose the Welch procedure only if there is overwhelming evidence that $\sigma_x \neq \sigma_y$, especially for small samples. For large samples, it does not make a lot of difference what you do, for all the procedures converge to a normal z-test or confidence interval in that case anyway.

Problems

In the following problems, assume $\sigma_x = \sigma_y$, unless specifically told otherwise.

20. Does the use of programmed materials increase computational skills in statistics? Two small classes were used to test this claim, the control group (x) being taught by conventional methods and the experimental group (y) with programmed materials. The results of a common test yielded scores as follows.

x: 82 87 91 54 97 76 64 98 92 57 80 53 64
y: 91 62 94 92 87 79 86 75 90 73 83 93 65 89 68 52

Test the null hypothesis against $H_1{:}\mu_x < \mu_y$ at the 5% level.

Ans: $P = .27$; Hence, accept H_0 ($ts = -.63$). 95% CI: $(-\infty, 5.8)$

21. To test whether or not a certain type of coating made metal more corrosion resistant, 5 metal rods were treated (the coating is expensive) (y), and 15 were not (x). After 10 weeks, weight losses were recorded in grams, as,

x: .194 .200 .199 .209 .198 .195 .203 .207 .203 .208 .196 .198 .195 .203 .196
y: .196 .192 .200 .199 .194

Test $H_0{:}\mu_x = \mu_y$ vs. $H_1{:}\mu_x > \mu_y$ at level .10.

Ans: $P = .054$; Reject H_0($ts = 1.69$). 90% CI: $(.0009, \infty)$

22. Of two technicians working a laboratory, one (x) is suspected of always reading a balance on the "high side," compared to the other (y). Each is asked to give 10 readings of a 2-gram weight, and the following results were reported.

x: 2.046 2.062 2.035 2.041 2.044 2.058 2.049 2.052 2.088 2.067
y: 2.025 2.023 2.048 2.033 2.051 2.089 2.061 2.082 2.029 2.041

Test the null hypothesis against $\mu_x > \mu_y$ at level $\alpha = .05$.

Ans: $P = .25$. Accept H_0($ts = .68$). 95% CI: $(-.009, \infty)$

23. Two different types of fertilizers are used on 10 specimens of corn. After several weeks the growth of the plants is carefully noted with the following results.

x:	2.4	2.4	1.1	0.9	3.9	2.5	2.3	3.8	1.6	2.9
y:	1.9	1.8	2.3	2.1	3.3	4.1	0.7	3.6	0.8	2.9

Is there any significant difference in the fertilizer effects? Let $\alpha = .01$.

Ans: With $H_1: \mu_x \neq \mu_y$, $P = .95$ ($ts = .06$). Accept H_0. 99% CI: $(-1.35, 1.41)$

24. In a study of the difference of attitudes toward current social issues between men (x) and women (y), a researcher reported the following data taken from test scores.

$\bar{x} = 79 \quad s_x = 6 \quad \bar{y} = 82 \quad s_y = 8 \quad n_x = 20 \quad n_y = 15$

Find a 90% confidence interval for the difference $\mu_x - \mu_y$. Would the null hypothesis have been rejected?

Ans: $(-6.89, .89)$ No, 0 is in the interval.

25. A professor notes that one class of 23 students achieved an average score of 67 on a test with a standard deviation of 10. Another section of the same subject having 21 students achieved an average of 63 with a standard deviation of 9. Find a 90% confidence interval for the difference in means. Are the two classes essentially the same?

Ans: $(-.74, 8.74)$ Yes, 0 is in the interval.

26. Two classes of respective sizes 29 and 35 always seem to score about the same on the average, but the scores in the first class (x) are consistently more variable than in the second one (y). On a recent exam, the class averages and sample standard deviations were $\bar{x} = 73$, $s_x^2 = 9.7$, while $\bar{y} = 75$ and $s_y^2 = 14.1$. Test the hypothesis $\mu_x = \mu_y$ at level .05.

Ans: $P = .02$ ($ts = -2.29$) Reject H_0. 95% CI: $(-3.7, -.29)$ using INA

$P = .02$ ($ts = -2.33$) Reject H_0. 95% CI: $(-3.7, -.32)$ using LSN

27. A random sample of size 40 was drawn from a normal distribution, yielding sample values $\bar{x} = 6.2$ and $s_x = 1.02$. Another random sample of size 40 from an allegedly different normal population resulted in sample values $\bar{y} = 6$ and $s_y = .72$. Is it safe to assume that the means are equal? Let $\alpha = .01$.

Ans: $P = .31$ ($ts = 1.01$), Reject H_0. 99% CI:$(-.31, .71)$ using LSN

28. Two bolt machines are supposed to be set to produce 2-inch bolts. Variability would not change in any case. Samples of 10 each were taken off the first run and resulted in observations:

Machine X: 2.067 2.016 2.011 2.020 2.066 2.052 2.014 2.069 2.053 2.054

Machine Y: 2.032 2.074 2.077 2.095 2.001 2.008 2.029 2.025 2.056 2.066

Is there a significant difference in the output? Let $\alpha = .01$.

Ans: $P = .75$ ($ts = -.33$). Do not reject H_0. 99% CI:(−.04, .032)

29. It is claimed that the following random samples were independently drawn from a $N(0,1)$ distribution. Test the hypothesis that $\mu_x = \mu_y$, assuming that $\sigma_x^2 = 1 = \sigma_y^2$ at level .05.

X: −1.752 −0.291 −0.933 −0.450 0.512 −0.702 0.284 −0.509 −1.776

Y: −0.044 0.263 0.986 −0.441 −0.866 −1.215

Ans: Using [Σ+], $\bar{x} = -.624$, $\bar{y} = -.2195$, $n_x = 9$, $n_y = 6$, $P = .44$ ($ts = -.77$), and 95% CI: (−1.44, 0.63), using LSN. In any case, do not reject H_0.

30. An analyst in the Office of the Secretary of Defense is interested in the relative ages of Operations Research (OR) analysts who work for the Department of the Army and the Department of the Navy. The OSD analyst takes a survey and finds the following:

Dept. of Army:	28	35	56	54	31	38	34
Dept. of Navy:	24	28	51	37	29		

Find a 95% CI for the difference in ages.

Ans: (−8.65, 19.91)

31. A mathematics professor has two classes taking Calculus I. He wishes to see if there is any difference in the two classes, so he gives both of them the same examination. The first class, composed of 29 students, scores 78 on the average, with a standard deviation of 11. The second class, which has 27 students, has an average score of 75, with a standard deviation of 9. Construct a 95% confidence interval for the difference in means to help the professor decide if the classes are essentially the same.

Ans: (−2.29, 8.29) The classes cannot be distinguished.

32. The effects of a certain drug on distance runners are being studied. A test measuring performance on a continuous scale was given to two groups with the following results.

I:	85.6	88.9	91.5	90.2	89.1	87.0
II:	80.1	79.0	76.3	88.5	90.3	

Analyze the data for significant difference.

Ans: $P = .06$ ($ts = 2.20$). Reject at any level $\alpha > .06$.

33. The lieutenant in charge of traffic control at a metropolitan police department is responsible for evaluating the performance of two divisions. He decides to take a random sample of the number of tickets issued daily during the last year by the two divisions as one measure of performance. He samples, with the following results.

Division A:	85	91	160	93	112	96	99	114	101
Division B:	98	86	130	148	180	56	78	91	109

Find a 95% confidence interval for the difference in means of the two divisions.

Ans: (−34.15, 28.60)

34. The owner of an NFL football team has decided to evaluate the diet used in his training camp. In so doing, he wants to compare the number of calories consumed at the morning meal to the number consumed at the dinner meal. He takes two random samples, with the following results.

Breakfast:	2000	1885	1900	3150	2500	2830	2150	2140	1980
Dinner:	2100	1895	1920	3010	2900	2740	2200	2700	

Determine a 99% confidence interval for the difference in mean calories consumed at breakfast and dinner.

Ans: (−796.91, 493.99)

35. Two different production runs on the same brand of golf ball produced the following results concerning the diameter of the balls.

	FIRST RUN	SECOND RUN
Mean:	1.65 inches	1.66 inches
Standard deviation:	.04 inches	.05 inches
Sample size:	100	100

At the .10 level of significance, did the runs produce the same size golf ball?

Ans: Using LSN, $P = .12$ ($ts = -1.56$). Do not reject equality of means.

36. Two random samples were taken of members of two different nationwide social clubs, The Squares and The Swingers, to determine their annual incomes. The following results were determined from the samples. Do the two clubs differ in incomes?

THE SQUARES		THE SWINGERS	
Mean income	$7123	Mean income	$7289
Standard deviation	298	Standard deviation	325
Sample size	221	Sample size	198

Ans: Using LSN, $P = 5.71 \times 10^{-8}$. ($ts = -5.43$); Reject H_0.

In the remaining problems, it is not safe to assume either σ_x, σ_y known, or $\sigma_x = \sigma_y$.

37. A distributor of light bulbs is considering adding one of two new types of bulbs to his line of products. Tests of the life of the bulbs resulted in the following time in days for each type.

Type I:	360	180	395	420	501	175	380	395	312	378	390
Type II:	420	401	376	388	491	401	412	495	400	399	

Find a 95% confidence interval for the mean difference in life time, and test for equality of means.

Ans: $\nu = 13$, CI: $(-134.74, 4.69)$; $P = .065$ ($ts = -2.01$). Do not reject H_0.

38. The following are ten independent measures of the density of a compound by two different analysts.

Analyst A:	.180	.182	.179	.184	.187	.177	.178	.179	.181	.180
Analyst B:	.181	.181	.180	.179	.181	.183	.181	.182	.181	.179

Test for equality of means for the two analysts. Use $\alpha = .05$.

Ans: $P = .92$ ($ts = -.098$), $\nu = 11$. 95% CI: $(-.002, .002)$. Do not reject H_0.

39. A random sample of 12 graduate students at a large public university had a mean age of 28 years, with a standard deviation of 1.8 years. A random sample of 14 graduate students at a small private eastern school had a mean age of 26 years, with a standard deviation of 1.6 years. Find a 95% confidence interval for the difference in mean age between graduate students at the two schools. Assess the difference in means.

Ans: $\nu = 22$, CI: $(0.60, 3.40)$. $P = .007$($ts = 2.97$). Reject equality of means.

40. The weights of 20 Texas high school football players have a sample mean of 220 pounds and a standard deviation of 12 pounds. A sample of 20 Florida high school footballers resulted in a mean weight of 215 pounds, with a standard deviation of 11 pounds. Construct a 95% confidence interval estimate for the difference in the mean weights of all Texas and all Florida high school football players. How do the 2 states compare?

Ans: $\nu = 37$, CI: $(-2.13, 12.13)$. They are not distinguishable.

Section 6.5: Equality of Variances

Of course, we can never know for certain if two variances are equal, as required by the t-procedure of the last section. In the end, it has to be an assumption, whether tenable or not. But we do have at our disposal a method for testing the hypothesis that two normal variances are equal, given random samples from each population. You know that that does not either prove or disprove the state of nature, but at least it might make you feel more (or less) comfortable about making the assumption $\sigma_x^2 = \sigma_y^2$. The procedure is based on yet another probability distribution we have not encountered before, the F-family.

You remember that the variance estimator S^2, apart from a constant, has a Chi-square distribution. If S_x^2 and S_y^2 are two such estimators from independent populations, then their ratio gives rise to a particular F-distribution. Such distributions are discussed in ASM starting on page 6–17 as the last program in the statistics module, ST–22. As you see, the F-family arises as the ratio of 2 independent Chi-square variables, each divided by its respective degrees of freedom. So, both degrees of freedom are required to keep track of a particular F-distribution. That is, the F-family is indexed by 2 degrees of freedom, the first one, ν_1, called the *numerator* degrees of freedom, and the second, ν_2, called the *denominator* degrees of freedom for obvious reasons. You may think of ν_1 and ν_2 as a first and last name, respectively, for a particular member of the F-family, and we can conveniently keep track of these names in ordered pair notation, (ν_1, ν_2). Thus, 2 members of the F-family are the same, if and only if, their first *and* second name is the same. So the F with d.f. = (2,4), for example, is different from the one with d.f. = (4,2).

The usual unpleasant formula for the $F(\nu_1, \nu_2)$ density function is displayed for you at the bottom of page 6–17. Three members are graphed for you in Figure 6.8 just above it. Basically, F–densities resemble Chi-square densities somewhat, and you might have expected that. But the double index certainly makes them different, as well as the formula, which unfortunately cannot be integrated in a convenient closed form. Once again, we have to be content with approximations. Program ST–22 allows us to find the CCDF at an arbitrary value F, that is, $Q(F)$, after entering the degrees of freedom *in order* with [A] and [B]. After entering F (please correct your copy of the user instructions on page 6–18, if it shows x under "enter") and pressing [C], $Q(F)$ will be displayed. As noted, neither ν_1 nor ν_2 ought to exceed about 120 (they will not for most applications), and execution time increases with the size of degrees of freedom. You may practice with Example 1 on page 6–19 of ASM, but please hold off on the other examples until we treat AOV in Chapter 8.

As for our current interest in the F-family, if S_x^2 and S_y^2 are sample variances from independent normal populations, based on respective sample sizes n_x and n_y, then $(n_x - 1)\, S_x^2/\sigma_x^2$ has a Chi-square distribution with $n_x - 1$ d.f., $(n_y - 1)S_y^2/\sigma_y^2$ has a Chi-square distribution with $n_y - 1$ d.f. Accordingly,

$$\frac{(n_x - 1)\, S_x^2/(n_x - 1)\, \sigma_x^2}{(n_y - 1)\, S_y^2/(n_y - 1)\, \sigma_y^2} = \frac{\sigma_y^2}{\sigma_x^2}\,\frac{S_x^2}{S_y^2} \tag{6.6}$$

has an F-distribution with d.f. = $(n_x - 1, n_y - 1)$. Or,

$$\frac{\sigma_x^2}{\sigma_y^2}\,\frac{S_y^2}{S_x^2} \tag{6.7}$$

has an F with d.f. = $(n_y - 1, n_x - 1)$, reversed degrees of freedom.

If we start with Equation 6.7, we may generate confidence intervals for the ratio σ_x^2/σ_y^2 from the fact that

$$1 - \alpha = \Pr\left(F_{1-\alpha/2} < \frac{\sigma_x^2}{\sigma_y^2}\frac{S_y^2}{S_x^2} < F_{\alpha/2}\right) \tag{6.8}$$

In terms of data, the nonsymmetric confidence limits would then be given by

$$l = \frac{s_x^2}{s_y^2} F_{1-\alpha/2} \qquad \text{and} \qquad u = \frac{s_x^2}{s_y^2} F_{\alpha/2} \tag{6.9}$$

where, keep in mind, d.f. $= (n_y - 1, n_x - 1)$. To find confidence limits for σ_y^2 / σ_x^2, we have only to invert these limits to obtain a lower limit of $1/u$ and an upper limit of $1/l$. Alternatively, we could start with Equation 6.6 and obtain a lower limit of

$$\frac{s_y^2}{s_x^2} F_{1-\alpha/2}$$

while the upper limit would be

$$\frac{s_y^2}{s_x^2} F_{\alpha/2}$$

only this time d.f. $= (n_x - 1, n_y - 1)$. Since these limits must be the same, it brings up an interesting relationship between percentile points and the *order* of the degrees of freedom: namely, for any proportion P, $F_P(\nu_1, \nu_2) = 1/F_{1-P}(\nu_2, \nu_1)$. In other words, the lower percentile point with one set of names is the reciprocal of the corresponding upper percentile point with names (d.f.) reversed. Thus, for example, $F_{.95}(2,4) = 1/F_{.05}(4,2)$.

Of course, we need these percentile points if we wish to compute confidence intervals, and, since ST–22 does not make these available except by trial and error, we have provided the usual table at several locations in the book where they are used—page 161 in the user instructions for ZS–6, for instance. For various choices of ν_1 and ν_2, entries for the typical upper percentile points are given. No lower percentile points are shown, nor are they necessary, because of the foregoing reciprocal relationship. You may notice that the table is rather skimpy in degrees of freedom after 10 or so. If accuracy is really important, you interpolate in the table, or else use ST–22 to advantage to find a more precise value. But in the applications here where the F-percentile is used, such extreme accuracy is rarely called for, and usually it suffices to select d.f. as close as possible to tabled ones. That is the procedure we will use in our solutions.

The confidence interval program for the ratio σ_x^2/σ_y^2 is presented in ZS–6 under the code NV for normal variances. You may use the usual data entry in Step 1. At Step 3, you press [D] to see a first name ν_1 for an F-distribution and [R/S] to see the second name. (These happen to be $n_x - 1$ and $n_y - 1$, in that order.) Now store the upper percentile $F_{\alpha/2}$ in R_{31}, reverse the degrees of freedom, and enter *that* upper percentile point in R_{41}. With a press of [R/S], you will find l and u as usual.

EXAMPLE 6.8

In the color perception test of Example 6.5, it was assumed that $\sigma_x^2 = \sigma_y^2$. The sample data revealed $s_x = 1.885$, with $n_x = 7$, while $s_y = 1.434$, with $n_y = 6$. To find a 95% confidence for the ratio σ_x^2/σ_y^2, proceed in ZS–6 to store these statistics in the appropriate registers, as required by Step 1 of NV. Then press [D] to see $\nu_1 = 6$ displayed and [R/S] to see $\nu_2 = 5$ in R_D. From the F-table, enter $F_{.025} = 6.98$ in R_{31}. Now reverse the degrees of freedom to (5,6), and enter $F_{.025} = 5.99$ in R_{41}. Press [R/S] to see $l = .25$, and find 10.35 in R_T. You can be 95% confident that σ_x^2/σ_y^2 is between .25 and 10.35. By the way, since 1 is one of the plausible values for σ_x^2/σ_y^2, it is quite tenable that $\sigma_x^2/\sigma_y^2 = 1$ or $\sigma_x^2 = \sigma_y^2$.

You see how a fixed size α test can be carried out by observing whether or not 1 is located in the confidence interval. But we can utilize ST–22 to carry out significance tests also. We have provided for this at Step 2 of the NV program. With the same data entry, you enter the usual H_1-code. Then you must set flag 4 to signal the calculator that a significance test is required, rather than a confidence interval. The P-value is found by pressing [D] once the flag is set. Remember, if flag 4 is *not* set, the program will proceed to the confidence interval routine. With the data for Example 6.8 still stored, it is only necessary to do this (using an H_1-code of 0) to find a P-value of .56 for the example. Thus the hypothesis of equality of variances is quite acceptable. Remember that if you ever press [D] to test H_0 and find ν_1 displayed instead of P, flag 4 was not set. Conversely, if you want to compute a CI, and, instead of seeing ν_1, you see a proportion P in R_D, flag 4 was not cleared. Press [RST] and try again if that ever happens.

With that list of complications aside, you should be able to manage the problems that follow. In all cases, assume that the data are from independent normal populations. Please take special note of the special procedures to be followed for one-sided confidence intervals—necessitated by the lack of symmetry in the density function, as with the Chi-square routines.

41. Find the 90th percentile of the F-random variable whose numerator degrees of freedom are 10 and whose denominator degrees of freedom are 15. What is the 5th percentile of this distribution?

Ans: $F_{.10}(10,15) = 2.06 \quad F_{.95}(10,15) = 1/F_{.05}(15,10) = 1/2.84 = .35$

42. Use the F-table to find the values of $F_{.025}(6,8)$, $F_{.05}(7,10)$, $F_{.995}(8,8)$, $F_{.05}(36,51)$, $F_{.10}(350,110)$.

Ans: 4.65 3.14 $1/7.50 = .13$ 1.69 1.21

43. Two nail machines seem to put out an average 1-inch nail alright, but machine X always seems to have more extreme values than machine Y. A sample of 1 day's output yielded the following data. At level 5%, test $H_o{:}\sigma_x^2 = \sigma_y^2$ against the alternative $H_1{:}\sigma_x^2 > \sigma_y^2$.

X:	1.09	1.02	1.02	0.95	1.05	1.05	0.67	1.09	0.88	1.04	1.04
	1.02	1.03	1.09	1.06							
Y:	1.08	1.01	1.02	1.01	1.03	1.03	1.01	1.08	1.05	1.04	1.03
	0.82	1.02	1.01	1.03							

Ans: $P = .016$ ($ts = 3.317$); lower 95% CI:(1.38, ∞); Reject H_0.

44. Two samples of 10 and 16 observations, respectively, yielded variances of $s_x^2 = .3888$ and $s_y^2 = 1.25$. It would appear that σ_x^2 is considerably less than σ_y^2 based on these estimates, but what would a 90% CI for σ_x^2/σ_y^2 be?

Ans: $\nu_1 = 9$, $\nu_2 = 15$; CI (.12, .94). Fails to include 1, so conclude $\sigma_x^2 \neq \sigma_y^2$. An upper one-sided limit would be $u = .73$, so we are 90% confident that $\sigma_x^2 < .73\sigma_y^2$.

45. A control laboratory is interested in comparing two methods of analysis of a specimen. There is no evidence that the variances in the two methods vary at all, but some old data contained the following.

Aqueous:	4.27	4.32	4.29	4.30	4.31	4.30	4.30	4.32	4.28	4.32
Alcoholic:	4.28	4.32	4.29	4.31	4.35	4.29	4.32	4.33	4.28	4.37
	4.38	4.28	4.32							

Does one method appear to be more variable than the other?

Ans: Test $H_o{:}\sigma_x^2 = \sigma_y^2$ vs. $H_1{:}\sigma_x^2 \neq \sigma_y^2$ to find $P = .052$ ($ts = .26$). A 95% CI for σ_x^2/σ_y^2 runs from .08 to 1.01 and just includes 1. At level $\alpha = .05$, we would not have enough evidence to reject equality and conclude that one process is no more variable than the other.

46. Two different brands of coffee-filling machines are supposed to be equivalent as far as the amount of liquid delivered. A 30-day record revealed, however, that, while both machines yielded about the same amount on the average, machine X seemed to be more variable than

machine Y. This was based on the fact that the sample variances reported were $s_x^2 = 2.96$ and $s_y^2 = 1.56$. Is the conclusion valid? Test $H_o: \sigma_x^2 = \sigma_y^2$ against $H_1: \sigma_x^2 > \sigma_y^2$ at level .10.

Ans: $P = .04$ ($ts = 1.90$). The lower 90% CI for σ_x^2/σ_y^2 is $(1.18, \infty)$. Either way, H_0 should be rejected in favor of H_1.

47. When the salaries of two departments in a firm were compared recently, it was found that they both received the same average raise. But the 30 members of department X had a standard deviation of \$190, while the 10 members of department Y were found to have a standard deviation of \$134. Find a 95% confidence interval for σ_x/σ_y. Would the hypothesis of equality have been accepted or rejected?

Ans: (.75, 2.27). Accepted, since 1 is in the CI.

48. A new type of voltmeter (X) is asserted to be more precise in terms of smaller variances than one that is presently in use (Y). But, unless σ_x is significantly less than σ_y, it would be better to remain with the present one, since the new one is more expensive. To test the new machine, 20 resistors were measured (in ohms) by both machines and yielded sample variances $s_x^2 = .621$, while $s_y^2 = .732$. Is the difference enough to warrant a change?

Ans: Let $H_1: \sigma_x < \sigma_y$, so that the burden is on data to reject H_0 and change. Then $P = .36$ ($ts = .85$); no change is warranted.

49. An experiment involving two normal random variables, X and Y, yielded the information $s_y^2 = 21.1$, $n_y = 17$ and $s_x^2 = 2.1$, $n_x = 23$. Find a 90% upper one-sided CI for the ratio σ_y^2/σ_x^2.

Ans: (0,19); $19 = 1/l$ where l is the lower bound on σ_x^2/σ_y^2.

50. The comptroller of a large company decided to determine if there was any difference in the expense accounts of the company's two sales divisions. A random sample of expense vouchers was conducted for the two divisions with the following results.

Division 1:	28.30	15.70	28.65	19.34	42.50	27.28	21.95	32.80
	31.75	39.40						
Division 2:	38.94	33.26	31.88	22.40	26.35	41.40	18.43	26.85
	16.25	29.40						

At the .05 level of significance, does the standard deviation of expense vouchers differ for the two divisions?

Ans: $P = .95$ ($ts = 1.05$). They do not appear to differ.

51. Is the assumption of equality of variances tenable for Problem 24?

Ans: $P = .24$ ($ts = .56$); 95% CI: (.20, 1.45). Data do support equality.

52. Repeat Problem 51 for Problem 25.

Ans: $P = .64$ ($ts = 1.23$); 95% CI: (.50, 3.04). Data support equality of variances assumption.

Ratios of variances is not the only setting in which the F-family arises, by any means. Since $2n\lambda\bar{X}$ has a Chi-square distribution whenever $\bar{X}$ is a random sample from an exponential population, it is not surprising that the ratio of sample means from two independent such populations would give rise to an F-distribution. Indeed, if X and Y represent two such populations with parameters λ_x and λ_y, and samples of respective sizes n_x and n_y are presented, then

$$\lambda_x\bar{X}/\lambda_y\bar{Y} = \frac{\mu_y\bar{X}}{\mu_x\bar{Y}} \tag{6.10}$$

has an F-distribution with d.f. $= (2n_x, 2n_y)$.

This is the basis for the inference regarding the ratio μ_y/μ_x, or μ_x/μ_y, if we prefer. The details are so much like those of the ratio σ_x^2/σ_y^2 from normal populations that essentially the only difference lies in the choice of degrees of freedom. These details are summarized for you under the code *Exp* in program ZS–6, with almost identical instructions. An example should serve to illustrate the very slight changes.

EXAMPLE 6.10

A certain component of an earth satellite is being tested for (exponential) mean time to failure. In a test laboratory, a sample of 60 items averaged 1,306 hours before failure. In actual space tests, 6 of these components lasted a total of 7,482 hours. Is it reasonable to assume that mean time to failure in space is at least as long as in earth environment?

Solution Letting X stand for earth environment, $n_x = 60$ and $\bar{x} = 1{,}306$, while $n_y = 6$ with $\bar{y} = 1{,}247$. Store these, as required by *Exp*, in ZS–6. For this situation, let us test $H_o{:}\mu_x \le \mu_y$ against $H_1{:}\mu_x > \mu_y$, putting the burden on the data to show that the apparent drop is real. Using an H_1-code of $+1$, set flag 4 and press [E] to discover a significance level of .50. It is not likely that the smaller observed average failure in space is a real one. A lower 95% one-sided confidence limit on μ_x/μ_y may be found by first pressing [E] to see $\nu_1 = 120$ displayed, then [R/S], to find $\nu_2 = 12$. From the F-table, $F_{.05} = 2.35$ for d.f. $= (100{,}12)$. Entering this value in R_{31} and 1 in R_{41}, a press of [R/S] displays $l = .4$. With 1 included in the interval $(.4, \infty)$, we cannot safely conclude $\mu_x > \mu_y$, as we would if l were greater than 1.

Recalling that $\mu_x = 1/\lambda_x$ and $\mu_y = 1/\lambda_y$, you will observe that a test of equality of means here is equivalent to a test of equality of failure rates. You do have to exercise care about one-sided situations, however. In the last example, the alternative $H_1{:}\mu_x > \mu_y$ would be equivalent to $H_1{:}\lambda_x < \lambda_y$, just the opposite. So if you have a problem calling for a test of $H_o:\lambda_x \leq \lambda_y$ versus $H_1:\lambda_x > \lambda_y$, you should test, equivalently, $H_o:\mu_x \geq \mu_y$ against $H_1:\mu_x < \mu_y$. Once this translation is made, no further changes are required. The *P*-value obtained will be the correct one.

In the matter of confidence intervals, a CI on μ_x/μ_y is clearly a CI on λ_y/λ_x, since the latter is just equivalent to the former. If you need a CI on μ_y/μ_x or λ_x/λ_y (the same thing), the simplest thing to do is to compute a CI for μ_x/μ_y and invert the limits. Thus, in our example, an upper one-sided CI on μ_y/μ_x would be given by $l = 1/u = 2.2$, leading to the same conclusion, of course. Had we wanted a lower one-sided CI on μ_y/μ_x, we would go back and compute an upper one-sided limit for μ_x/μ_y by entering 1 in R_{31}, and then 1.85 [$F_{.05}(12,100)$] in R_{41}. Press [E], followed by two [R/S]'s to find $l = 1.05$, which you ignore. The value $u = 1.94$ is found in R_T, and a press of [1/*x*] converts this to .52, a lower limit on μ_y/μ_x.

We offer a few exponential exercises to close this section and chapter. This will provide you with still more practice utilizing the *F*-distribution.

Problems

53. An item designed to operate under 2 different stress levels is assumed to have an exponential lifetime. Is it likely that the failure rates are different based on the following observed times to failure?

Level 1:	34	15	151	77	97	300		
Level 2:	102	322	21	82	99	8	18	42

Ans: Testing $H_0:\lambda_x = \lambda_y$ or $\mu_x = \mu_y$, $P = .62$ based on $ts = 1.29$. A 95% CI for λ_y/λ_x extends from .44 to 4.12 and comfortably includes the value 1. There is no basis for any difference in failure rates.

54. Arrival times for orders placed on 2 related items from a warehouse vary, and sampling revealed the following data in hours.

Item 1:	41	32	29	19	6
Item 2:	51	71	80	42	93

Is there any basis for rescheduling maintenance schedules because mean arrival times are different?

Ans: A test of $H_0:\mu_x = \mu_y$ yields a significance level of $P = .14$, so there is little evidence that the apparent difference of $\bar{x} - y = -42$ hr is real. A 95% CI for μ_x/μ_y runs from as low as .10 hr to 1.4 hr, so it could easily be true that $\mu_x = \mu_y$.

55. Failure data in a missile program revealed that 56 replacements of guidance package failed after a total operating time of 56,631 hr when used in one missile. When used in different type missile, 17 replacements lasted a total of 34,320 hr. How likely is it that the missile type makes a difference?

Ans: A test of $H_0:\mu_x = \mu_y$ gives $P = .007$, too small to attribute the average difference to chance.

ZS–6: User Instructions

STEP	PROCEDURE	ENTER	PRESS	DISPLAY
DE	*Original Data Entry*			
1	Enter data using ST–04.			
2	Exit ST–04.		[RST]	n
3	Process data for storage.		[D′]	0
	Note: Step 3 is not repeatable.			
PN	*Paired Data:* $\mu_x - \mu_y$			
1	Enter data using DE *or*:			
	a. Enter sample size.	n	[STO]03	n
	b. Enter sample means.			
	1. Original means	$\bar{x}$	[STO]47	$\bar{x}$
	OR:	$\bar{y}$	[STO]37	$\bar{y}$
	2. Mean difference	$\bar{d}$	[STO]47	$\bar{d}$
		0	[STO]37	0
	c. Enter sample standard deviation.	s_d	[STO]27	s_d
2	*Test* $H_o{:}\mu_x - \mu_y = \theta_o$.			
	a. Enter H_1-code.*	H_1-code	[B′]	H_1-code
	b. Enter θ_0, and compute P-value.	θ_o	[R/S]	P
3	*CI for* $\mu_x - \mu_y$			
	a. Calculate degrees of freedom.		[B]	ν
	b. Enter $t_{\alpha/2}$ with d.f. $= \nu$,	$t_{\alpha/2}$	[R/S]	l
	and calculate limits.		[$x \rightleftarrows t$]	u
	Note: For one-sided intervals, enter t_α at 3b and ignore l or u, as the case may be.			
INA ($\sigma_x = \sigma_y$)	Independent Data: $\mu_x - \mu_y$			
1	Enter data using DE *or*:			
	a. Clear memory and enter sample sizes.		[CMs]	
		n_x	[STO]13	n_x
		n_y	[STO]03	n_y
	b. Enter sample averages.	$\bar{x}$	[STO]47	$\bar{x}$
		$\bar{y}$	[STO]37	$\bar{y}$

$$* \ H_1\text{-code} = \begin{cases} 1 \text{ if } H_1{:}\theta > \theta_o \\ 0 \text{ if } H_1{:}\theta \neq \theta_o \\ -1 \text{ if } H_1{:}\theta < \theta_o \end{cases}$$

STEP	PROCEDURE	ENTER	PRESS	DISPLAY
	c. Enter standard deviations.			
	1. Pooled estimate available OR:	s_p	[STO]33	s_p
	2. Original S.D.'s available	0 s_x s_y	[STO]33 [STO]48 [STO]38	0 s_x s_y
2	*Test* $H_o{:}\mu_x - \mu_y = \theta_o$. a. Enter H_1-code. b. Enter θ_o, and compute P-value.	H_1-code θ_o	[C′] [R/S]	H_1-code P
3	*CI for* $\mu_x - \mu_y$ a. Calculate degrees of freedom. b. Enter $t_{\alpha/2}$ with d.f. $= \nu$, and calculate limits.	$t_{\alpha/2}$	[C] [R/S] [$x \gtrless t$]	ν l u
	(See previous note for one-sided limits.)			
INB ($\sigma_x \neq \sigma_y$) 1	*Independent data:* $\mu_x - \mu_y$ (Welch Approximate t) Enter data using DE *or*: a. Enter sample sizes.	n_x n_y	[STO]13 [STO]03	n_x n_y
	b. Enter sample means. c. Enter sample standard deviations.	$\bar{x}$ $\bar{y}$ s_x s_y	[STO]47 [STO]37 [STO]48 [STO]38	$\bar{x}$ $\bar{y}$ s_x s_y
2	*Test* $H_o{:}\mu_x - \mu_y = \theta_o$. a. Enter H_1-code. b. Enter θ_o, and compute P-value.	H_1-code θ_o	[A′] [R/S]	H_1-code P
3	*CI for* $\mu_x - \mu_y$ a. Calculate degrees of freedom. b. Enter $t_{\alpha/2}$ with d.f. $= \nu$, and calculate limits.	$t_{\alpha/2}$	[A] [R/S] [$x \gtrless t$]	ν l u
	(See previous note for one-sided limits.)			

STEP	PROCEDURE	ENTER	PRESS	DISPLAY
LSN	*Large sample normal:* *or* σ_x, σ_y known: $\mu_x - \mu_y$			
1	Enter summary data only:			
	a. Enter sample sizes.	n_x	[STO]13	n_x
		n_y	[STO]03	n_y
	b. Enter sample means.	$\bar{x}$	[STO]47	$\bar{x}$
		$\bar{y}$	[STO]37	$\bar{y}$
	c. Enter standard deviations.	σ_x or s_x	[STO]48	σ_x or s_x
		σ_y or s_y	[STO]38	σ_y or s_y
2	*Test* H_o:$\mu_x - \mu_y = \theta_o$			
	a. Enter H_1-code.	H_1-code	[A′]	H_1-code
	b. Set flag 5.		[St flg]5	H_1-code
	c. Enter θ_o, and calculate *P*-value.	θ_o	[R/S]	P
3	*CI for* $\mu_x - \mu_y$			
	a. Initialize (ignore output).		[A]	*xxx*
	b. Enter $z_{\alpha/2}$,	$z_{\alpha/2}$	[R/S]	l
	and calculate limits.		[$x \leftrightarrows t$]	u
	(See previous note for one-sided limits.)			

TABLE T. STUDENT −t PERCENTILES

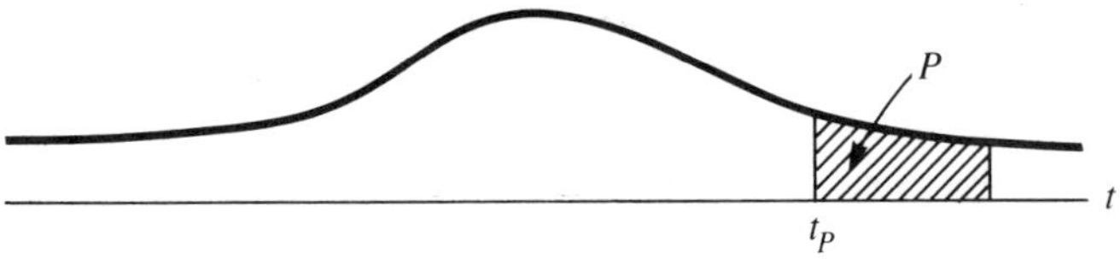

	P				
ν	.10	.05	.025	.01	.005
1	3.078	6.314	12.71	31.82	63.66
2	1.886	2.920	4.303	6.965	9.925
3	1.638	2.353	3.182	4.541	5.841
4	1.533	2.132	2.776	3.747	4.604
5	1.476	2.015	2.571	3.365	4.032
6	1.440	1.943	2.447	3.143	3.707
7	1.415	1.895	2.365	2.998	3.499
8	1.397	1.860	2.306	2.896	3.355
9	1.383	1.833	2.262	2.821	3.250
10	1.372	1.812	2.228	2.764	3.169
11	1.363	1.796	2.201	2.718	3.106
12	1.356	1.782	2.179	2.681	3.054
13	1.350	1.771	2.160	2.650	3.012
14	1.345	1.761	2.145	2.624	2.977
15	1.341	1.753	2.131	2.602	2.947
16	1.337	1.746	2.120	2.583	2.921
17	1.333	1.740	2.110	2.567	2.898
18	1.330	1.734	2.101	2.552	2.878
19	1.328	1.729	2.093	2.539	2.861
20	1.325	1.725	2.086	2.528	2.845
21	1.323	1.721	2.080	2.518	2.831
22	1.321	1.717	2.074	2.508	2.819
23	1.320	1.714	2.069	2.500	2.807
24	1.318	1.711	2.064	2.492	2.797
25	1.316	1.708	2.060	2.485	2.787
26	1.315	1.706	2.056	2.479	2.779
27	1.314	1.703	2.052	2.473	2.771
28	1.313	1.701	2.048	2.467	2.763
29	1.311	1.699	2.045	2.462	2.756
30	1.310	1.697	2.042	2.457	2.750
> 30	1.282	1.645	1.960	2.326	2.576

STEP	PROCEDURE	ENTER	PRESS	DISPLAY
NV 1	*Independent Data:* σ_x^2/σ_y^2 Enter data using DE *or*: a. Enter sample sizes. b. Enter sample standard deviations.	 n_x n_y s_x s_y	 [STO]13 [STO]03 [STO]48 [STO]38	 n_x n_y s_x s_y
2	*Test* $H_o{:}\sigma_x^2 = \sigma_y^2$. a. Enter H_1-code; set flag 4. b. Calculate P-value.	 H_1-code	 [St flg]4 [D]	 H_1-code P
3	*CI for* σ_x^2/σ_y^2 a. Compute degrees of freedom From accompanying table: b. Enter F-value with d.f. = (ν_1, ν_2). c. Enter F-value with d.f. = (ν_2, ν_1). d. Calculate limits.	 $F_{\alpha/2}$ $F_{\alpha/2}$	 [D] [R/S] [STO]31 [STO]41 [R/S] $[x \leftrightarrows t]$	 ν_1 ν_2 $F_{\alpha/2}$ $F_{\alpha/2}$ l u
	Note: a. For lower one-sided interval, enter F_α at Step 3b, 1 at Step 3c, and ignore u. b. For upper one-sided interval, enter 1 at Step 3b, F_α at Step 3c, and ignore l.			
Exp 1	*Independent Exponential*: μ_x/μ_y Enter data using DE *or*: a. Enter sample sizes. b. Enter sample means.	 n_x n_y $\bar{x}$ $\bar{y}$	 [STO]13 [STO]03 [STO]47 [STO]37	 n_x n_y $\bar{x}$ $\bar{y}$

STEP	PROCEDURE	ENTER	PRESS	DISPLAY
2	*Test* $H_o{:}\mu_x = \mu_y$ a. Enter H_1-code; set flag 4. b. Calculate P-value.	H_1-code	[St flg]4 [E]	H_1-code P
3	*CI for* μ_x/μ_y a. Compute degrees of freedom From accompanying table: b. Enter F-value with d.f. $= (\nu_1, \nu_2)$. c. Enter F-value with d.f. $= (\nu_2, \nu_1)$. d. Calcualte limits. (See note under NV3 for one-sided limits.)	 $F_{\alpha/2}$ $F_{\alpha/2}$	[E] [R/S] [STO]31 [STO]41 [R/S] [R/S]	ν_1 $F_{\alpha/2}$ $F_{\alpha/2}$ l u

REGISTER CONTENTS							
00	Used by ST–04 (00–06)	10		20	Used (20–26)	30 ts	40 $\hat{\theta}$
01		11		21		31 $t_{\alpha/2}, F_{\alpha/2}$	41 $F_{\alpha/2}$
02		12		22		32 SE	42 $P(ts)$
03 n_y		13 n_x		23		33 s_p	43
04		14		24		34 θ_o	44
05		15 ν (ν_1)		25		35	45
06		16 ν_2		26		36 H_1-code	46
07 Used		17	Used (17–19)	27 s_d		37 $\bar{y}$	47 $\bar{x},\bar{d}$
08 Used		18		28		38 s_y	48 s_x
09 Used		19		29		39 e $(\hat{\theta} \pm e)$	49

TABLE F. F-PERCENTILES

P	$\downarrow \nu_2$	$\nu_1 \rightarrow$ 1	2	3	4	5	6	7	8	9	10	12	15	20	30	50	100	200	> 200
.10		39.9	49.5	53.6	55.8	57.2	58.2	58.9	59.4	59.9	60.2	60.7	61.2	61.7	62.3	62.7	6.30	63.2	63.3
.05		161	200	216	225	230	234	237	239	241	242	244	246	248	250	252	253	254	254
.025	1	648	800	864	900	922	937	948	957	963	969	977	985	993	1000	1008	1013	1016	1020
.01		4050	5000	5400	5620	5760	5860	5930	5980	6020	6060	6110	6160	6210	6260	6303	6334	6350	6370
.005		16200	20000	21600	22500	23100	23400	23700	23900	24100	24200	24400	24600	24800	25000	25200	25300	25400	25500
.10		8.53	9.00	9.16	9.24	9.29	9.33	9.35	9.37	9.38	9.39	9.41	9.42	9.44	9.46	9.47	9.48	9.49	9.49
.05		18.5	19.0	19.2	19.2	19.3	19.3	19.4	19.4	19.4	19.4	19.4	19.4	19.5	19.5	19.5	19.5	19.5	19.5
.025	2	38.5	39.0	39.2	39.2	39.3	39.3	39.4	39.4	39.4	39.4	39.4	39.4	39.4	39.5	39.5	39.5	39.5	39.5
.01		98.5	99.0	99.2	99.2	99.3	99.3	99.4	99.4	99.4	99.4	99.4	99.4	99.4	99.5	99.5	99.5	99.5	99.5
.005		199	199	199	199	199	199	199	199	199	199	199	199	199	199	199	199	199	199
.10		5.54	5.46	5.39	5.34	5.31	5.28	5.27	5.25	5.24	5.23	5.22	5.20	5.118	5.17	5.15	5.14	5.14	5.13
.05		10.1	9.55	9.28	9.12	9.01	8.94	8.89	8.85	8.81	8.79	8.74	8.70	8.66	8.62	8.58	8.55	8.54	8.53
.025	3	17.4	16.0	15.4	15.1	14.9	14.7	14.6	14.5	14.5	14.4	14.3	14.3	14.2	14.1	14.0	14.0	13.9	13.9
.01		34.1	30.8	29.5	28.7	28.2	27.9	27.7	27.5	27.3	27.2	27.1	26.9	26.7	26.5	26.3	26.2	26.2	26.1
.005		55.6	49.8	47.5	46.2	45.4	44.8	44.4	44.1	43.9	43.7	43.4	43.1	42.8	42.5	42.2	42.0	41.9	41.8
.10		4.54	4.32	4.19	4.11	4.05	4.01	3.98	3.95	3.93	3.92	3.90	3.87	3.84	3.82	3.80	3.78	3.77	3.76
.05		7.71	6.94	6.59	6.39	6.26	6.16	6.09	6.04	6.00	5.96	5.91	5.86	5.80	5.75	5.70	5.66	5.65	5.63
.025	4	12.2	10.6	9.98	9.60	9.36	9.20	9.07	8.98	8.90	8.84	8.75	8.66	8.56	8.46	8.38	8.32	8.29	8.26
.01		21.2	18.0	16.7	16.0	15.5	15.2	15.0	14.8	14.7	14.5	14.4	14.2	14.0	13.8	13.7	13.6	13.5	13.5
.005		31.3	26.3	24.3	23.2	22.5	22.0	21.6	21.4	21.1	21.0	20.7	20.4	20.2	19.9	19.7	19.5	19.4	19.3
.10		4.06	3.78	3.62	3.52	3.45	3.40	3.37	3.34	3.32	3.30	3.27	3.24	3.21	3.17	3.15	3.13	3.12	3.11
.05		6.61	5.79	5.41	5.19	5.05	4.95	4.88	4.82	4.77	4.74	4.68	4.62	4.56	4.50	4.44	4.41	4.39	4.37
.025	5	10.0	8.43	7.76	7.39	7.15	6.98	6.85	6.76	6.68	6.62	6.52	6.43	6.33	6.23	6.14	6.08	6.05	6.02
.01		16.3	13.3	12.1	11.4	11.0	10.7	10.5	10.3	10.2	10.1	9.89	9.72	9.55	9.38	9.24	9.13	9.08	9.02
.005		22.8	18.3	16.5	15.6	14.9	14.5	14.2	14.0	13.8	13.6	13.4	13.1	12.9	12.7	12.4	12.3	12.2	12.1
.10		3.78	3.46	3.29	3.18	3.11	3.05	3.01	2.98	2.96	2.94	2.90	2.87	2.84	2.80	2.77	2.75	2.73	2.72
.05		5.99	5.14	4.76	4.53	4.39	4.28	4.21	4.15	4.10	4.06	4.00	3.94	3.87	3.81	3.75	3.71	3.69	3.67
.025	6	8.81	7.26	6.60	6.23	5.99	5.82	5.70	5.60	5.52	5.46	5.37	5.27	5.17	5.07	4.98	4.92	4.88	4.85
.01		13.7	10.9	9.78	9.15	8.75	8.47	8.26	8.10	7.98	7.87	7.72	7.56	7.40	7.23	7.09	6.99	6.93	6.88
.005		18.6	14.5	12.9	12.0	11.5	11.1	10.8	10.6	10.4	10.2	10.0	9.81	9.59	9.36	9.17	9.03	8.95	8.88

TABLE F. F-PERCENTILES

P	$\downarrow\nu_2$	$\nu_1 \rightarrow$ 1	2	3	4	5	6	7	8	9	10	12	15	20	30	50	100	200	> 200
.10		3.59	3.26	3.07	2.96	2.88	2.83	2.78	2.75	2.72	2.70	2.67	2.63	2.59	2.56	2.52	2.50	2.48	2.47
.05		5.59	4.74	4.35	4.12	3.97	3.87	3.79	3.73	3.68	3.64	3.57	3.51	3.44	3.38	3.32	3.27	3.25	3.23
.025	7	8.07	6.54	5.89	5.52	5.29	5.12	4.99	4.90	4.82	4.76	4.67	4.57	4.47	4.36	4.28	4.21	4.18	4.14
.01		12.2	9.55	8.45	7.85	7.46	7.19	6.99	6.84	6.72	6.62	6.47	6.31	6.16	5.99	5.86	5.75	5.70	5.65
.005		16.2	12.4	10.9	10.1	9.52	9.16	8.89	8.68	8.51	8.38	8.18	7.97	7.75	7.53	7.35	7.22	7.15	7.08
.10		3.46	3.11	2.92	2.81	2.73	2.67	2.62	2.59	2.56	2.54	2.50	2.46	2.42	2.38	2.35	2.32	2.31	2.29
.05		5.32	4.46	4.07	3.84	3.69	3.58	3.50	3.44	3.39	3.35	3.28	3.22	3.15	3.08	3.02	2.97	2.95	2.93
.025	8	7.57	6.06	5.42	5.05	4.82	4.65	4.53	4.43	4.36	4.30	4.20	4.10	4.00	3.89	3.81	3.74	3.71	3.67
.01		11.3	8.65	7.59	7.01	6.63	6.37	6.18	6.03	5.91	5.81	5.67	5.52	5.36	5.20	5.07	4.96	4.91	4.86
.005		14.7	11.0	9.60	8.81	8.30	7.95	7.69	7.50	7.34	7.21	7.01	6.81	6.61	6.40	6.22	6.09	6.02	5.95
.10		3.36	3.01	2.81	2.69	2.61	2.55	2.51	2.47	2.44	2.42	2.38	2.34	2.30	2.25	2.22	2.19	2.17	2.16
.05		5.12	4.26	3.86	3.63	3.48	3.37	3.29	3.23	3.18	3.14	3.07	3.01	2.94	2.86	2.80	2.76	2.73	2.71
.025	9	7.21	5.71	5.08	4.72	4.48	4.32	4.20	4.10	4.03	3.96	3.87	3.77	3.67	3.56	3.47	3.40	3.37	3.33
.01		10.6	8.02	6.99	6.42	6.06	5.80	5.61	5.47	5.35	5.26	5.11	4.96	4.81	4.65	4.52	4.42	4.36	4.31
.005		13.6	10.1	8.72	7.96	7.47	7.13	6.88	6.69	6.54	6.42	6.23	6.03	5.83	5.62	5.45	5.32	5.26	5.19
.10		3.29	2.92	2.73	2.61	2.52	2.46	2.41	2.38	2.35	2.32	2.28	2.24	2.20	2.15	2.12	2.09	2.07	2.06
.05		4.96	4.10	3.71	3.48	3.33	3.22	3.14	3.07	3.02	2.98	2.91	2.84	2.77	2.70	2.64	2.59	2.56	2.54
.025	10	6.94	5.46	4.83	4.47	4.24	4.07	3.95	3.85	3.78	3.72	3.62	3.52	3.42	3.31	3.22	3.15	3.12	3.08
.01		10.0	7.56	6.55	5.99	5.64	5.39	5.20	5.06	4.94	4.85	4.71	4.56	4.41	4.25	4.12	4.01	3.96	3.91
.005		12.8	9.43	8.08	7.34	6.87	6.54	6.30	6.12	5.97	5.85	5.66	5.47	5.27	5.07	4.90	4.77	4.71	4.64
.10		3.18	2.81	2.61	2.48	2.39	2.33	2.28	2.24	2.21	2.19	2.15	2.10	2.06	2.01	1.97	1.94	1.92	1.90
.05		4.75	3.89	3.49	3.26	3.11	3.00	2.91	2.85	2.80	2.75	2.69	2.62	2.54	2.47	2.40	2.35	2.32	2.30
.025	12	6.55	5.10	4.47	4.12	3.89	3.73	3.61	3.51	3.44	3.37	3.28	3.18	3.07	2.96	2.87	2.80	2.76	2.72
.01		9.33	6.93	5.95	5.41	5.06	4.82	4.64	4.50	4.39	4.30	4.16	4.01	3.86	3.70	3.57	3.49	3.41	3.36
.005		11.8	8.51	7.23	6.52	6.07	5.76	5.52	5.35	5.20	5.09	4.91	4.72	4.53	4.33	4.17	4.04	3.97	3.90
.10		3.07	2.70	2.49	2.36	2.27	2.21	2.16	2.12	2.09	2.06	2.02	1.97	1.92	1.87	1.83	1.79	1.79	1.76
.05		4.54	3.68	3.29	3.06	2.90	2.79	2.71	2.64	2.59	2.54	2.48	2.40	2.33	2.25	2.18	2.12	2.10	2.07
.025	15	6.20	4.77	4.15	3.80	3.58	3.41	3.29	3.20	3.12	3.06	2.96	2.86	2.76	2.64	2.55	2.47	2.44	2.40
.01		8.68	6.36	5.42	4.89	4.56	4.32	4.14	4.00	3.89	3.80	3.67	3.52	3.37	3.21	3.08	2.98	2.92	2.87
.005		10.8	7.70	6.48	5.80	5.37	5.07	4.85	4.67	4.54	4.42	4.25	4.07	3.88	3.69	3.52	3.39	3.33	3.26

.10		2.97	2.59	2.38	2.25	2.16	2.09	2.04	2.00	1.96	1.94	1.89	1.84	1.79	1.74	1.69	1.65	1.63	1.61
.05		4.35	3.49	3.10	2.87	2.71	2.60	2.51	2.45	2.39	2.35	2.28	2.20	2.12	2.04	1.97	1.91	1.88	1.84
.025	20	5.87	4.46	3.86	3.51	3.29	3.13	3.01	2.91	2.84	2.77	2.68	2.57	2.46	2.35	2.25	2.17	2.13	2.09
.01		8.10	5.85	4.94	4.43	4.10	3.87	3.70	3.56	3.46	3.37	3.23	3.09	2.94	2.78	2.64	2.54	2.48	2.42
.005		9.94	6.99	5.82	5.17	4.76	4.47	4.26	4.09	3.96	3.85	3.68	3.50	3.32	3.12	2.96	2.83	2.76	2.69
.10		2.88	2.49	2.28	2.14	2.05	1.98	1.93	1.88	1.85	1.82	1.77	1.72	1.67	1.61	1.55	1.51	1.48	1.46
.05		4.17	3.32	2.92	2.69	2.53	2.42	2.33	2.27	2.21	2.16	2.09	2.01	1.93	1.84	1.76	1.70	1.66	1.62
.025	30	5.57	4.18	3.59	3.25	3.03	2.87	2.75	2.65	2.57	2.51	2.41	2.31	2.20	2.07	1.97	1.88	1.84	1.79
.01		7.56	5.39	4.51	4.02	3.70	3.47	3.30	3.17	3.07	2.98	2.84	2.70	2.55	2.39	2.25	2.13	2.07	2.01
.005		9.18	6.35	5.24	4.62	4.23	3.95	3.74	3.58	3.45	3.34	3.18	3.01	2.82	2.63	2.46	2.32	2.25	2.18
.10		2.81	2.41	2.20	2.06	1.97	1.90	1.84	1.80	1.76	1.73	1.68	1.63	1.57	1.50	1.44	1.39	1.36	1.33
.05		4.03	3.18	2.79	2.56	2.40	2.29	2.20	2.13	2.07	2.03	1.95	1.87	1.78	1.69	1.60	1.52	1.48	1.44
.025	50	5.34	3.97	3.39	3.05	2.83	2.67	2.55	2.46	2.38	2.32	2.22	2.11	1.99	1.87	1.75	1.65	1.60	1.54
.01		7.17	5.06	4.20	3.72	3.41	3.19	3.02	2.89	2.79	2.70	2.56	2.42	2.27	2.10	1.95	1.82	1.75	1.68
.005		8.63	5.90	4.83	4.23	3.85	3.58	3.38	3.22	3.09	2.99	2.82	2.65	2.47	2.27	2.10	1.95	1.87	1.78
.10		2.76	2.36	2.14	2.00	1.91	1.83	1.78	1.73	1.69	1.66	1.61	1.56	1.49	1.42	1.35	1.29	1.26	1.21
.05		3.94	3.09	2.70	2.46	2.31	2.19	2.10	2.03	1.97	1.93	1.85	1.77	1.68	1.59	1.48	1.39	1.34	1.28
.025	100	5.18	3.83	3.25	2.92	2.70	2.54	2.42	2.32	2.24	2.18	2.08	1.97	1.85	1.71	1.59	1.48	1.42	1.35
.01		6.90	4.82	3.98	3.51	3.21	2.99	2.82	2.69	2.59	2.50	2.38	2.22	2.07	1.89	1.74	1.60	1.52	1.43
.005		8.24	5.59	4.54	3.96	3.59	3.33	3.13	2.97	2.85	2.74	2.58	2.41	2.23	2.02	1.84	1.68	1.59	1.48
.10		2.73	2.33	2.11	1.97	1.88	1.80	1.75	1.70	1.66	1.63	1.58	1.52	1.46	1.38	1.31	1.24	1.20	1.14
.05		3.89	3.04	2.65	2.42	2.26	2.14	2.06	1.98	1.93	1.88	1.80	1.72	1.62	1.52	1.41	1.32	1.26	1.19
.025	200	5.18	3.76	3.18	2.85	2.63	2.47	2.35	2.26	2.18	2.11	2.01	1.90	1.78	1.64	1.51	1.39	1.32	1.23
.01		6.76	4.71	3.88	3.41	3.11	2.89	2.73	2.60	2.50	2.41	2.27	2.13	1.97	1.79	1.63	1.48	1.39	1.28
.005		8.06	5.44	4.41	3.84	3.47	3.21	3.01	2.86	2.73	2.63	2.47	2.30	2.11	1.91	1.71	1.54	1.44	1.31
.10		2.71	2.30	2.08	1.94	1.85	1.77	1.72	1.67	1.63	1.60	1.55	1.49	1.42	1.34	1.26	1.18	1.13	1.00
.05		3.84	3.00	2.60	2.37	2.21	2.10	2.01	1.94	1.88	1.83	1.75	1.67	1.57	1.46	1.35	1.24	1.17	1.00
.025	> 200	5.02	3.69	3.12	2.79	2.57	2.41	2.29	2.19	2.11	2.05	1.94	1.83	1.71	1.57	1.43	1.29	1.21	1.00
.01		6.63	4.61	3.78	3.32	3.02	2.80	2.64	2.51	2.41	2.32	2.18	2.04	1.88	1.70	1.52	1.35	1.25	1.00
.005		7.88	5.30	4.28	3.72	3.35	3.09	2.90	2.74	2.62	2.52	2.36	2.19	2.00	1.79	1.59	1.40	1.28	1.00

Bivariate Populations

1. (Example 6.3) Before (x) and after (y) weights were recorded in pounds after two weeks of dieting. Find a 95% CI for the mean difference $\mu_x - \mu_y$, and conduct a significance test of equality of means. Test for a weight loss of at least 2 pounds.

x:	119	122	136	130	129	136	134	133	119	115
y:	114	119	134	126	119	137	124	127	119	107

Solution

ZS STEP	ENTER	PRESS	DISPLAY	COMMENTS
PN		[Pgm]04		Initialize module program
1		[E′]	0.	ST–04 for raw data entry.
DE	119	[A]	1.	Enter first x-value.
	114	[B]	1.	Follow with first y-value.
	122	[A]	2.	Enter second x-value.
	119	[B]	2.	Follow with second y-value.
	136	[A]	3.	Enter succeeding pairs.
	134	[B]	3.	
	130	[A]	4.	
	126	[B]	4.	
	129	[A]	5.	
	119	[B]	5.	
	136	[A]	6.	
	137	[B]	6.	
	134	[A]	7.	
	124	[B]	7.	
	133	[A]	8.	
	127	[B]	8.	
	119	[A]	9.	
	119	[B]	9.	
	115	[A]	10.	Enter last x-value.
	107	[B]	10.	Follow with last y-value (n = 10).
2		[RST]	10.	Exit ST–04 and lower all flags.
3		[D′]	0.	ZS program processes ST–04 data.
PN				
3a		[B]	9.	Calculates and displays $\nu = 9 =$ d.f.
3b	2.262	[R/S]	1.94	Enter $t_{.025}$ from t-table and display $l = 1.94$.
		[$x \rightleftarrows t$]	7.46	Display u, so CI is (1.94, 7.46).
2a	0	[B′]	0.	Enter H_1-code = 0 for two-sided test.
	0	[R/S]	.0039	Use $\theta_0 = 0$ for this case and find $P = .0039$; reject at usual levels.
		[RCL]30	3.85	Display value of ts.

ZS STEP	ENTER	PRESS	DISPLAY	COMMENTS
2a	1	[B′]	1.	Use H_1-code of 1 making H_o:$\mu_x - \mu_y \leq 2$ the disclaimer.
	2	[R/S]	.027	With $\theta_o = 2$, P-value is enough to reject at $\alpha = 5\%$.

2. (Example 6.5) A test of color perception was administered to a control group (x) and an experimental group (y) with results:

x: 16.3 14.7 12.3 13.5 16.0 17.1 17.3
y: 14.0 16.5 17.7 15.9 18.0 16.3

Analyze the two groups for differences. Also test for equality of variances.

Solution (*assuming* $\sigma_x = \sigma_y$)

ZS STEP	ENTER	PRESS	DISPLAY	COMMENTS
INA		[Pgm04]		Initialize ST–04 for data entry
1		[E′]	0.	using above raw scores.
DE	16.3	[A]	1.	Enter first x-value.
1	14.7	[A]	2.	Continue x-values, assuming
	12.3	[A]	3.	data are independent.
	13.5	[A]	4.	
	16.0	[A]	5.	
	17.1	[A]	6.	
	17.3	[A]	7.	Last x-value entered; $n_x = 7$.
	14	[B]	1.	Begin entering y-value with
	16.5	[B]	2.	label B in ST–04.
	17.7	[B]	3.	
	15.9	[B]	4.	
	18	[B]	5.	
	16.3	[B]	6.	Conclude y-entries; $n_y = 6$.
2		[RST]	6.	Exit ST–04, and enter ZS–6 program.
3				
INA		[D′]	0.	Process data.
2a	0	[C′]	0.	Enter H_1-code for two-sided test.
	0	[R/S]	.274	Display P-value (for $\theta_o = 0$) of .27.
2a		[RCL]30	−1.15	Accept H_o($ts = -1.15$).
3a		[C]	11.	Reveal d.f. $= n_x + n_y - 2 = 11$ for this case.
3b	2.201	[R/S]	−3.16	Entering $t_{.025} = 2.201$, CI
		[$x \rightleftharpoons t$]	.99	runs from $l = -3.16$ to $u = .99$, which does include 0.

Solution (assuming $\sigma_x \neq \sigma_y$)

INB				
1	Same as INA, so data are already entered and processed.			
3a		[A]	10.	Calculates degrees of freedom
3b	2.228	[R/S]	−3.14	for approximate CI based on Welch t.
		[$x \rightleftarrows t$]	.97	Comes close to preceding solution.
2a	0	[A′]	0.	Begins Welch t-test with H_1-code,
2b	0	[R/S]	.266	followed by $\theta_o = 0$, to yield about the same P-value.

Solution (to test for $\sigma_x^2 = \sigma_y^2$)

NV				
1	Same as INA			
2	0	[St flg]4	0.	H_1-code for two-sided test. Flag 4 signals NV a test is being called for.
		[D]	.56	Large P-value; accept $\sigma_x^2 = \sigma_y^2$
		[RCL]30	1.73	with $ts = 1.73$.
3a		D	6	To compute CI display $\nu_1 = n_x - 1$,
		[R/S]	5	and $\nu_2 = n_y - 1$.
3b	6.98	[STO]31	6.98	Enter $F_{.025}$ with d.f. = (6,5).
3c	5.99	[STO]41	5.99	Enter $F_{.025}$ with reversed d.f. = (5,6).
3d		[R/S]	.25	Shows a 95% CI that in-
		[$x \rightleftarrows t$]	10.36	clude the value 1. Accept $\sigma_x^2 = \sigma_x^2$.

3. (Example 6.10) A sample of 60 exponential times to failure averaged $\bar{x}$ = 1,306 hours. Six independent times averaged $\bar{y}$ = 1,247 hours. Test $H_o{:}\mu_x \leq \mu_y$.

ZS STEP	ENTER	PRESS	DISPLAY	COMMENTS
Exp				
1	60	[STO]13	60.	Enter summary data.
	6	[STO]03	6.	
	1306	[STO]47	1306.	
	1247	[STO]37	1247.	Data entry concluded
2a	1	[St flg]4	1.	Enter H_1-code for $H_1{:}\mu_x > \mu_y$, and set flag 4 to signal H-test.
2b		[E]	.50	P-value of .50 obtained; do not reject.
3a		[E]	120	$\nu_1 = 2n_x$ displayed
		[R/S]	12	$\nu_2 = 2n_y$ displayed
3b	2.35	[STO]31	2.35	$F_{.05}$ for $\nu_1 = 100$, $\nu_2 = 12$ entered
3c	1	[STO]41	1.	Store 1 in R_{41} to compute lower CI.
3d		[R/S]	.45	Lower bound on μ_x/μ_y displayed

7

Proportions

Section 7.1: Introduction

All of the procedures of the last chapter required either normal or exponential assumptions. In any case, continuous measurements are certainly required, and, whatever the assumptions are in that case, you know that the sample mean is approximately normal for large sample sizes. Indeed, that is true whether or not the underlying population is assumed to be continuous. This places inference regarding population means within reach of the normal procedure for many cases. As a matter of fact, the t-tests are rather robust in that regard, meaning that even for small sample sizes, the procedures can tolerate reasonable departures from normality. In any event, there are few options open to you (the exponential case is certainly a very special one), so it reduces to either assuming normality or not solving the problem.

You need to be especially cautious when data are discrete, and there is no way even to justify assuming that the underlying scale is continuous. Counting data (*enumeration*, also occasionally called *attributes data*) fall into this category, and one case, the Bernoulli, we can handle reasonably well. You recall from Chapter 2 that a Bernoulli random variable arises whenever one encounters a dichotomized situation (like pass-fail, go-no go, success-failure), in which we arbitrarily let 1 stand for one of the categories to be observed and 0 the other. A random variable having such a range, then, has a probability distribution indexed by a single parameter $p = \Pr(X = 1)$, since $q = \Pr(X = 0) = 1 - p$ is just a function of p. As with any parameter in real life, we seldom know its value and need to estimate it with data.

The data in the Bernoulli case, the random sample, $X_1, X_2, \ldots, X_n$, will result in a string of 0's and 1's and is almost always summarized by the sum

$$Y = \sum_{i=1}^{n} X_i$$

which represents a count of the total number of 1's, whatever they stand for. You learned in Chapter 2 that Y has a binomial distribution with parameters n (known and fixed) and p, where p is the unknown "success" probability. It is this distribution that forms the basis for statistical inference concerning the unknown proportion p.

Since the parameter p is itself a population mean, (the common mean of each Bernoulli X_i), it is quite in keeping with general estimation theory that the sample mean $\bar{X} = Y/n$, in this case, is an appropriate estimator for p. In fact, it is unbiased, and we know that, since $V(X_i) \equiv pq$, $V(\bar{X}) = pq/n$. Moreover, by the Central Limit Theorem, $\bar{X}$ is approximately normal for large sample sizes. As usual, we have no problems to speak of when n is large, and, as you will see, we can generate confidence intervals and tests of hypotheses for such cases using the normal distribution. Lacking large samples sizes, there are definite problems created by the fact that Y, and hence $\bar{X}$, is discrete and can only assume a finite number of values. We will address these in what follows.

Section 7.2: Confidence Intervals

First, the matter of confidence intervals. If n is large, we use the fact that $\bar{X}$ is approximately normal. In this case, our parameter is $\theta = p$, and $\hat{\theta} = \hat{p} = \bar{x}$ is its estimate. Also, since $V(\bar{X}) = pq/n$, the standard error of the mean is (estimated by) SE $= \sqrt{\hat{p}\hat{q}/n} = \sqrt{\bar{x}(1-\bar{x})/n}$. The usual normal symmetric confidence limits for p would then be given by

$$\hat{\theta} \pm z_{\alpha/2}\,\mathrm{SE} \equiv \hat{p} \pm z_{\alpha/2}\sqrt{\hat{p}\hat{q}/n} \equiv \bar{x} \pm z_{\alpha/2}\sqrt{\bar{x}(1 - \bar{x})/n} \qquad (7.1)$$

These are the formulas you will usually find presented in standard textbooks for this case, and we have programmed it under the code B(p) in program ZS–7 at Step 3.

You see from the user instructions for B(p) that data entry is particularly simple, requiring merely that you store the sample size n in R_{03}

and the parameter estimate $\hat{p} = \bar{x} = y/n$ (where y in this section always stands for the number of 1's in the sample) in R_{40}. For the two-sided confidence interval, you have only to enter half the desired risk, $\alpha/2$, and press [C]. The limits will be found in the usual way. For one-sided limits, merely enter the entire risk α and ignore l or u, as the case may be. Although you would rarely consider risks larger than $^1/_2$, you should be aware that in this program you may only enter values of α up to .5.

EXAMPLE 7.1

The reliability of a switch is being tested, and in 200 cycles, there were only 165 successes. Analyze the probability of success for this switch.

Solution Model a cycle as a Bernoulli trial with unknown probability p for success. Now in ZS–7, store 200 in R_{03} and $^{165}/_{200} = .825$ in R_{40}. Taking a nominal $\alpha = .05$, enter .025 and press [C] to find $l = .77$, and [$x \rightleftharpoons t$] to find .88. Thus, you can be 95% confident that $.77 < p < .88$. It is not likely that p is as high as 95%, say. A lower one-sided interval on p may be found by entering .05 and pressing [C] to find $l = .78$. We would report (.78, 1) as the lower one-sided interval for p. The point estimate .825 is compatible with all values of p larger than .78, but not smaller. Put another way, all choices of p_0 larger than .78 would be acceptable hypotheses in a test against alternatives of the form $H_1{:}p > p_0$ at the 5% level of significance. If $p_0 < .78$, on the other hand, $H_0{:}p = p_0$ (or $H_0{:}p \leq p_0$) would be rejected in favor of $H_1{:}p > p_0$ at level 5%.

EXAMPLE 7.2

A manufacturer sells items in large lots and wishes to advertise that his lots contain no more than 5% defective items. He has a sample of 100 items drawn off a production line and tested to find only 2 defectives. What is an upper bound on p, the probability of a defective item?

Solution Here, imagine each test as a Bernoulli trial where 1 now stands for *failure* to pass the test—p is this unknown probability. From the data, $\hat{p} = .02$. With $n = 100$ and $\alpha = .01$, ZS–7 yields a 99% upper one-sided limit of .053, or .05 to 2 decimal places. The manufacturer can be pretty confident that p does not exceed 5%. In reporting $p \leq .05$, there is only a 1% chance that he is in error.

When the sample size is small (say, less than 20 or so), the use of Equation 7.1 may not be so appropriate, being based on the Central Limit Theorem. There is a rather difficult analysis that may be applied directly to the binomial distribution, and it yields formulas for small sample sizes. Curiously enough, the confidence limits are given in terms of F-percentiles, where the degrees of freedom depend on the sample outcome (reminiscent of the Welch t) and differ for l and u. Another matter worthy of note is that the confidence in reporting $l < p < u$ is not exactly the specified $1 - \alpha$ (it was not for the normal approximation either), but rather is bounded below by $1 - \alpha$. So the result is conservative in the sense that the confidence is *at least* $1 - \alpha$ that $l < p < u$.

The formulas for l and u are best given in terms of $y = n\hat{p} = n\bar{x}$, the total number of 1's in the sample. For the record they are

$$l = \frac{y}{y + (n - y + 1)\,F_{\alpha/2}} \quad \text{where} \quad \begin{aligned} \nu_1 &= 2(n - y + 1) \\ \nu_2 &= 2y \end{aligned}$$

$$u = \frac{(y + 1)F_{\alpha/2}}{(n - y) + (y + 1)F_{\alpha/2}} \quad \text{where} \quad \begin{aligned} \nu_1 &= 2(y + 1) \\ \nu_2 &= 2(n - y) \end{aligned} \tag{7.2}$$

You see that the degrees of freedom for the F-percentile depend on y in different ways for the two limits. These formulas have been programmed for you at Step 4 in B(p). You first enter the data as n and $\hat{p}$ in Step 1 (this covers all the cases). Then, you press [B] to find ν_1 for the lower limit l, that is, $2(n - y + 1)$. A press of [R/S] will then reveal $\nu_2 = 2y$, whereupon you should look up $F_{\alpha/2}$ for this pair of d.f. and store the result in R_{31}. This takes care of the lower limit l. Now you press [R/S] twice to find the *new* pair of d.f. for the upper limit, so the last number you see is $2(n - y)$. Look up $F_{\alpha/2}$ for this new pair and store the result in R_{41} for computing u. Then l may be displayed by yet another press of [R/S], with the value of u located, as usual, in R_T. The procedure is a little complicated, to be sure, but it is a small price to pay to handle small sample sizes.

If it should happen that $y = 0$ (certainly a possibility) then you see from Equation 7.2 that $\nu_2 = 0$ for the lower limit, and there is no F-percentile for such a case. Thus, l cannot be computed. But u can, and the upper one-sided confidence interval (CI) is from $(0, u)$. If, on the other hand, $y = n$, then the limit for u suffers from the same problem, but l does not, and so a lower one-sided interval can be computed as $(l, 1)$, but not an upper one. If you stop and think about what it means to have $y = 0$ or n, then that all makes perfect sense. It does mean that, in either of these cases, it is not possible to compute a two-sided CI for p. The user instructions have a built-in safeguard against this, for you simply will not be able to effect Step 3b or Step 3d when that happens. But you see from the (typical) note that one-sided instructions are not so affected.

EXAMPLE 7.3

Nine Bernoulli trials resulted in exactly 4 successes. Find a 95% CI for the probability p of success.

Solution Storing 9 in R_{03} and $4/9$ in R_{40} as per Step 1 of B(p), a press of [B] shows 12 in R_D. This is $\nu_1 = 2(n - y + 1)$. A press of [R/S] reveals $\nu_2 = 8$, and from the F-table $F_{.025} = 4.20$ for d.f. $= (12, 8)$. This is stored in R_{31}. Two additional presses of [R/S] yield $(\nu_1, \nu_2) = (10, 10)$

and $F_{.025} = 3.72$ for d.f. $= (10, 10)$. This is stored in R_{41} and [R/S] shows $l = .14$ in R_D. A press of [$x \leftrightarrows t$] displays $u = .79$. Thus, our confidence is *at least* 95% that $.14 < p < .79$. That is quite a spread of plausible values, of course, but not much more information should be anticipated with only 9 trials.

EXAMPLE 7.4

In studying the reliability of a device, 25 trials resulted in 23 successes. Find a lower one-sided confidence bound on p, the probability of success.

Solution Enter 25 in R_{03} and $23/25 = .92$ in R_{40}. Now press [B] and [R/S] to see $(\nu_1, \nu_2) = (6, 46)$. As you see from the note, you should store $F_{.05} = 2.29$ in R_{31}. Now press [R/S] twice to find new d.f. $(\nu_1, \nu_2) = (48, 4)$. This is to be ignored and 1 entered in R_{41}. Now press [R/S] to find $l = .77$. The confidence is *at least* 95% that p is *at least* 77%! Had an upper one-sided interval been called for, we would have entered 1 in R_{31} after the first display of d.f. $= (6, 46)$ at Step 4a. Following that entry, the next stage would show new d.f. $(\nu_1, \nu_2) = (48, 4)$, and $F_{.05} = 5.70$ would be stored in R_{41}. The last [R/S] shows .88, which we ignore, and retrieve the desired $u = .99$ from R_T. Thus, the confidence is at least 95% that p is *no more* than 99%.

In this last example, the decision as to whether $n = 25$ is large or small is marginal. We might have opted for the large sample procedure here. In fact, entering .05 and pressing [C] reveals a normal lower confidence bound of .83, not terribly different from the small sample bound. Had the observed proportion of .92 been based on a sample of size 30, this bound would increase only slightly to .84. The bound based on Equation 7.2 would not change significantly, since ν_1 is still 6, and ν_2 only increases to 56, which we would probably approximate with 50 anyway. So the solution is about the same. If $n = 50$, the normal lower bound is .86 and the binomial lower bound is .83. There is much closer agreement there.

Practice with the following problems and those following Section 7.3 should help you to implement the CI portions of the B(p) program. We will turn to hypothesis testing in the next section.

Problems

1. A poll of 600 California voters was taken to see if they favored a proposition banning smoking in public buildings. If 55% of the voters polled were in favor of the proposition, what is a 95% confidence interval for the proportion of all California voters who will favor the proposition?

Ans: (.51, .59)

2. In a poll of 1,500 cowboys, it was found that 800 of them believed that white stallions are the most spirited of all of the American horses. Assuming the poll was truly random, what is the 95% confidence interval for the population proportion, p, of cowboys who believe the white stallion to be the most spirited?

   Ans: (.51, .56)

3. In a random sample of children from urban schools, 43% were found to need immediate dental treatment for cavities. Construct a 95% confidence interval for the proportion, p, of children from urban schools who need treatment, when the sample size is 50, and again when the sample size is 100.

   Ans: $n = 50$ (.29, .57); $n = 100$ (.33, .53)

4. In a recent survey of high school principals, 91 of the 100 interviewed indicated that they had Master's degrees or higher. Assuming that the survey was random, construct a 95% confidence interval for the proportion of all high school principals who have at least a Master's degree.

   Ans: (.85, .97)

5. In a public opinion poll, 480 out of 600 persons interviewed approved of the administration's welfare policies. Establish a 98% confidence interval for the population proportion who approve of the policies.

   Ans: (.76, .84)

6. 100 graduates of a training program were selected to evaluate whether or not they passed the Certified Public Accountant Examination on the first try. If 30% were found to be successful, construct a 99% confidence interval for the percentage of all 10,000 graduates of the training program who passed on the first try.

   Ans: (.18, .42)

7. Out of a random sample of 12 cups of coffee from a vending machine, 3 were overfills. Find a 90% CI for the proportion of such overfills.

   Ans: (.07, .52)

8. Out of a sample of 25 people polled, 20% were in favor of pay TV. Find an upper one-sided 95% confidence bound on p, the proportion of the population favoring pay TV.

   Ans: Using $\alpha = .05$ and $F_{.05} = 2.02$ for d.f. $= (12, 40)$, $p < .38$. (The normal routine gives $p < .33$).

9. An engineer wants to modify the design of a highly reliable ($p = .95$) item. Being concerned over degradation of reliability, management allows him to modify 10 such items (they are expensive) and test them. Nine out of the ten tests were successful, and the engineer claims that reliability cannot have dropped below 90%. Test his claim with confidence intervals.

Ans: Letting $H_1{:}p > .90$, a lower one-sided bound is required. For $\alpha = .05$ this gives an interval of (.6,1) for the plausible values of p. Since .9 is in the interval, the hypothesis $H_0{:}p \leq .90$ cannot be rejected.

10. In a sample of 40 parents, 26 were in favor of an open dormitory policy on campus. Find a 99% confidence interval for the true proportion, p, in favor of such a policy, and use this interval to decide if $p \neq .5$.

Ans: The binomial interval extends from .43 to .83 and the normal from .46 to .84. In either case, $H_0{:}p = .5$ cannot be rejected.

Section 7.3: Tests for Bernoulli p

Tests of hypotheses about the parameter p of a Bernoulli population are carried out in much the same manner as described earlier for testing μ in a normal population. However, since the binomial distribution is discrete, there are some technical problems associated with selection of test size and computation of significance levels. These problems arise because only a few α values can actually be achieved in any given problem. The usual practice is to set a *bound* on the Type I error probability that is acceptable to the decision maker. This is automatically done when you use a confidence interval or limit to test a hypothesis $H_0{:}p = p_0$ about p (as you have seen). The only change necessary in the hypothesis testing procedure that you are used to is simply one of stating the α value as a *bound*. For example, instead of stating that "H_0 is rejected at the $\alpha =$ .05 level" for the binomial situation, you should state "H_0 is rejected at a level not exceeding $\alpha = .05$."

For tests of significance, an associated problem arises for two-sided alternatives. It is essentially the same problem as the one discussed earlier for any nonsymmetric probability distribution, as most of the binomials will be. In any event, the logical test statistic to use is the observed number of "successes" $y = n\hat{p} = n\bar{x}$, as noted earlier. The random variable Y then has a binomial distribution with parameters n and p_0 whenever the hypothesis $H_0{:}p = p_0$ is true.

For the one-sided alternative $H_1{:}p < p_0$, we would want to reject H_0 whenever the observed value is too small, so we logically set the observed significance level to $P(ts) = P(y)$, the binomial CDF at y. When the alternative is $H_1{:}p > p_0$, we have to be careful for we would like the significance level to be the probability of observing a value *as large or larger* than the observed y, to be consistent with previous methods. When TS is continuous, this is just $Q(ts)$, but here we cannot ignore the equality, and so the significance level will have to be $Q(y - 1) = \Pr(Y \geq y)$ for integer y.

For the two-sided alternative $H_1{:}p \neq p_0$, there is a problem in even defining the significance level, as we have remarked.* Lacking such a de-

* Again, see J. D. Gibbons and J. W. Pratt, *op. cit.*

finitive procedure, we have elected to take the particular one-sided value $P(y)$ and report $2P(y)$, if $P(y) < .5$; otherwise, we report $2[1 - P(y)]$, computed from a binomial distribution with parameters n and p_0. At least this will be consistent with the method used whenever the distribution is symmetric and follows the modus operandi for the F-tests and Chi-square tests of previous chapters. Beyond that, we make no claim for its validity or preference over other proposed methods, none of which appear to be universally acceptable. This means that you should be somewhat conservative in your reliance on P-values computed this (or any other) way. As you will see, most of the problems are posed as one-sided problems, which is the way many of them naturally arise. For these, there is no problem.

As to the computations themselves, matters have been programmed for you in ZS–7 under B(p) Step 2. After entering the same two pieces of data, n and $y = n\bar{x}$, you proceed to Step 2a and enter the usual H_1-code and press [B′]. At this point you have a choice. For *small* sample sizes, you enter the selected p_0 and press [R/S] to see a display of P. This is Step 2b, and the P-value here is based on the binomial distribution, as we just discussed it, and clearly utilizes program ST–20. You recall that program may be slow running for large values of n and may not run at all for some choices of n and p_0, even with n not so large. So, for even moderately large sample sizes, you may elect to perform a normal test based on CLT by entering p_0 and pressing [C′], instead of [R/S]. This is Step 2b′.

EXAMPLE 7.5

A certain baseball player scored hits 30% of the times at bat last season. So far this season, he has been at bat 18 times and has scored only 3 hits. Is this a significantly lower batting average?

Solution First, enter the data as per Step 1 of B(p). Letting $H_0: p \geq .3$ be the disclaimer, the H_1-code is -1. Enter this and press [B′]. Now enter the hypothetical value .3, and press [R/S] to discover $P = .16$. This must be $P(3)$ for the binomial with parameters $n = 18$, $p_0 = .3$. The evidence for rejecting H_0 is weak.

EXAMPLE 7.6

Conduct a significance test for $H_0: p = .3$ against $H_1: p \neq .3$ for the situation of Example 7.5 and compare with the 95% CI for p.

Solution With the data still stored, enter an H_1-code of 0 and press [B′]. Now enter the value .3 and press [R/S] to find $P = .33$. For a 95% CI, press [B] and [R/S] to find d.f. = (32, 6). Store $F_{.025} = 5.07$ in R_{31}, as required. Now find new d.f. = (8, 30) and store $F_{.025} = 2.65$ in R_{41}. The 95% CI for p is thus (.04, .41), which includes the value .3. Either way, we cannot reject H_0.

EXAMPLE 7.7

In reliability testing on a new item being developed, preliminary tests revealed no successes out of 5 trials. Test the hypothesis that p is at least 25% at level $\alpha = .05$.

Solution Enter $n = 5$ and $\bar{x} = 0$ in R_{03} and R_{40}, respectively. With $H_1:p < .25$, enter an H_1-code of -1 and press [B′]. Now enter .25 and press [R/S] to find $P = .24$. The conclusion is that $p \geq .25$, even with 0 successes. If you press [B] and [R/S], you will find d.f. = (12, 0). Store 1 in R_{31} since l cannot be computed. The next 2 [R/S]'s will yield new d.f. = (2, 10), and you store $F_{.05} = 4.10$ in R_{41}. The upper one-sided 95% confidence interval is (0, .45). At the other extreme, if y had been 5, then a lower one-sided 95% confidence interval for p could be computed, but neither a two-sided one nor an upper one-sided one. With $\hat{p} = 1$ in R_{40} in place of 0, try this to see the first d.f. = (2, 10). For this choice, you enter $F_{.05} = 4.10$ in R_{31}. The second d.f. = (12, 0) so u cannot be computed. Store 1 in R_{41}. Press [R/S], and you will see .55 as the lower bound on p.

EXAMPLE 7.8

A certain professor has always given 10% A's in his course. But this term, in a class of 85 students, 12 received A's. Is the professor getting easier? Conduct a test with size not greater than 5%.

Solution With $n = 85$ in R_{03} and $^{12}/_{85}$ in R_{40}, we wish to test $H_0:p \leq .10$ against $H_1:p > .10$. Use an H_1-code of $+ 1$ and press [B′]. Here it is clear we want the large sample test, so enter the value of .1 and press [C′]. Since $P = .10$, and this exceeds 5%, H_0 should not be rejected. The sample estimate $\hat{p} = .14$ can be explained by chance and does not reflect a real change in p.

In the problems that follow, you may often elect to present a confidence interval solution. We show both when appropriate.

Problems

11. A manufacturer of camera flash bulbs claims that his product is highly reliable, that p, the probability of success, is greater than .99. In a random sample of 100 such bulbs, only three failed to work. Is his claim justified?

Ans: With $H_1:p > .99$, $P = .98$. A lower 95% CI for p is (.94,1). Either way $H_0:p \leq .99$ may not be rejected.

12. A certain piece of very expensive support equipment for an outer space vehicle is said to have reliability at least 95%. And in 8 tests, there was only 1 failure. Does this evidence support $H_0{:}p \geq .95$, where p is the probability of success?

Ans: With $H_1{:}p < .95$, $P = .34$, supporting H_0. An upper 95% to CI is (0, .99), and this includes the hypothesized value.

13. A certain sensing device is supposed to withstand heats of up to 100° centigrade, with probability at least .99. Ten such devices were tested, and 2 of them burned out before 100°. Is there reasonable ($\alpha =$ .01) doubt about the claim?

Ans: $H_1{:}p > .99$, $P = 1$. A lower 99% CI is (.38,1). Indeed, there is doubt about the claim; the data supports $H_0{:}p \leq .99$.

14. A switchboard operator believes that at least 40% of the calls received are long distance. If a random sample of 200 calls produced 70 long distance calls, do you agree with the operator? Set $\alpha = .10$.

Ans: If $H_0{:}p \geq .4$, then $P = .07$. A 90% upper bound on p is .39. For $\alpha = .10$ then, H_0 should be rejected.

15. An internal auditor claims that at least 90% of all discrepancies are corrected. If a random sample of 200 discrepancies yielded 175 corrections, is the claim supported? Set $\alpha = .05$.

Ans: Let $H_1{:}p > .90$, in which case $P = .88$, supporting $H_0{:}p \leq$ .90. A lower 95% bound on p is .84, making $p \geq .90$ plausible.

16. A company plans to market nationally a new product only if more than 70% of the test market responds favorably to the new product. If a random sample of 1,000 persons from the test market yielded 730 favorable responses, at the .05 level of significance, what can you conclude?

Ans: Let $H_1{:}p > .70$, putting the burden on the data to reject H_0 in favor of H_1 for a marketing signal. Then $P = .02$. A 95% lower bound on p is .71. Either way, H_0 should be rejected, and marketing should proceed.

17. It is claimed that 70% of the income tax returns from a certain county have 1 or more arithmetic errors. A spot check of 15 returns from this county revealed only 2 were without any errors. Can the claim be justified at the 10% level of significance?

Ans: With $H_1{:}p > .7$ the claim ($p =$ the probability of error), $P = .13 > .10$. Hence, do not reject H_0. Claim is not supported.

18. According to last year's "statistics," 70% of the population were in favor of stricter smog control laws. This year, only 65% of 1,000 respondents favored stricter laws. Does this represent a significant decrease?

Ans: The *null* hypothesis would be $H_0{:}p \geq .70$. With $H_1{:}p < .7$, $P = .00028$, so H_0 should be rejected. Conclude the decrease is significant.

19. Of 694 respondents, 364 were in favor of more dissemination of birth control information. Test the hypothesis that in the population the proportion so in favor is .5.

Ans: With $H_1{:}p \neq .5$, $P = .20$ supporting, $H_0{:}p = .5$. The 95% CI for p runs from .49 to .56.

20. According to genetic theory, one-fourth of the offspring of a certain pair of mice should be white, the remainder should be black or spotted. In 3 litters, these mice have produced 21 mice, of which 9 were white. Does this show that the genetic model is wrong?

Ans: Letting $H_0{:}p = .25$, $H_1{:}p \neq .25$, $P = .04$. A 95% CI for p runs from .2 to .7. Evidence is a little weak either way. $P = .04$ would have us reject H_0, and the CI would have us accept H_0; however, the two-sided P-value is subject to doubt as to its validity.

21. To see if one particular type of microcomputer dominated all the rest in one area, a survey of 86 owners was conducted, and it was found that 53 owned this type of computer. Can it be concluded that the true proportion is greater than .5?

Ans: Let $H_0{:}p \leq .5$ be the disclaimer. Then $P = .02$. H_0 should be rejected for any size $\alpha \geq .02$.

22. Test the null hypothesis $p = .70$ against the alternative $p \neq .70$, if a sample proportion of .74 was obtained from a sample of size 300. Let $\alpha = .05$.

Ans: P = .13, 95% CI for p is (.69, .79). Do not reject H_0.

23. A company uses a rule not to market a new product unless there is greater than 75% acceptability demonstrated. If the latest product received 2,023 favorable responses from 2,550 interviews, should they market the product?

Ans: Using $H_1{:}p > .75$, P $= 2 \times 10^{-7}$. A lower 95% CI on p is given by (.78, 1). The company should market the product.

24. A sample of 100 items from a shipment produced 6 defective parts. The company guarantees 97% nondefective shipments. Is the guarantee suspect?

Ans: Using $H_1{:}p < .97$, $P = .04$. The 95% upper bound on p is .98. The evidence is weak either for or against the guarantee.

Section 7.4: Two Proportions

Comparing two independent Bernoulli populations, X and Y, reduces to comparing their only parameters, p_x and p_y. Such a situation often arises in practice. For instance, does the proportion of men favoring a given political candidate really differ from the proportion of women who favor that candidate? Is the proportion of one ethnic group who applies for welfare really different from another group, or are the apparent differences just due to chance? In each case, the question can be reduced to asking whether or not $p_x = p_y$. If random samples of sizes n_x from X and n_y from Y are taken, then $\hat{p}_x = \bar{x}$ and $\hat{p}_y = \bar{y}$ are the usual estimates, and analysis would be based on the difference of means, $\bar{x} - \bar{y}$.

Unless both n_x and n_y are large, there is nothing to offer you for either confidence intervals or hypothesis testing. When both sample sizes are large, we can appeal to the Central Limit Theorem for both, since $\bar{X}$ and $\bar{Y}$ are then independent sample means, and, hence, $\bar{X} - \bar{Y}$ is approximately normal with mean $E(\bar{X} - \bar{Y}) = p_x - p_y$ and variance

$$\sigma_{\bar{x}}^2 + \sigma_{\bar{y}}^2 = \frac{p_x q_x}{n_x} + \frac{p_y q_y}{n_y}$$

Put another way,

$$\frac{\bar{X} - \bar{Y} - (p_x - p_y)}{\sqrt{\dfrac{p_x q_x}{n_x} + \dfrac{p_y q_y}{n_y}}} \tag{7.3}$$

has a standard normal distribution, and this is the basis for the usual normal confidence intervals and tests.

As for confidence intervals, it is natural to estimate the standard error by

$$\text{SE} = \sqrt{\frac{\hat{p}_x(1 - \hat{p}_x)}{n_x} + \frac{\hat{p}_y(1 - \hat{p}_y)}{n_y}}$$

and then form confidence limits for $p_x - p_y$ in the usual way, as,

$$\bar{x} - \bar{y} \pm z_{\alpha/2}\,\text{SE} \tag{7.4}$$

This formula has been programmed for you in ZS–7 under the code B($p_x - p_y$). Data entry is quite like that for independent means in ZS–6, with sample sizes stored in R_{13} and R_{03} and sample means $\hat{p}_x = \bar{x}$, $\hat{p}_y = \bar{y}$ in R_{47} and R_{37}, respectively. Enter any $\alpha/2$, and press [A] to find l and u in the usual way.

For testing hypotheses, we consider only $H_0{:}p_x - p_y = 0$, the hypothesis of equality of proportions (extended in the usual way to $H_0{:}p_x - p_y \geq 0$ and $H_0{:}p_x - p_y \leq 0$). When H_0 is true, p_x and p_y are both equal to some common value p. In that case, the standard error of the difference $\bar{X} - \bar{Y}$ may be expressed further as

$$\sqrt{\frac{p_x(1 - p_x)}{n_x} + \frac{p_y(1 - p_y)}{n_y}} = \sqrt{p\,(1 - p)} \cdot \sqrt{\frac{1}{n_x} + \frac{1}{n_y}}$$

Since $\hat{p}_x$ and $\hat{p}_y$ both estimate the same parameter, it makes sense to "pool" the total observations (compare the way s_p^2 was formed in estimating the assumed common variance of two normal populations). In that event,

$$\hat{p} = \frac{\sum_{i=1}^{n_x} x_i + \sum_{j=1}^{n_y} y_j}{n_x + n_y}$$

is just the overall average number of "successes" from the 2 samples. Then the standard error is estimated by

$$\text{SE} = \sqrt{\hat{p}(1 - \hat{p})}\sqrt{\frac{1}{n_x} + \frac{1}{n_y}}$$

and the distribution of

$$\text{TS} = \frac{\overline{X} - \overline{Y}}{\text{SE}} \tag{7.5}$$

is approximately standard normal. The usual Z—test can then be conducted, and the significance level computed from the normal distribution. Again, this is all programmed for you on ZS–7 at Step 2 of B($p_x - p_y$). Since the hypothesized value is always 0, it is only necessary to enter the H_1-code and press [A′] to have the P-value displayed.

EXAMPLE 7.9

In a random sample of 500 men (x), 350 were found to favor a certain political issue. In a similar sample of 300 women (y), 200 were so inclined. Is there any real difference between the sexes on this issue?

Solution Enter the data $n_x = 500$, $\hat{p}_x = .7$, $n_y = 300$, and $\hat{p}_y = 2/3$ in B($p_x - p_y$), according to the instructions for Step 1. Using an H_1-code of 0, enter this and press [A′] to find $P = .32$. The data support $H_0: p_x = p_y$. To find a 95% CI for the difference $p_x - p_y$, enter .025 and press [A] to find $l = -.03$ and $u = .10$. Since 0 is included in the interval, H_0 is accepted at level $\alpha = .05$.

Problems

25. In a random sample of 100 in one part of a city, 41 favored a certain presidential candidate, while only 38% of 150 polled in a different part of the city favored this candidate. Test for a significant difference in these proportions with size $\alpha = .01$.

Ans: If $H_1: p_x \neq p_y$, $P = .63$. A 99% CI for $p_x - p_y$ is $(-.13, .19)$. Either way, conclude that the proportions are equal.

26. In 1953, 58% ($\hat{p}_x$) of 1,000 people polled favored the death penalty.

Ten years later, only 51% ($\hat{p}_y$) of 1,000 people polled were so inclined. Is this a significant decrease?

Ans: With H_0:$p_x \leq p_y$ the disclaimer, $P = .0008$. The 95% lower one-sided CI for $p_x - p_y$ is (.03,1). H_0 should be rejected.

27. Two years ago, only 60% of 30 service personnel in a large appliance service center were in favor of a closed union shop. Last year, 39 out of 56 such service personnel were in favor of such conditions for a nominal 70%. Is this a real increase?

Ans: With H_0:$p_x \geq p_y$ the disclaimer, (x representing 2 years ago), $P = .18$. The 95% upper bound on $p_x - p_y$ is .08. Hence, H_0 should be accepted, if the size α is no more than 18%.

28. The government is trying to decide between two suppliers of electronic parts. Out of a sample of 100 parts from each source, 5 defective parts were found for supplier A and 7 for supplier B. Is there any significant difference between the 2 suppliers?

Ans: For H_1:$p_x \neq p_y$, $P = .55$. It is doubtful that there is any real difference.

29. To decide which of two used car salespersons should receive the salesperson of the year award, past records were examined. Salesperson A sold an automobile to 19% of 835 customers contacted, while salesperson B was able to sell 25% of only 483 customers contacted. Is B a better salesperson?

Ans: H_0:$p_A \geq p_B$, $P = .005$, so the data lend fairly strong support to $p_A < p_B$, and B should be judged superior.

30. A mathematics proficiency test was administered to 2 schools in the same district. Out of 250 students in one school, 86% passed, while only 83% of the 198 students in the other school passed. Is this difference significant?

Ans: Taking H_1 to be $p_x \neq p_y$, $P = .38$, indicating no significant difference.

STEP	PROCEDURE	ENTER	PRESS	DISPLAY
B(p)	*Bernoulli Parameter*			
1	Enter data.			
	a. Enter sample size.	n	[STO]03	n
	b. Enter proportion estimate.	$\hat{p}$	[STO]40	$\hat{p}$
2	*Test* $H_0{:}p = p_0$.			
	a. Enter H_1-code.*	H_1-code	[B′]	H_1-code
	b. Enter p_0, and compute P-value *or*:	p_0	[R/S]	P
	b′. For n large, normal test.	p_0	[C′]	P
3	*CI for p for n large*			
	Enter risk and calculate limits.	$\alpha/2$	[C]	l
	Note: For one-sided limits, enter α and ignore l or u.		[$x \leftrightarrows t$]	u
4	*CI for p for n small*			
	a. Find first d.f. for F.		[B]	ν_1
			[R/S]	ν_2
	b. Enter $F_{\alpha/2}$ with d.f. = (ν_1, ν_2).	$F_{\alpha/2}$	[STO]31	$F_{\alpha/2}$
	c. Find second d.f. for F.		[R/S]	ν_1
			[R/S]	ν_2
	d. Enter $F_{\alpha/2}$ with new d.f. = (ν_1, ν_2).	$F_{\alpha/2}$	[STO]41	$F_{\alpha/2}$
	e. Calculate Limits.		[R/S]	l
			[$x \leftrightarrows t$]	u
	Note: For lower one-sided limit, enter F_α at Step 3b, 1 at Step 3d, and ignore u. For upper one-sided limit, enter 1 at Step 3b, F_α at Step 3d, and ignore l.			

$$*H_1\text{-code} = \begin{cases} 1 \text{ if } H_1{:}\theta > \theta_0 \\ 0 \text{ if } H_1{:}\theta \neq \theta_0 \\ -1 \text{ if } H_1{:}\theta < \theta_0 \end{cases}$$

STEP	PROCEDURE	ENTER	PRESS	DISPLAY
$B(p_x - p_y)$	*Two Bernoulli Parameters*			
1	Enter data.			
	a. Enter sample sizes.	n_x	[STO]13	n_x
		n_y	[STO]03	n_y
	b. Enter proportion estimates.	$\hat{p}_x$	[STO]47	$\hat{p}_x$
		$\hat{p}_y$	[STO]37	$\hat{p}_y$
2	*Test* $H_0: p_x = p_y$			
	Enter H_1-code, and calculate P-value.	H_1-code	[A′]	P
3	*CI for* $p_x - p_y$			
	Enter risk, and calculate limits.	$\alpha/2$	[A]	l
			[$x \rightleftarrows t$]	u
	Note: For one-sided limits, enter α and ignore l or u.			

REGISTER CONTENTS				
00	10	20	30 ts	40 $\hat{\theta}$
01	11	21 Used	31 $z_{\alpha/2}$, $F_{\alpha/2}$	41 $F_{\alpha/2}$
02	12	22 Used	32 SE	42 $P(ts)$
03 n_y	13 n_x	23 Used	33 Used	43
04 u	14	24 Used	34 θ_0	44
05 l	15	25 Used	35	45
06	16	26 Used	36 H_1-code	46
07	17	27	37 $\hat{p}_y$	47 $\hat{p}_x$
08	18	28	38	48
09	19	29	39 $e(\hat{\theta} + e)$	49

TABLE F. F-PERCENTILES

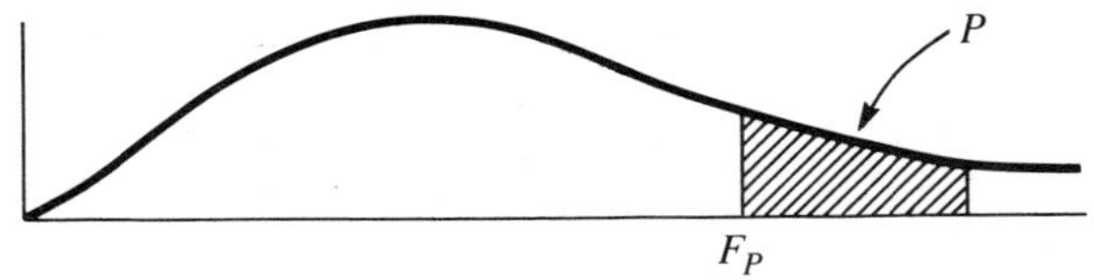

P	$\downarrow \nu_2$ \ $\nu_1 \rightarrow$ 1	2	3	4	5	6	7	8	9	10	12	15	20	30	50	100	200	> 200
.10		39.9	49.5	53.6	55.8	57.2	58.2	58.9	59.4	59.9	60.2	60.7	61.2	61.7	62.3	62.7	6.30	63.2	63.3
.05		161	200	216	225	230	234	237	239	241	242	244	246	248	250	252	253	254	254
.025	1	648	800	864	900	922	937	948	957	963	969	977	985	993	1000	1008	1013	1016	1020
.01		4050	5000	5400	5620	5760	5860	5930	5980	6020	6060	6110	6160	6210	6260	6303	6334	6350	6370
.005		16200	20000	21600	22500	23100	23400	23700	23900	24100	24200	24400	24600	24800	25000	25200	25300	25400	25500
.10		8.53	9.00	9.16	9.24	9.29	9.33	9.35	9.37	9.38	9.39	9.41	9.42	9.44	9.46	9.47	9.48	9.49	9.49
.05		18.5	19.0	19.2	19.2	19.3	19.3	19.4	19.4	19.4	19.4	19.4	19.4	19.5	19.5	19.5	19.5	19.5	19.5
.025	2	38.5	39.0	39.2	39.2	39.3	39.3	39.4	39.4	39.4	39.4	39.4	39.4	39.4	39.5	39.5	39.5	39.5	39.5
.01		98.5	99.0	99.2	99.2	99.3	99.3	99.4	99.4	99.4	99.4	99.4	99.4	99.4	99.5	99.5	99.5	99.5	99.5
.005		199	199	199	199	199	199	199	199	199	199	199	199	199	199	199	199	199	199
.10		5.54	5.46	5.39	5.34	5.31	5.28	5.27	5.25	5.24	5.23	5.22	5.20	5.118	5.17	5.15	5.14	5.14	5.13
.05		10.1	9.55	9.28	9.12	9.01	8.94	8.89	8.85	8.81	8.79	8.74	8.70	8.66	8.62	8.58	8.55	8.54	8.53
.025	3	17.4	16.0	15.4	15.1	14.9	14.7	14.6	14.5	14.5	14.4	14.3	14.3	14.2	14.1	14.0	14.0	13.9	13.9
.01		34.1	30.8	29.5	28.7	28.2	27.9	27.7	27.5	27.3	27.2	27.1	26.9	26.7	26.5	26.3	26.2	26.2	26.1
.005		55.6	49.8	47.5	46.2	45.4	44.8	44.4	44.1	43.9	43.7	43.4	43.1	42.8	42.5	42.2	42.0	41.9	41.8
.10		4.54	4.32	4.19	4.11	4.05	4.01	3.98	3.95	3.93	3.92	3.90	3.87	3.84	3.82	3.80	3.78	3.77	3.76
.05		7.71	6.94	6.59	6.39	6.26	6.16	6.09	6.04	6.00	5.96	5.91	5.86	5.80	5.75	5.70	5.66	5.65	5.63
.025	4	12.2	10.6	9.98	9.60	9.36	9.20	9.07	8.98	8.90	8.84	8.75	8.66	8.56	8.46	8.38	8.32	8.29	8.26
.01		21.2	18.0	16.7	16.0	15.5	15.2	15.0	14.8	14.7	14.5	14.4	14.2	14.0	13.8	13.7	13.6	13.5	13.5
.005		31.3	26.3	24.3	23.2	22.5	22.0	21.6	21.4	21.1	21.0	20.7	20.4	20.2	19.9	19.7	19.5	19.4	19.3
.10		4.06	3.78	3.62	3.52	3.45	3.40	3.37	3.34	3.32	3.30	3.27	3.24	3.21	3.17	3.15	3.13	3.12	3.11
.05		6.61	5.79	5.41	5.19	5.05	4.95	4.88	4.82	4.77	4.74	4.68	4.62	4.56	4.50	4.44	4.41	4.39	4.37
.025	5	10.0	8.43	7.76	7.39	7.15	6.98	6.85	6.76	6.68	6.62	6.52	6.43	6.33	6.23	6.14	6.08	6.05	6.02
.01		16.3	13.3	12.1	11.4	11.0	10.7	10.5	10.3	10.2	10.1	9.89	9.72	9.55	9.38	9.24	9.13	9.08	9.02
.005		22.8	18.3	16.5	15.6	14.9	14.5	14.2	14.0	13.8	13.6	13.4	13.1	12.9	12.7	12.4	12.3	12.2	12.1
.10		3.78	3.46	3.29	3.18	3.11	3.05	3.01	2.98	2.96	2.94	2.90	2.87	2.84	2.80	2.77	2.75	2.73	2.72
.05		5.99	5.14	4.76	4.53	4.39	4.28	4.21	4.15	4.10	4.06	4.00	3.94	3.87	3.81	3.75	3.71	3.69	3.67
.025	6	8.81	7.26	6.60	6.23	5.99	5.82	5.70	5.60	5.52	5.46	5.37	5.27	5.17	5.07	4.98	4.92	4.88	4.85
.01		13.7	10.9	9.78	9.15	8.75	8.47	8.26	8.10	7.98	7.87	7.72	7.56	7.40	7.23	7.09	6.99	6.93	6.88
.005		18.6	14.5	12.9	12.0	11.5	11.1	10.8	10.6	10.4	10.2	10.0	9.81	9.59	9.36	9.17	9.03	8.95	8.88

.10		3.59	3.26	3.07	2.96	2.88	2.83	2.78	2.75	2.72	2.70	2.67	2.63	2.59	2.56	2.52	2.50	2.48	2.47
.05		5.59	4.74	4.35	4.12	3.97	3.87	3.79	3.73	3.68	3.64	3.57	3.51	3.44	3.38	3.32	3.27	3.25	3.23
.025	7	8.07	6.54	5.89	5.52	5.29	5.12	4.99	4.90	4.82	4.76	4.67	4.57	4.47	4.36	4.28	4.21	4.18	4.14
.01		12.2	9.55	8.45	7.85	7.46	7.19	6.99	6.84	6.72	6.62	6.47	6.31	6.16	5.99	5.86	5.75	5.70	5.65
.005		16.2	12.4	10.9	10.1	9.52	9.16	8.89	8.68	8.51	8.38	8.18	7.97	7.75	7.53	7.35	7.22	7.15	7.08
.10		3.46	3.11	2.92	2.81	2.73	2.67	2.62	2.59	2.56	2.54	2.50	2.46	2.42	2.38	2.35	2.32	2.31	2.29
.05		5.32	4.46	4.07	3.84	3.69	3.58	3.50	3.44	3.39	3.35	3.28	3.22	3.15	3.08	3.02	2.97	2.95	2.93
.025	8	7.57	6.06	5.42	5.05	4.82	4.65	4.53	4.43	4.36	4.30	4.20	4.10	4.00	3.89	3.81	3.74	3.71	3.67
.01		11.3	8.65	7.59	7.01	6.63	6.37	6.18	6.03	5.91	5.81	5.67	5.52	5.36	5.20	5.07	4.96	4.91	4.86
.005		14.7	11.0	9.60	8.81	8.30	7.95	7.69	7.50	7.34	7.21	7.01	6.81	6.61	6.40	6.22	6.09	6.02	5.95
.10		3.36	3.01	2.81	2.69	2.61	2.55	2.51	2.47	2.44	2.42	2.38	2.34	2.30	2.25	2.22	2.19	2.17	2.16
.05		5.12	4.26	3.86	3.63	3.48	3.37	3.29	3.23	3.18	3.14	3.07	3.01	2.94	2.86	2.80	2.76	2.73	2.71
.025	9	7.21	5.71	5.08	4.72	4.48	4.32	4.20	4.10	4.03	3.96	3.87	3.77	3.67	3.56	3.47	3.40	3.37	3.33
.01		10.6	8.02	6.99	6.42	6.06	5.80	5.61	5.47	5.35	5.26	5.11	4.96	4.81	4.65	4.52	4.42	4.36	4.31
.005		13.6	10.1	8.72	7.96	7.47	7.13	6.88	6.69	6.54	6.42	6.23	6.03	5.83	5.62	5.45	5.32	5.26	5.19
.10		3.29	2.92	2.73	2.61	2.52	2.46	2.41	2.38	2.35	2.32	2.28	2.24	2.20	2.15	2.12	2.09	2.07	2.06
.05		4.96	4.10	3.71	3.48	3.33	3.22	3.14	3.07	3.02	2.98	2.91	2.84	2.77	2.70	2.64	2.59	2.56	2.54
.025	10	6.94	5.46	4.83	4.47	4.24	4.07	3.95	3.85	3.78	3.72	3.62	3.52	3.42	3.31	3.22	3.15	3.12	3.08
.01		10.0	7.56	6.55	5.99	5.64	5.39	5.20	5.06	4.94	4.85	4.71	4.56	4.41	4.25	4.12	4.01	3.96	3.91
.005		12.8	9.43	8.08	7.34	6.87	6.54	6.30	6.12	5.97	5.85	5.66	5.47	5.27	5.07	4.90	4.77	4.71	4.64
.10		3.18	2.81	2.61	2.48	2.39	2.33	2.28	2.24	2.21	2.19	2.15	2.10	2.06	2.01	1.97	1.94	1.92	1.90
.05		4.75	3.89	3.49	3.26	3.11	3.00	2.91	2.85	2.80	2.75	2.69	2.62	2.54	2.47	2.40	2.35	2.32	2.30
.025	12	6.55	5.10	4.47	4.12	3.89	3.73	3.61	3.51	3.44	3.37	3.28	3.18	3.07	2.96	2.87	2.80	2.76	2.72
.01		9.33	6.93	5.95	5.41	5.06	4.82	4.64	4.50	4.39	4.30	4.16	4.01	3.86	3.70	3.57	3.49	3.41	3.36
.005		11.8	8.51	7.23	6.52	6.07	5.76	5.52	5.35	5.20	5.09	4.91	4.72	4.53	4.33	4.17	4.04	3.97	3.90
.10		3.07	2.70	2.49	2.36	2.27	2.21	2.16	2.12	2.09	2.06	2.02	1.97	1.92	1.87	1.83	1.79	1.79	1.76
.05		4.54	3.68	3.29	3.06	2.90	2.79	2.71	2.64	2.59	2.54	2.48	2.40	2.33	2.25	2.18	2.12	2.10	2.07
.025	15	6.20	4.77	4.15	3.80	3.58	3.41	3.29	3.20	3.12	3.06	2.96	2.86	2.76	2.64	2.55	2.47	2.44	2.40
.01		8.68	6.36	5.42	4.89	4.56	4.32	4.14	4.00	3.89	3.80	3.67	3.52	3.37	3.21	3.08	2.98	2.92	2.87
.005		10.8	7.70	6.48	5.80	5.37	5.07	4.85	4.67	4.54	4.42	4.25	4.07	3.88	3.69	3.52	3.39	3.33	3.26
.10		2.97	2.59	2.38	2.25	2.16	2.09	2.04	2.00	1.96	1.94	1.89	1.84	1.79	1.74	1.69	1.65	1.63	1.61
.05		4.35	3.49	3.10	2.87	2.71	2.60	2.51	2.45	2.39	2.35	2.28	2.20	2.12	2.04	1.97	1.91	1.88	1.84
.025	20	5.87	4.46	3.86	3.51	3.29	3.13	3.01	2.91	2.84	2.77	2.68	2.57	2.46	2.35	2.25	2.17	2.13	2.09
.01		8.10	5.85	4.94	4.43	4.10	3.87	3.70	3.56	3.46	3.37	3.23	3.09	2.94	2.78	2.64	2.54	2.48	2.42
.005		9.94	6.99	5.82	5.17	4.76	4.47	4.26	4.09	3.96	3.85	3.68	3.50	3.32	3.12	2.96	2.83	2.76	2.69

TABLE F. F-PERCENTILES

P	$\downarrow \nu_2$	$\nu_1 \rightarrow$ 1	2	3	4	5	6	7	8	9	10	12	15	20	30	50	100	200	> 200
.10	30	2.88	2.49	2.28	2.14	2.05	1.98	1.93	1.88	1.85	1.82	1.77	1.72	1.67	1.61	1.55	1.51	1.48	1.46
.05	30	4.17	3.32	2.92	2.69	2.53	2.42	2.33	2.27	2.21	2.16	2.09	2.01	1.93	1.84	1.76	1.70	1.66	1.62
.025	30	5.57	4.18	3.59	3.25	3.03	2.87	2.75	2.65	2.57	2.51	2.41	2.31	2.20	2.07	1.97	1.88	1.84	1.79
.01	30	7.56	5.39	4.51	4.02	3.70	3.47	3.30	3.17	3.07	2.98	2.84	2.70	2.55	2.39	2.25	2.13	2.07	2.01
.005	30	9.18	6.35	5.24	4.62	4.23	3.95	3.74	3.58	3.45	3.34	3.18	3.01	2.82	2.63	2.46	2.32	2.25	2.18
.10	50	2.81	2.41	2.20	2.06	1.97	1.90	1.84	1.80	1.76	1.73	1.68	1.63	1.57	1.50	1.44	1.39	1.36	1.33
.05	50	4.03	3.18	2.79	2.56	2.40	2.29	2.20	2.13	2.07	2.03	1.95	1.87	1.78	1.69	1.60	1.52	1.48	1.44
.025	50	5.34	3.97	3.39	3.05	2.83	2.67	2.55	2.46	2.38	2.32	2.22	2.11	1.99	1.87	1.75	1.65	1.60	1.54
.01	50	7.17	5.06	4.20	3.72	3.41	3.19	3.02	2.89	2.79	2.70	2.56	2.42	2.27	2.10	1.95	1.82	1.75	1.68
.005	50	8.63	5.90	4.83	4.23	3.85	3.58	3.38	3.22	3.09	2.99	2.82	2.65	2.47	2.27	2.10	1.95	1.87	1.78
.10	100	2.76	2.36	2.14	2.00	1.91	1.83	1.78	1.73	1.69	1.66	1.61	1.56	1.49	1.42	1.35	1.29	1.26	1.21
.05	100	3.94	3.09	2.70	2.46	2.31	2.19	2.10	2.03	1.97	1.93	1.85	1.77	1.68	1.59	1.48	1.39	1.34	1.28
.025	100	5.18	3.83	3.25	2.92	2.70	2.54	2.42	2.32	2.24	2.18	2.08	1.97	1.85	1.71	1.59	1.48	1.42	1.35
.01	100	6.90	4.82	3.98	3.51	3.21	2.99	2.82	2.69	2.59	2.50	2.38	2.22	2.07	1.89	1.74	1.60	1.52	1.43
.005	100	8.24	5.59	4.54	3.96	3.59	3.33	3.13	2.97	2.85	2.74	2.58	2.41	2.23	2.02	1.84	1.68	1.59	1.48
.10	200	2.73	2.33	2.11	1.97	1.88	1.80	1.75	1.70	1.66	1.63	1.58	1.52	1.46	1.38	1.31	1.24	1.20	1.14
.05	200	3.89	3.04	2.65	2.42	2.26	2.14	2.06	1.98	1.93	1.88	1.80	1.72	1.62	1.52	1.41	1.32	1.26	1.19
.025	200	5.18	3.76	3.18	2.85	2.63	2.47	2.35	2.26	2.18	2.11	2.01	1.90	1.78	1.64	1.51	1.39	1.32	1.23
.01	200	6.76	4.71	3.88	3.41	3.11	2.89	2.73	2.60	2.50	2.41	2.27	2.13	1.97	1.79	1.63	1.48	1.39	1.28
.005	200	8.06	5.44	4.41	3.84	3.47	3.21	3.01	2.86	2.73	2.63	2.47	2.30	2.11	1.91	1.71	1.54	1.44	1.31
.10	> 200	2.71	2.30	2.08	1.94	1.85	1.77	1.72	1.67	1.63	1.60	1.55	1.49	1.42	1.34	1.26	1.18	1.13	1.00
.05	> 200	3.84	3.00	2.60	2.37	2.21	2.10	2.01	1.94	1.88	1.83	1.75	1.67	1.57	1.46	1.35	1.24	1.17	1.00
.025	> 200	5.02	3.69	3.12	2.79	2.57	2.41	2.29	2.19	2.11	2.05	1.94	1.83	1.71	1.57	1.43	1.29	1.21	1.00
.01	> 200	6.63	4.61	3.78	3.32	3.02	2.80	2.64	2.51	2.41	2.32	2.18	2.04	1.88	1.70	1.52	1.35	1.25	1.00
.005	> 200	7.88	5.30	4.28	3.72	3.35	3.09	2.90	2.74	2.62	2.52	2.36	2.19	2.00	1.79	1.59	1.40	1.28	1.00

Proportions

1. In 9 independent Bernoulli trials, there were exactly 4 successes. Find a 95% CI for p and test $H_0{:}p = .5$.

Solution

ZS STEP	ENTER	PRESS	DISPLAY	COMMENTS
B(p)				
1a	9.	[STO]03	9.	Enter sample size.
1b	.44 ...	[STO]40	.44 ...	Enter $\hat{p} = 4/9$, the estimate of p.
4a		[B]	12.	Since n is small, find first
		[R/S]	8.	pair of d.f. = (12,8).
4b	4.20	[STO]31	4.20	Enter first $F_{.025}$ percentile.
4c		[R/S]	10.	Discover revised d.f.
		[R/S]	10.	= (10,10).
4d	3.72	[STO]41	3.72	Store second $F_{.025}$ percentile.
4e		[R/S]	.14	Lower confidence limit displayed
		[$x \leftrightarrows t$]	.79	and u found in R_T
3a	0.	[B′]	0.	Enter H_1-code for $H_1{:}p \neq .5$.
3b	.5	[R/S]	1.	Significance level 1; accept H_0.

2. A device was tested 25 times and passed 23 times. Find a lower one-sided CI on p, the probability of passing. Test $H_0{:}p \geq .95$.

Solution

ZS STEP	ENTER	PRESS	DISPLAY	COMMENTS
B(p)				
1a	25	[STO]03	25.	Enter data as above.
1b	23÷ 25	[STO]40	.92	$\hat{p} = 23/25 = .92$
2a	−1	[B′]	−1.	Enter H_1-code for $H_1{:}p < .95$.
2b	.95	[R/S]	.36	Significance level $P = .36$. Note somewhat lengthy running time.
2b′	.95	[C′]	.25	Comparing with large sample test, $P = .25$
4a		[B]	6.	Intitialize d.f. for small n CI to
		[R/S]	46.	determine percentile.
4b	2.29	[STO]31	2.29	Following instructions, store $F_{.05}$ in R_{31}
4c		[R/S]	48.	Calculate new d.f $\nu_1 = 48$ and
		[R/S]	4.	$\nu_2 = 4$.
4d	1	[STO]41	1.	Enter 1 in R_{41}, and calculate lower confidence bound.
4e		[R/S]	.77	

3. A sample of size 100 was taken from a lot with replacement, and 2 defective items were found. Test the manufacturer's claim that $p < .05$ at $\alpha = .01$.

Solution

ZS STEP	ENTER	PRESS	DISPLAY	COMMENTS
B(p)				
1a	100.	[STO]03	100.	Enter data as usual.
1b	.02	[STO]40	.02	
2a	−1.	[B′]	−1.	H_1-code for H_1:p < .05, the null hypothesis being H_0:$p \geq$ .05, the disclaimer.
2b	.05	[R/S]	.12	Enter p_0, and find P = .12, supporting H_0, not H_1.
2b′	.05	[C′]	.08	Compare normal test and find P = .08.

4. In a random sample of 500 men (X), 350 were found to favor a certain political issue. In a similar sample of 300 women (Y), 200 were so inclined. Is there any real difference between sexes on this issue?

Solution To test H_0:$p_x = p_y$,

ZS STEP	ENTER	PRESS	DISPLAY	COMMENTS
B($p_x - p_y$)				
1a	500	[STO]13	500.	Enter first sample size.
	300	[STO]03	300.	Enter second sample size.
1b	.7	[STO]47	.7	Enter first proportion estimate $\hat{p}_x$.
	2/3	[STO]37	0.66...	Enter second proportion estimate $\hat{p}_y$
2	0.	[A′]	.32	Enter H_1-code for H_1:$p_x \neq p_y$ and find P
		[RCL]30	.98	= .32 with ts = .98. Data supports H_0.
3	.025	[A]	−.03	A 95% CI for the difference $u_x - \mu_y$ extends from −.03 to
		[$x \rightleftarrows t$]	.10	+ .10; includes 0.

8

Analysis of Variance

Section 8.1: Introduction

Many times our statistical investigations lead us to the problem of comparing more than two populations. To start with, let's take a simple example that we will follow through: suppose a school system has three schools located in different parts of town. There may be some real concern about whether or not the same quality of education is being offered to all the residents of this town. One possible way of investigating this situation would be to draw a random sample from each of the 3 schools and compare achievement, as measured by scores on some type of standardized test. Recognizing that individual differences occur, we might use the school mean as a criterion for measuring achievement, so that we are interested in testing the hypothesis that these 3 population means are all the same against the alternative that they are not (which means, then, that at least some two of them differ).

It would be very easy to criticize the assumption that mean achievement on a standardized test adequately measures quality of education in the first place. Such parameters may actually measure something else, like heterogeneity of the population as regards location in the town boundaries—the reader may think of many other possibilities.

Typical of many introductory treatments of a subject, we are going to bypass any lengthy discussion of such nuances and get on with the mechanics of carrying out procedures.

Before turning to the procedures, it may be instructive to return momentarily to the problem of testing for equality of just two means. You recall from Chapter 6 that the only nice solution for small samples occurs when it is reasonable to assume the variances are equal. Then, an exact test based on the t-distribution, having $\nu = n_x + n_y - 2$, may be formed. A test of size α will be completed, if $H_o{:}\mu_x = \mu_y$ is rejected whenever the test statistic

$$ts = \frac{\bar{x} - \bar{y}}{s_p} \qquad (s_p \text{ is the pooled standard deviation}) \tag{8.1}$$

either exceeds $t_{\alpha/2}$ or is less than $-t_{\alpha/2}$, or, equivalently, whenever $|ts| > t_{\alpha/2}$. A significance test is conducted by computing the significance level as $P = 2Q(|ts|) = \Pr(|TS| > |ts|)$.

Now, it happens to be a theoretical fact that if T has a t-distribution with d.f. $= \nu$, then T^2 has an F-distribution with d.f. $= (1, \nu)$. But then, $|TS| > t_{\alpha/2}$ is equivalent to $(TS)^2 > t^2_{\alpha/2}$, which is the same as examining an upper one-sided F-test. In fact, if you will check your t-table entries, you will find close agreement between $t^2_{\alpha/2}$ and the F_α-point in the column headed $\nu_1 = 1$ in the F-table. So, the two-sided t-test of $H_o{:}\mu_x = \mu_y$ can be converted to an upper one-sided F-test, with d.f. $= (1, n_x + n_y - 2)$. Also, the significance level for the two-sided t-test may be equivalently expressed

$$P = \Pr(|TS| > |ts|) = \Pr((TS)^2 > (ts)^2) = \Pr(F > (ts)^2)$$

as a one-sided significance level from the F-distribution.

The reason for bringing all this up is that when more than two populations are involved, the t statistic is no longer adequate, but an F-statistic can handle the more general situation. Seeing the equivalence for two populations should make the more general case plausible. The scheme of testing for equality of means 2-by-2, using the t-test, is a possibility, but it should be avoided, since performing several such tests, each of size α, for example, will not preserve the overall size α. Just as with the 2-sample t-test, we will have to assume that the population variances are all equal in the general case, as we shall see. Like the t-test, the corresponding F-tests we construct will be rather robust with regard to normality assumptions, but are rather sensitive to the assumption of equality of variances, reminiscent of the Behrens–Fisher problem. The reason for the name *Analysis of Variance* (AOV for short) is that the F-tests, as you shall see, are constructed on the basis of comparing the ratio of two independent *variance* estimators (hence Chi-square random variables) under conditions when the means are all equal and when they are not.

To keep things simple at first, we will carry through our school system example, using fictitious data based on small samples. This should give a suitable basis for the general idea, which will then be outlined. Practice with the exercises that follow will then tie the loose ends together. The first thing that should strike you, is that we need a more extensive notation for keeping track of observations. For three populations, things are not so bad, and we could just use X, Y, and Z. But for the general case of K such means, using distinct letters is an inefficient method of housekeeping. Instead we turn to double subscripts. We will use X_1, X_2 through X_K to stand for our K population random variables with respective means $\mu_1, \mu_2, \ldots, \mu_K$. Now, from each of these random variables, we want to consider random samples of possibly different sizes. So, we will represent a random sample of X_1 of size n_1, say, as a collection of n_1 independent, identically distributed random variables. Keeping the 1 fixed, we use a second subscript to do this and present the array as

$$X_{11}, X_{12}, X_{13}, \ldots, X_{1n_1}$$

(We occasionally use a comma to separate the subscripts, so as to avoid confusion. Thus, we would represent the 12th random variable as $X_{1,12}$, not X_{112}, in case there is a possibility of confusing this with, say, the second random variable of the 11th population random variable $X_{11,2}$. It is usually quite clear by context what is intended.) The second sample of size n_2 would then be denoted

$$X_{21}, X_{22}, X_{23}, \ldots, X_{2n_2}$$

and so on through the Kth population represented by the random sample

$$X_{K1}, X_{K2}, X_{K3}, \ldots, X_{Kn_K}$$

The total sample size is then

$$N = \sum_{i=1}^{K} n_i$$

At first, it will be helpful if we suppose that the sample sizes are all equal, say equal to n, in which case $N = Kn$. After we clarify the model with this case and follow through the example, we will generalize the results and display formulas for arbitrary sample sizes.

As for assumptions, a model that leads to the analysis of variance is one in which the population random variables $X_1, X_2, \ldots, X_K$ are themselves to be independent normal random variables with respective means

 $\mu_1, \mu_2, \ldots, \mu_K$. Like the 2-sample case, it is assumed that the population random variables all have a *common variance* σ^2. In that case, we may write for $i = 1,2$ through K

$$X_{ij} = \mu_i + E_{ij} \qquad j = 1, 2,\ldots,n \tag{8.2}$$

where $E_{i1}, E_{i2},\ldots,E_{in}$ are independent normal random variables with mean 0 and common variance σ^2.

There is an alternate form of the model, which actually lends itself better to an intuitive analysis (one we must be content with here). Suppose we average the means, letting

$$\mu = \frac{1}{K} \sum_{i=1}^{K} \mu_i$$

Then

$$\delta_i = \mu_i - \mu \qquad i = 1, 2, \ldots, K$$

is the amount by which μ_i differs from this *grand* or *overall* mean μ. We may think of δ_i as a measure of achievement that is peculiar to the *i*-th population—school, in our example. Now the model may be rewritten, conveniently, as

$$X_{ij} = \mu + \delta_i + E_{ij} \qquad \begin{array}{l} i = 1, 2,\ldots,K \\ j = 1, 2,\ldots,n \end{array} \tag{8.3}$$

In this way, a random observation X_{ij} is viewed as "split" into a portion, μ, that is a general characteristic common to all K populations (schools), a portion, δ_i, that is peculiar to the *i*-th population (school), and a random amount E_{ij}, also called *error*.

As regards the differences $\delta_1, \delta_2,\ldots,\delta_K$, as any sum of deviations from an average, the sum must vanish, that is,

$$\sum_{i=1}^{K} \delta_i = 0$$

That being the case, we could write

$$\delta_K = -\delta_1 - \delta_2 - \ldots - \delta_{K-1}$$

as a combination of δ_1 through δ_{K-1}, so that among $\delta_1, \delta_2,\ldots,\delta_K$, there are only $K - 1$ distinct parameters. In this form, we see that, if, indeed,

$$\mu_1 = \mu_2 = \mu_3 = \ldots = \mu_K \qquad \text{then } \delta_1 = 0, \delta_2 = 0,\ldots,\delta_K = 0$$

Conversely, if $\delta_1 = \delta_2 = ... = \delta_K = 0$, this says that

$$\mu_1 = \mu, \mu_2 = \mu, ..., \mu_K = \mu$$

Consequently, a test of $H_0: \mu_1 = \mu_2 = ... = \mu_K = \mu$ is exactly equivalent to a test of $H_0: \delta_1 = \delta_2 = ... = \delta_K = 0$.

This brings us to the data. For this kind of setup, the data, the observations, and summary statistics may be kept track of in an array consisting of K rows. The general format would be

$$\left.\begin{array}{l} \text{Row (population) 1: } x_{11}, x_{12}, x_{13}, ..., x_{1n} \rightarrow \bar{x}_1, s_1^2 \\ \text{Row (population) 2: } x_{21}, x_{22}, x_{23}, ..., x_{2n} \rightarrow \bar{x}_2, s_2^2 \\ \text{Row (population) 3: } x_{31}, x_{32}, x_{33}, ..., x_{3n} \rightarrow \bar{x}_3, s_3^2 \\ \quad \vdots \\ \text{Row (population) } K\text{: } x_{K1}, x_{K2}, x_{K3}, ... x_{Kn} \rightarrow \bar{x}_K, s_K^2 \end{array}\right\} \begin{array}{l} \bar{\bar{x}} = \frac{1}{K} \sum_{i=1}^{K} \bar{x}_i \\ s_{\bar{x}}^2 = \frac{1}{K-1} \sum_{i=1}^{K} (\bar{x}_i - \bar{\bar{x}})^2 \\ s_p^2 = \frac{1}{K} \sum_{i=1}^{K} s_i^2 \end{array} \qquad (8.4)$$

For our school example, $K = 3$, and we are going to take $n = 10$, supposing that actual scores for each school are

$$\left.\begin{array}{l} \text{Row (school) 1: 60 62 78 87 61 78 77 84 73 84} \rightarrow 74.4, 102.04... \\ \text{Row (school) 2: 85 90 72 70 75 81 60 81 76 86} \rightarrow 77.6, 78.93... \\ \text{Row (school) 3: 62 67 61 75 63 74 86 67 89 65} \rightarrow 70.9, 98.54... \end{array}\right\} \begin{array}{l} \bar{\bar{x}} = 74.3 \\ s_{\bar{x}}^2 = 11.23 \\ s_p^2 = 93.174 \end{array} \qquad (8.5)$$

For each row, we have processed the data to produce a row mean and sample variance, and then processed these K data values to produce a mean of K sample means, $\bar{\bar{x}}$, a sample variance for these K sample means, $s_{\bar{x}}^2$, and finally, a pooled variance, s_p^2, found by averaging the K sample variances.

Now let us put together an intuitive analysis that will generate the F-test of H_o. You know the data in each row constitute a random sample from a normal distribution having the same variance σ^2 under the AOV assumption. Hence, each of the K s_i^2's is an estimate of σ^2, and their average is a natural pooled estimate, s_p^2, of this common variance σ^2. And, of course, this is so whether H_o is true or false.

On the other hand, each row mean, $\bar{x}_i$, comes from a normal distribution with mean μ_i and variance σ^2/n, you will recall. But, when H_o is true, each μ_i is just μ, so that the row means, $\bar{x}_1, \bar{x}_2, ..., \bar{x}_K$, then constitute a random sample under our assumption of a single normal distribution having mean μ and variance σ^2/n. As usual the mean of *this* sample, $\bar{\bar{x}}$, is an estimate of μ, and the sample variance of *this* sample, $s_{\bar{x}}^2$, is the usual estimate of σ^2/n, or, $ns_{\bar{x}}^2$ is an estimate of σ^2. So, *if* H_0 *is true,* s_p^2 and $ns_{\bar{x}}^2$ ought to be "about" the same, since they are two estimates of the same parameter. That is, the ratio $ns_{\bar{x}}^2/s_p^2$ ought to be "close" to the value 1. Moreover, the values $\bar{x}_1, \bar{x}_2, \ldots, \bar{x}_K$ tend to be within 3 SE's, that is, $\pm 3\sigma/\sqrt{n}$, of their common mean μ, and the quantity

$$\sum_{i-1}^{K} (\bar{x}_i - \bar{\bar{x}})^2$$

should reflect this. (See Figure 8.1.)

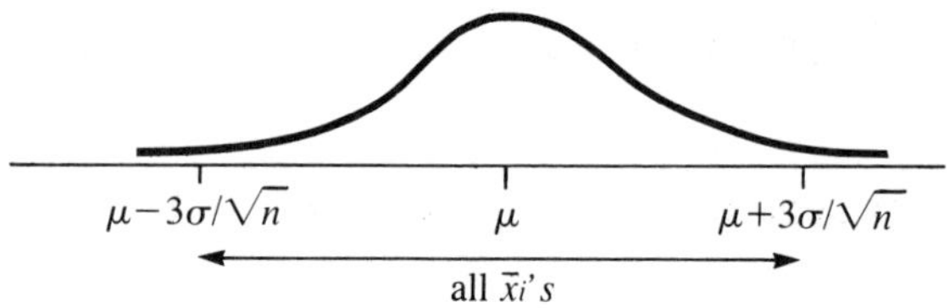

Figure 8.1. When H_0 is true

On the other hand, *if* H_o *is false,* then at least two population means are different. Each $\bar{x}_i$ will still tend to be within 3 SE's of each respective mean μ_i, at least two of which are not the same now. We might have a situation like that depicted in Figure 8.2, where $\mu_1 \neq \mu_3$, while $\mu_2 = \mu_4$. Still, the overall center of gravity, $\bar{\bar{x}}$, will be roughly in the center of the $\bar{x}_i$'s. Accordingly,

$$\sum_{i=1}^{K} (\bar{x}_i - \bar{\bar{x}})^2$$

will tend to be larger than its value when H_o is true.

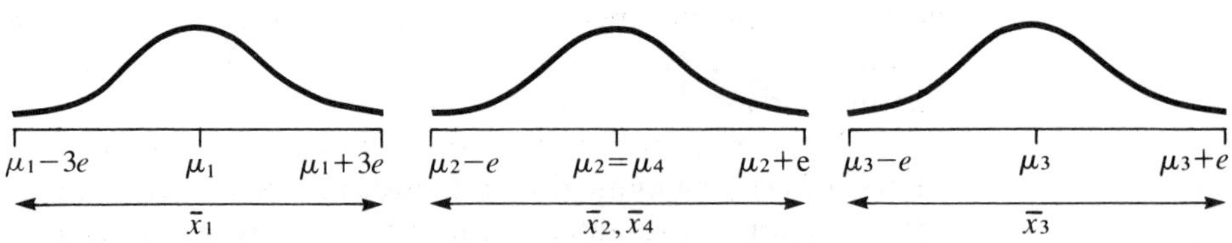

Figure 8.2. When H_O is false ($e = 3\sigma/\sqrt{n}$)

Now, s_p^2 is not affected one iota by all of this, so only the numerator of $ns_{\bar{x}}^2/s_p^2$ is affected by the status of H_o, and will tend to be larger when H_o is false, than when it is true. Intuition tells us to reject H_o whenever this ratio is "too large." To be explicit about it, $(K-1)(ns_{\bar{x}}^2)/\sigma^2$ has a Chi-square distribution with d.f. $= K - 1$, while (it may be shown) $(N - K)s_p^2/\sigma^2$ has a Chi-square distribution with d.f. $= N - K$. Dividing each by the respective degrees of freedom, it is the ratio

$$\frac{\dfrac{(K-1)(ns_{\bar{x}}^2)}{\sigma^2(K-1)}}{\dfrac{(N-K)s_p^2}{\sigma^2(N-k)}} = \frac{ns_{\bar{x}}^2}{s_p^2} \tag{8.6}$$

which has an F-distribution with d.f. $= (K - 1, N - K)$, when H_o is true. In our example, this ratio is 1.21, and we will have to decide whether this is "large" or not by consulting the F-distribution with d.f. $= (2,27)$. But, utilizing ST–22, we find $\Pr(F > 1.21) = .31$, so there is really insufficent evidence for rejecting H_o for the example.

To see the changes that are necessary for the more general case, we may rewrite the numerator of the F-ratio (apart from constants) as

$$(K-1)(ns_{\bar{x}}^2) = \sum_{i=1}^{K} n\,(\bar{x}_i - \bar{\bar{x}})^2$$

This expresses the result as a sum of squared deviations, and we see that when the sample sizes are *not* all equal, the common factor n should be replaced by n_i, the i-th sample size. Because the squared deviations are from row to row, this sum of squares is often called *row sum of squares* and denoted S_R^2 or RSS or some variation of that.

The ASM manual discusses this model on page 4–20, where this is called *treatment* sum of squares and is denoted TSS as part of their AOV development. But in ST–16, an extension of this model, this will be changed to RSS for *row* sum of squares. We prefer to remain consistent with that model (to be discussed shortly), so we would ask you to change that in ST–15, or note the equivalence to our choice of RSS. Thus, in general,

$$\text{RSS} = \sum_{i=1}^{K} n_i(\bar{x}_i - \bar{\bar{x}})^2 \qquad \bar{\bar{x}} = \frac{1}{K}\sum_{i=1}^{K} \bar{x}_i \qquad \bar{x}_i = \frac{1}{n_i}\sum_{j=1}^{n_i} x_{ij} \tag{8.7}$$

It is the sum of squares RSS, divided by $K - 1$, the degrees of freedom, that forms the estimate of σ^2, based on the variation from row to row. This quantity, RSS$/(K - 1)$ is sometimes called the *mean* (for average) row sum of squares and denoted MRSS (or some variation of that). Often you will also find it called the *explained* variance or *between row variation,* meaning an accounting for variability based on differences from row to row.

By contrast, the denominator of the F-ratio for $n_i = n$ can be expressed (apart from constants) as

$$Ks_p^2 = \sum_{i=1}^{K} s_i^2 = \sum_{i=1}^{K} \sum_{j=1}^{n} \frac{(x_{ij} - \bar{x}_i)^2}{(n-1)}$$

or

$$(N - K)s_p^2 \equiv (Kn - K)\, s_p^2 = K(n-1)\, s_p^2 = \sum_{i=1}^{K} \sum_{j=1}^{n} (x_{ij} - \bar{x}_i^2)$$

When sample sizes are not all equal, this sum of squares, denoted ESS in ASM, would be written

$$\text{ESS} = \sum_{i=1}^{K} \sum_{j=1}^{n_i} (x_{ij} - \bar{x}_i)^2 \tag{8.8}$$

and is called *error* sum of squares. The reason for this terminology is that, since deviations from the row means are taken into account in computing ESS, it accounts for variability due simply to chance fluctuations or to the *error* term in the model (see Equation 8.3). As an estimate of the variance σ^2 of that term, ESS/$(N - K)$ is sometimes called *mean* error sum of squares and is denoted MESS (or some variation thereof). You will also find it called the *unexplained* variance, unexplained in the sense of not being accounted for by any differences from row to row. The quantity ESS is also referred to as the *residual* sum of squares, because if we define the total sum of squared deviations from the overall mean by

$$\text{SS} = \sum_{i=1}^{K} \sum_{j=1}^{n_i} (x_{ij} - \bar{\bar{x}})^2 \tag{8.9}$$

then it may be shown that SS = RSS + ESS, so that ESS is what *remains* when the row sum of squares is subtracted from the total.

Finally, with these terms in mind, you see that the F-ratio used to test H_o is simply the ratio of these two mean sum of squares, that is,

$$F = \frac{\text{MRSS}}{\text{MESS}} = \frac{\text{Explained variance}}{\text{Unexplained variance}} \tag{8.10}$$

When H_o is true, this ratio has an F-distribution with d.f. = $(K - 1, N - K)$. That much being established, the matter of coordinating the various sums of squares is one primarily of housekeeping, and this is usually summarized in a table form known as an AOV table (or ANOVA table). The format we are going to follow is displayed in Table 8.1. For the school example, using the quantities we already computed in Equation 8.5, the following AOV table summarizes the procedure.

AOV TABLE

Source	Sum of squares	d.f.	MSS	F	P
Rows	224.6	2	112.3	1.21	.31
Error	2515.7	27	93.174		
Total	2740.3	29			

Naturally, we want to utilize the TI–59 programs for processing the row data, as well as carrying out the necessary F-test. You see from the user instructions that program ST–06 is the program required for data entry. Once that is accomplished, program ST–15, as you see, will output the various quantities we want for the AOV table, and then ST–22 may be used to compute the P-value. Actually, if that is all we wanted to do—and we only needed to do it once—then we could rely entirely on these three programs and carry out the procedure as in our school example. Unfortunately, programs ST–15 and ST–22 share some registers in the calculator, and you cannot rely on ST–15 or its published register contents to output the same results should you wish to rework the problem for any reason. Thus, it is wise to store the various key outputs in an AOV table *before* performing an F-test.

TABLE 8.1. ONE-WAY AOV TABLE

Source	Sum of squares	d.f.	MSS	F	PV
Rows	SS_r	$r-1$	MSS_r	MSS_r/Mss_u	
Residual = Error	SS_u	$N-r$	MSS_u		
Total	SS_t	$N-1$			

One of the reasons you may wish to store those key elements is to study the problem further, particularly if H_o is rejected, for then you would naturally want to know which means are causing the significant difference. There is a confidence interval method to do just that, as well as to provide intervals for other quantities of interest. Consequently, we have programmed the requirements in program ZS–8 under Code AOV–1 to provide for all of these contingencies. As you can see from the user instructions, the first step is to enter the data using program ST–06.

Program ST–06 is discussed on page 3 9 of ASM. After initializing the program by pressing [E′], you enter the data row by row using label A. A couple of matters need to be brought to your attention. First of all, you must not repartition the calculator, due to the number of steps in ZS–8. A total of $N = 29$ data points are allowed in standard partition. If you exceed these requirements, just press [D] to reset the pointer and continue entering the data. The second thing to note is that you must press [B′] and [C′] *at the conclusion of each row entry* to develop the necessary IDB. The result of pressing [B′] is $\bar{x}_i$, the row mean for that row. You need to record this somewhere near the data for possible future use. The output of [C′] is not s^2, as published, but MSD; in any event you do *not* need to record this, although you must press [C′] before moving to the next row. This is illustrated in the examples following ZS–8 instructions and is summarized below.

EXAMPLE 8.1

Three types of solvents are tested on grease-soaked material, and the amount of grease removed in milligrams is noted as follows:

Solvent A:	11	12	12	
Solvent B:	13	15		
Solvent C:	12	10	11	11

Enter the data using ST–06, and verify that the 3 row means are: $\bar{x}_1 =$ 11.67, $\bar{x}_2 = 14$, and $\bar{x}_3 = 11$.

Solution Having accomplished that much, you should now press [RST] to exit ST–06 and proceed to Step 2 of AOV–1, which calls for a simple press of [A]. If you will do that, you will see 12.22 in the display, which is the value of RSS to be entered in the AOV table. You may now fill the first block of the table by rows using successive [R/S] presses, and you can see under the Display column of the user instructions just what goes where in the table. The eighth output is $N - 1 = 8$, the d.f. for SS, and that much completes the sum of squares portion of the table. Just go on with [R/S] presses to get the remaining 2 values of interest, namely, the *F*-ratio and its *P*-value. Your completed table should appear as follows.

AOV TABLE

Source	Sum of squares	d.f.	MSS	*F*	*P*
Rows	12.22	2	6.11	7.86	.02
Error	4.67	6	0.78		
Total	16.89	8			

With that small a *P*-value, you would reject the hypothesis that the solvents were indistinguishable at the usual 5% level of significance, but you would not reject at the 1% level. If you do reject, it is not hard to guess (here) which one is causing the difference. But more of that in the next section.

As noted in the user instructions, if you lose your way at any point in these outputs, you may begin anew by just pressing [A] again. In addition, you may check the register contents for the location of various quantities that are entered in the table. Practice with the problems that follow. You will have additional opportunities after we examine related confidence intervals in the following section.

Problems

1. The data below are coded data for 4 different treatments of dyed material. Test the hypothesis that there is no difference in treatments. Use $\alpha = .01$.

Treatment A:	5	4	2	8		$\bar{x}_1 = 4.75$
Treatment B:	2	5	1			$\bar{x}_2 = 2.67$
Treatment C:	9	3				$\bar{x}_3 = 6$
Treatment D:	1	4	6	7	3	$\bar{x}_4 = 4.2$

Ans: $F = .72$, $P = .56$. Do not reject H_o.

2. Does the quarter in which a beginning statistics course is taken make a difference? The final grades for such a course for the past 3 quarters revealed the following data.

Fall quarter:	71	71	76	78	80	79	85	78	72	73	$\bar{x}_1 = 77.88$
	80	83	81	77	83	79	78				
Winter quarter:	81	87	91	92	84	93	94	86	73	71	$\bar{x}_2 = 84.75$
	81	84									
Spring quarter:	73	76	75	71	75	72	75	78	75	76	
	76	77	79	74	77	77	78	87	86	86	$\bar{x}_3 = 77.62$
	87										

Test the null hypothesis at the 5% level of significance.

Ans: $F = 7.79$, $P = .001$. Reject H_0.

3. Five professors have been teaching different sections of Calculus I. A common examination was given to all five sections and resulted in final scores:

Professor A:	83	61	73	52	66	65	62	57	49	84	$\bar{x}_1 = 68.75$
	64	65	77	66	72	80	75	76	78	70	
Professor B:	65	62	57	71	59	64	59	58	71	82	
	71	84	82	68	77	51	72	92	60	73	$\bar{x}_2 = 67.77$
	64	49									
Professor C:	72	76	66	76	73	62	58	50	69	65	
	57	52	62	71	74	56	72	66	61	85	$\bar{x}_3 = 66.50$
	72	68									
Professor D:	73	75	90	56	78	54	71	85	69	69	$\bar{x}_4 = 73.05$
	52	73	82	86	71	89	69	76	70		
Professor E:	55	58	86	85	83	66	67	79	66	58	
	71	88	72	74	91	63	81	75	76	86	$\bar{x}_5 = 74.19$
	78										

Is it safe to assume that all five classes are on essentially the same footing for purposes of assigning students to Calculus II? Use $\alpha = .10$.

Ans: $F = 2.29$, $P = .06$. Reject H_0.

4. To see whether or not there was any significant difference in grading practices by departments at a university, a pilot study was carried out in which ten grade point averages for senior students in each department were selected at random from the files. The following results were reported.

Physics:	1.43	1.54	2.21	1.60	2.21	3.00	2.52	$\bar{x}_1 = 2.204$
	2.67	2.85	2.01					
Statistics:	2.01	1.54	1.19	1.74	2.17	0.98	1.40	$\bar{x}_2 = 1.426$
	0.92	1.53	0.78					
Chemistry:	3.20	2.15	3.39	3.27	1.84	2.80	0.91	$\bar{x}_3 = 2.370$
	2.23	1.04	2.87					
Mathematics:	1.38	1.44	2.18	2.59	1.66	2.60	2.31	$\bar{x}_4 = 1.816$
	1.20	0.84	1.96					
Biology:	2.09	2.47	1.60	2.21	1.65	1.15	3.51	$\bar{x}_5 = 2.098$
	2.22	1.80	2.28					

Do the data support the null hypothesis at the 5% level?

Ans: $F = 3.27$, $P = .02$. Reject H_o.

5. A supermarket manager has always stocked 3 highly competitive soaps on the basis that they were demanded with about equal frequency (on the average anyway). The record of monthly demand was carefully noted at seven different stores in the chain and produced the following results.

Soap A:	122	86	109	95	98	110	106	$\bar{x}_1 = 103.71$
Soap B:	102	85	96	97	84	95	85	$\bar{x}_2 = 92.00$
Soap C:	107	93	97	85	74	89	78	$\bar{x}_3 = 89.00$

Is the stocking policy justified? Use $\alpha = .05$.

AOV TABLE

Source	Sum of squares	d.f.	MSS	F	P
Rows	846.38	2	423.19	3.99	.04
Error	1907.43	18	105.97		
Total	2753.81	20			

Ans: Reject H_o, since $P < .05$.

Section 8.3: Scheffe's Confidence Intervals

When the null hypothesis of equality of means is rejected, as in Problem 5, for example, it is natural to want to probe further and discover which population means are reasonably different as indicated by the data. We could compute, say, 95% confidence intervals for each difference, $\mu_i - \mu_j$, and see whether or not 0 is included in the interval. But this is just equivalent to, and has the same inherent problems as, applying *t*-tests 2-by-2. An elegant solution has been developed by Scheffe,* which takes into account not only differences, but other combinations, that may be of interest.

First, a *contrast* (*linear* contrast to be more precise) among the K population means in an AOV model is defined to be any linear combination of the form,

$$\sum_{i=1}^{K} c_i\mu_i \qquad \text{where} \quad \sum_{i=1}^{K} c_i = 0 \tag{8.11}$$

You see that, for example, if $c_1 = 1$, $c_2 = -1$, and $c_i = 0$ for all other i, Equation 8.11 defines the simple difference $\mu_1 - \mu_2$. Other differences may be similarly defined, so that contrasts include all possible differences. But they include much more. For example, the μ's might represent the mean quality of three suppliers, and the contrast $\mu_1 - .5\mu_2 - .5\mu_3$ would allow the possibility of comparing (for equality) the quality of one supplier with an even mix of the other two.

There are, of course, many possibilities for a contrast among K means. Remarkable though it may seem, Scheffe provides a method for constructing confidence intervals for *all possible* contrasts in such a way that, for a given risk α, the probability is $1 - \alpha$ that *all* such contrasts are simultaneously covered by their respective interval limits. Moreover, the intervals require of the data only MESS from the AOV table, along with the row means, for computation. Indeed, it turns out that a size α test in AOV is simply equivalent to the Scheffe confidence intervals in the sense that H_o will be rejected by the *F*-test if, and only if, the CI for *some* contrast (not necessarily a simple difference) fails to include 0. The formula for the CI for $\sum_{i=1}^{K} c_i\mu_i$, expressed in the symmetric form of $\hat{\theta} \pm e$, is given by

$$\sum_{i=1}^{K} c_i\bar{x}_i \pm \sqrt{(K-1)F_\alpha \text{ MESS} \sum_{i=1}^{K} (c_i^2/n_i)} \qquad \text{d.f.} = (K-1,\ N-K) \tag{8.12}$$

* H. Scheffe, *The Analysis of Variance* (New York: John Wiley & Sons, 1959).

You see the F-percentile is retrieved from the F-table with d.f. exactly the same as used to compute the P-value in AOV.

As you may have already noticed in the user instructions for AOV–1, these confidence interval formulas have been programmed for you at Step 3. After the AOV outputs have been stored by Step 2, you simply initialize by pressing [E′]. Then you successively enter triplets in the form c_i, $\bar{x}_i$, n_i with presses of [R/S]. As you go along, you are given a running count to help you keep track of where you are. If you are distracted for any reason, you just start over with [E′]. Also, as you see from the note, you only need $K - 1$ stored in R_{14} and MESS from the AOV table stored in R_{48} to run this part of the program. These respective values for Problem 5 were 2 and 105.97, so if you will store these, we will illustrate the results with that problem.

EXAMPLE 8.2

Problem 5 had to do with three competitive brands of soap with MESS = 105.97 and $\nu_1 = 2$. The respective row means were $\bar{x}_1 = 103.71$, $\bar{x}_2 = 92$, and $\bar{x}_3 = 89$, and each $n_i = 7$. Examine 95% CI's for differences, as well as the contrast $\mu_1 - .2\mu_2 - .8\mu_3$.

Solution Press [E′] after storing 2 in R_{14} and 105.97 in R_{48}. There are only three distinct differences to study, $\mu_1 - \mu_2$, $\mu_1 - \mu_3$, and $\mu_2 - \mu_3$. For $\mu_1 - \mu_2$, choose $c_1 = 1$, $c_2 = -1$, $c_3 = 0$, and enter these with the instructions at Step 3b, using the sequence

1 [R/S] 103.71 [R/S] 7 [R/S]

−1 [R/S] 92.00 [R/S] 7 [R/S]

Since $c_3 = 0$, it may be ignored in the entry. Now use $F_{.05} = 3.49$ from the F-table with d.f. = (2,20). Enter it and press [A′]. You should discover l and u in the usual way, and the 95% CI for $\mu_1 - \mu_2$ is (−2.8, 26.2). In a similar way, you should discover that the interval for $\mu_1 - \mu_3$ is (.17, 29.2) and for $\mu_2 - \mu_3$ is (−11.5, 17.5). So, there is a significant difference between μ_1 and μ_3, while it is hard to distinguish between μ_1 and μ_2, as well as between μ_2 and μ_3.

For the contrast $\mu_1 - .2\mu_2 - .8\mu_3$, press [E′] again, and input the sequence

1 [R/S] 103.71 [R/S] 7 [R/S]

−0.2 [R/S] 92.00 [R/S] 7 [R/S]

−0.8 [R/S] 89.00 [R/S] 7 [R/S]

Entering 3.49 again, a press of [A′] produces the interval (.79, 27.43); such a contrast is significantly different from 0: that is, we can distinguish μ_1 from the convex mix $.2\mu_2 + .8\mu_3$.

As published in the register contents for ZS–8, the error term e, used in the CI formulas of Equation 8.12 when *differences* are considered, may be found in R_{49}. For simple differences in Example 8.2, this number is always* 14.54, so we could calculate that value and then just see whether or not $|\bar{x}_i - \bar{x}_j|$ exceeds 14.54 to judge whether or not there is a discernible difference between μ_i and μ_j. If the difference does exceed 14.54, then the CI could not include 0; otherwise, it does. Thus, since $\bar{x}_1 - \bar{x}_3 = 103.71 - 89 = 14.71$, we can see that μ_1 and μ_3 are discernible; $\bar{x}_2 - \bar{x}_3 = 3$, on the other hand, is not greater than 14.54, so μ_2 and μ_3 are not discernible. Likewise, $\bar{x}_1 - \bar{x}_2 = 11.71$ does not quite make it either. To compare only differences, then, when the sample sizes are all equal, it is only necessary to construct a single CI using the program. Then, retrieve the value of e from R_{49} and simply compare numerical differences $|\bar{x}_i - \bar{x}_j|$ to decide which are significant. The results may be conveniently summarized in a triangular table like Table 8.2, where numerical differences are entered in the appropriate cells and e is recorded for comparison.

TABLE 8.2. SUMMARY OF DIFFERENCES WHEN $n_i \equiv n$

	$\bar{x}_1$	$\bar{x}_2$	$\bar{x}_3$
$\bar{x}_1$	0	11.71	14.71
$\bar{x}_2$		0	3.00
$\bar{x}_3$			0

$e = 14.56$

Do not be surprised if AOV should reject H_o, and yet you are unable to find any significant differences. As we remarked, rejection of H_o merely means the CI for *some* contrast must fail to include 0, but that contrast need not be a simple difference. The data themselves may not be distinct enough to separate out such differences. But, as you know, this does not mean such differences do not exist, anymore than does establishing such differences prove that they do, much less are important detections. Here is an example that shows this apparent contradiction.

EXAMPLE 8.3

The following data were supplied by a consumer protection agency studying the average life of four brands of transistors (in 100 hours) with sample averages recorded.

* Since $c_i^2 = 1$ and $n_i \equiv 7$

Brand 1:	12	15	18	10	$\bar{x}_1 = 13.75$
Brand 2:	14	17	12	19	$\bar{x}_2 = 15.50$
Brand 3:	21	19	20	23	$\bar{x}_3 = 20.75$
Brand 4:	14	21	25	20	$\bar{x}_4 = 20.00$

Processing these data, first through ST–06, and then through ZS–8, produces the following AOV Table.

AOV TABLE

Source	Sum of squares	d.f.	MSS	F	P
Rows	139.5	3	46.5	4.09	.03
Error	136.5	12	11.375		
Total	276.0	15			

Thus, $H_o\!:\!\mu_1 = \mu_2 = \mu_3 = \mu_4$ should be rejected at the default level of 5%. Using $F_{.05} = 3.49$ for d.f. $= (3,12)$, a CI for $\mu_1 - \mu_2$ is $(-9.47, 5.97)$ with a factor of $e = 7.72$ found in R_{49}. The table of numerical (unsigned) differences is

	$\bar{x}_1$	$\bar{x}_2$	$\bar{x}_3$	$\bar{x}_4$
$\bar{x}_1$	0	1.75	7.00	6.25
$\bar{x}_2$		0	5.25	4.50
$\bar{x}_3$			0	0.75
$\bar{x}_4$				0

[e=7.72]

and you see that no CI would fail to include 0. There are no detectable significant differences. However, you will find that the CI for the contrast $\bar{x}_1 + \bar{x}_2 - \bar{x}_3 - \bar{x}_4$ does fail to include 0, that is, $\mu_1 + \mu_2$ is discernible from $\mu_3 + \mu_4$.

You should now be able to solve the following problems related to both of these last two sections.

Problems

For each of the following, test H_o, the hypothesis of equality of means, and examine CI's for differences, as well as any other contrasts that might be called for. The appropriate AOV tables are shown following problems 6–19.

6. Four different fertilizers all claim to be best for maximum plant growth. After subjecting a number of plants to the four different

types, plant growth in inches after a certain period of time was measured and is reported below.

										$\bar{x}$
Fertilizer 1:	4.3	4.2	4.1	2.9	3.6	3.8				3.82
Fertilizer 2:	3.9	4.1	4.2	4.6	3.2	3.5	3.4			3.84
Fertilizer 3:	4.1	4.2	4.6	3.3						4.05
Fertilizer 4:	3.8	4.1	4.4	3.8	3.9	3.4	3.7	4.4	4.3	3.98

Ans: $F = .32$, $P = .81$. No significant differences. All CI's include 0.

7. Samples were taken from the coffee machines at four different locations in a building to see if they were uniform in yielding 8 ounces of liquid.

						$\bar{x}$
Machine A:	8.05	7.72	7.42	7.52	8.02	7.746
Machine B:	7.81	7.52	7.62	8.01	8.09	7.810
Machine C:	8.04	8.02	7.93	7.86	8.01	7.972
Machine D:	7.71	7.91	7.81	8.06	8.02	7.902

Ans: $F = 1.18$, $P = .35$. No significant difference ($e = .40$, $max\Delta = .23$).

8. To see whether or not there is a significant difference in the type of material used in a kitchen counter top, three different types of material were used and subjected to light at six different locations. The following data are given in measured foot-candles on the surface.

							$\bar{x}$
Material #1:	41	48	48	43	41	41	43.67
Material #2:	44	37	39	37	44	45	41.00
Material #3:	49	45	40	47	44	37	43.67

Ans: $F = .93$, $P = .42$. No significant differences ($e = 6.12$, $max\Delta = 2.67$).

9. The following data represent crop yields in bushels corresponding to different soil treatments. Is there any significant ($\alpha = .05$) difference in treatments?

						$\bar{x}$
Treatment A:	28.7	30.6	47.8	27.2	42.3	35.32
Treatment B:	20.9	25.3	33.0	36.1	37.2	30.50
Treatment C:	25.9	28.6	37.5	26.2	37.8	31.20

Ans: $F = .60$, $P = .57$. No significant differences ($e = 13.3$, $max\Delta = 4.82$).

10. The recruiting division of a very large national company is divided into five regions. In 1979, the region's productivity was sampled with the following results in thousands of recruits.

						$\bar{x}$
Northeast:	4.0	4.6	7.0	5.1	3.0	4.74
Southeast:	4.3	3.4	6.0	5.0	5.1	4.76
Midwest:	3.2	3.7	3.9	4.2	3.3	3.66
Southwest:	4.5	5.0	4.9	7.0	4.2	5.12
Western:	5.0	5.1	5.3	5.0	5.1	5.10

Ans: $F = 1.96$, $P = .14$. No significant differences ($e = 2.04$, $max\Delta = 1.46$).

11. Real estate agents for three sectors of the country were sampled independently to determine annual income. The following are the sample results in thousands of dollars.

						$\bar{x}$
East:	25	20	28	35	29	27.4
Central:	27	19	26	27	28	25.4
West:	26	20	27	29	30	26.4

Test H_0 at the 5% level of significance.

Ans: Do not reject H_0. $F = .25$ $P = .78$ ($e = 7.82$, $max\Delta = 2$).

12. Samples were taken from the Army, Air Force, and Navy to measure the strength of members of the various services with the following results. Test the null hypothesis.

							$\bar{x}$
Army:	88	92	93	87	80	85	87.50
Air Force:	91	86	81	84	78	86	84.33
Navy:	89	88	90	85	79	86	86.17

Ans: $F = .77$, $P = .48$. ($e = 6.93$, $max\Delta = 3.17$). Do not reject H_0.

13. Ten students from each of three curriculums at a large university were randomly selected to measure grade point averages with the following results. Test the null hypothesis.

											$\bar{x}$
English:	3.5	3.6	3.1	3.2	3.0	2.9	3.6	3.7	3.4	3.1	3.31
Business:	3.8	3.7	3.6	4.0	3.9	3.5	3.8	3.4	3.6	3.2	3.65
Mathematics:	3.0	3.1	2.9	3.2	3.4	3.9	3.2	3.1	3.6	3.4	3.28

Ans: Reject at $\alpha \geq .01$. With 95% confidence, μ_1 and μ_2, as well as μ_2 and μ_3 differ, but not μ_1 and μ_3.

14. Three watchmakers were observed to determine how long it took them to assemble one component of a watch. The record of assembly times, in minutes, is as follows. Test H_0.

						$\bar{x}$
Watchmaker 1:	4.0	4.5	4.0	4.2	4.1	4.16
Watchmaker 2:	4.6	4.5	4.6	4.5	5.0	4.64
Watchmaker 3:	4.2	4.1	4.0	4.5	4.5	4.26

Ans: Reject H_0. $F = 6.92$, $P = .01$. With $e = .38$, discernible differences are $\mu_1 - \mu_2$, $\mu_2 - \mu_3$, but not $\mu_1 - \mu_3$.

15. The manager of a large production plant is concerned that machines in one division are producing products of a nonuniform weight. He samples products from three machines, with the following results (in ounces). Determine whether or not the machines are distinguishable in terms of their output.

						$\bar{x}$
Machine 1:	28.24	30.16	29.25	29.78	29.75	29.44
Machine 2:	29.45	29.65	29.70	29.30	29.00	29.42
Machine 3:	28.75	29.70	29.00	29.25	29.45	29.23

Ans: $F = .26$, $P = .78$. No significant differences with $e = .89$.

16. The superintendent of a large school district requested a sample from each of 5 schools within his district of fifth grade student performance on a standard English proficiency test. His independent sample produced the following scores. Do schools differ?

							$\bar{x}$
School 1:	88	87	89	70	96	85	85.83
School 2:	82	85	84	74	91	81	82.83
School 3:	87	89	88	72	94	83	85.50
School 4:	60	72	76	81	59	74	70.33
School 5:	75	78	79	82	81	87	80.33

Ans: Reject H_0. $F = 4.69$, $P = .006$. With $e = 13.82$, School 4 clearly differs from Schools 1 and 3. No discernible differences among the other possible pairs.

17. A toymaker whose toys all require batteries for operation wants to determine which brand of battery to recommend for use with his toys. A random sample of three popular brands yielded the following results (in hours). Is any outstanding?

						$\bar{x}$
Brand 1:	26	26	28	27	30	27.4
Brand 2:	25	24	29	23	25	25.2
Brand 3:	24	27	28	21	22	24.4

Ans: $F = 2.09$, $P = .17$. With $e = 4.24$, no discernible differences.

18. A test was conducted by a psychologist to determine if there was a difference in the cardiovascular fitness of individuals who ran less than 20 miles per week, those who didn't run, and those who ran more than 20 miles per week. His independent random sample yielded the following cardiovascular test scores. Are there any significant differences?

											$\bar{x}$
No running:	33.5	40.0	44.0	47.0	59.5	55.0	62.0	63.0	69.5	58.0	53.15
Less than 20 miles/week:	56.5	58.0	44.5	86.5	55.0	59.5	56.5	61.0	65.5	87.0	63.0
More than 20 miles/week:	91.5	96.5	93.5	82.0	96.5	79.0	85.0	99.5	86.5	80.0	89.0

Ans: Reject H_o. $F = 28$; $P = 3 \times 10^{-7}$. Those who run more than 20 miles per week differ from each of the other two groups, but they in turn are indiscernible.

19. Process the data for the example on page 4–22 of ASM to construct an AOV Table.

20. In a comparison of four balances to see if there were differences in their ability to weigh materials, five samples of material were tested on each machine with averages $\bar{x}_1 = 73.8$, $\bar{x}_2 = 74.4$, $\bar{x}_3 = 78.8$, and $\bar{x}_4 = 71.8$. A value of ESS = 10.4 was reported. Which mean differences are significant at level 5%?

Ans: d.f. = (3, 16), MESS = .65, $e = 1.6$. The discernible differences are $\mu_1 - \mu_3$, $\mu_1 - \mu_4$, $\mu_2 - \mu_3$, $\mu_2 - \mu_4$, and $\mu_3 - \mu_4$, using $F_{.05} = 3.3$.

AOV TABLES FOR PROBLEMS 6–19

6. AOV TABLE

Source	Sum of squares	d.f.	MSS	F	P
Rows	.2028	3	.0676	.32	.81
Error	4.691	22	.2132		
Total	4.894	25			

e will vary with each interval

7. AOV TABLE

Source	Sum of squares	d.f.	MSS	F	P
Rows	.1489	3	.0496	1.18	.35
Error	.6711	16	.0419		
Total	.8200	19			

$e = .40$
$F_{.05} = 3.20$

8. AOV TABLE

Source	Sum of squares	d.f.	MSS	F	P
Rows	28.44	2	14.22	.93	.42
Error	228.67	15	15.24		
Total	257.11	17			

$e = 6.12$

9. AOV TABLE

Source	Sum of squares	d.f.	MSS	F	P
Rows	67.83	2	33.91	.60	.57
Error	681.31	12	56.78		
Total	749.14	14			

$e = 13.29$

10. AOV TABLE

Source	Sum of squares	d.f.	MSS	F	P
Rows	7.102	4	1.775	1.95	.14
Error	18.184	20	.9092		
Total	25.29	24			

$e = 2.04$

11. AOV TABLE

Source	Sum of squares	d.f.	MSS	F	P
Rows	10	2	5.00	.25	.78
Error	235.6	12	19.63		
Total	245.6	14			

$e = 7.82$

12. AOV TABLE

Source	Sum of squares	d.f.	MSS	F	P
Rows	30.33	2	15.17	0.77	.48
Error	293.67	15	19.58		
Total	324.00	17			

$e = 6.93$

13. AOV TABLE

Source	Sum of squares	d.f.	MSS	F	P
Rows	.8447	2	.4223	5.51	.01
Error	2.07	27	.0767		
Total	2.9147	29			

$e = .32$
$F_{.05} = 3.3$

14. AOV TABLE

Source	Sum of squares	d.f.	MSS	F	P
Rows	.641	2	.32	6.92	.01
Error	.556	12	.046		
Total	1.197	14			

$e = .38$

15. AOV TABLE

Source	Sum of squares	d.f.	MSS	F	P
Rows	.1313	2	.066	.26	.78
Error	3.082	12	.257		
Total	3.213	14			

$e = .89$

16. AOV TABLE

Source	Sum of squares	d.f.	MSS	F	P
Rows	967.13	4	241.78	4.69	.006
Error	1287.83	25	51.51		
Total	2254.97	29			

$e = 13.82$
$F_{.05} = 2.78$

17. AOV TABLE

Source	Sum of squares	d.f.	MSS	F	P
Rows	24.13	2	12.07	2.09	.17
Error	69.20	12	5.77		
Total	93.33	14			

$e = 4.24$

18. AOV TABLE

Source	Sum of squares	d.f.	MSS	F	P
Rows	6860.82	2	3430.41	27.58	3×10^{-7}
Error	3358.53	27	124.39		
Total	10219.34	29			

$e = 12.91$
$F_{.05} = 3.35$

19. AOV TABLE

Source	Sum of squares	d.f.	MSS	F	P
Rows	47.5	2	23.75	2.63	.09
Error	216.5	24	9.02		
Total	264	26			

When data are classified in two ways, the housekeeping and analysis becomes a little more complicated, but only slightly, provided the data are in the right form. In addition to row effects as before, there are now column effects to take into account also. The data requirements are that there be just one observation per combination of row and column. Assuming R rows and C columns, the random variables involved have the following assumptions (similar to Equation 8.2.)

$$X_{ij} = \mu_{ij} + E_{ij} \qquad \begin{array}{l} i = 1, 2,\ldots,R \\ j = 1, 2,\ldots,C \end{array} \tag{8.13}$$

Here, μ_{ij} is the $(i - j)$-th mean, and, as before, the E_{ij}'s are independent random variables, each having a normal distribution with mean 0 and common variance σ^2. But, now,* if we let

$$\mu_{i.} = \frac{1}{C} \sum_{j=1}^{C} \mu_{ij}$$

be the average mean value across the i-th row,

$$\mu_{.j} = \frac{1}{R} \sum_{i=1}^{R} \mu_{ij}$$

be the average of the j-th column means, and

$$\mu_{..} \equiv \mu = \frac{1}{RC} \sum_{i=1}^{R} \sum_{j=1}^{C} \mu_{ij} = \frac{1}{R} \sum_{i=1}^{R} \mu_{i.} = \frac{1}{C} \sum_{j=1}^{C} \mu_{.j}$$

be the average of all the population means, then several things may be noted. First of all, these row and column means are estimated by the corresponding row and column data means, namely, $\hat{\mu}_{i.} = \bar{x}_{i.}$, $\hat{\mu}_{.j} = \bar{x}_{.j}$, and, not surprisingly,

$$\hat{\mu}_{..} = \bar{\bar{x}} = \frac{1}{RC} \sum_{i=1}^{R} \sum_{j=1}^{C} x_{ij} = \bar{x}_{..}$$

As in the last section, we may then rewrite the model (analogous to Equation 8.3) as

$$X_{ij} = \mu + (\mu_{i.} - \mu) + (\mu_{.j} - \mu) + E_{ij} \tag{8.14}$$

* It is convenient here to adopt the rather common device of replacing a subscript by a dot to indicate that the sum over that variable was taken to define the quantity in question.

displaying that portion of the observation, $\delta_i = \mu_{i.} - \mu$, that is peculiar to the i-th row, and that portion, $\gamma_j = \mu_{.j} - \mu$, that is peculiar to the j-th column. Ultimately, we will want confidence intervals and contrasts defined over rows and columns. But, for now, we are mainly interested in testing two sets of null hypotheses.

$$H_o{:}\mu_{1.} = \mu_{2.} = \ldots = \mu_{R.} \quad \text{or, equivalently,} \qquad \qquad (8.15)$$
$$H_o{:}\delta_i \equiv 0$$

the hypothesis that there are no differences in row means, and

$$H_o{:}\mu_{.1} = \mu_{.2} = \ldots = \mu_{.C} \quad \text{or, equivalently,} \qquad \qquad (8.16)$$
$$H_o{:}\gamma_j \equiv 0$$

the hypothesis that there are no differences in column means. Basically, these two tests are just versions of what we did in the last section. Only, now we will need an added sum of squares term, CSS, to account for column effects, thus leaving a smaller residual sum of squares, ESS, as a common basis for two separate F-tests. When both hypotheses are true, the random variables are just a random sample of size $N = RC$ from the same normal distribution.

Some of the details regarding these sums of squares are discussed very briefly in ASM on page 4–24, and we will try to stay within the notation adopted there. The companion program, ST–16, of the statistics module will be exploited also. You see that deviations are to be taken from an overall mean, $\bar{x}$, the average of all the RC data values observed. There will, accordingly, be R row means $\bar{x}_{1.}, \bar{x}_{2.}, \ldots, \bar{x}_{R.}$, and C column means $\bar{x}_{.1}, \bar{x}_{.2}, \ldots, \bar{x}_{.C}$. Then, for this array, the various sums of squares are

$$\text{RSS} = C\sum_{i=1}^{R}(\bar{x}_{i.} - \bar{x})^2 \qquad \text{MRSS} = \frac{\text{RSS}}{(R-1)}$$

$$\text{CSS} = R\sum_{j=1}^{C}(\bar{x}_{.j} - \bar{x})^2 \qquad \text{MCSS} = \frac{\text{CSS}}{(C-1)} \qquad \qquad (8.17)$$

$$\text{ESS} = \text{Res} = \sum_{i=1}^{R}\sum_{j=1}^{C}(x_{ij} - \bar{x}_{i.} - \bar{x}_{.j} + \bar{x})^2$$
$$\text{MESS} = \frac{\text{ESS}}{(R-1)(C-1)}$$

These are bases for a 2-way AOV table, which would appear as in Table 8.3. As you can see, there will now be two F-ratios to test

TABLE 8.3. TWO-WAY AOV TABLE

Source	Sum of squares	d.f.	MSS	F	PV
Rows	RSS	R-1	MRSS	F_R	PV_R
Columns	CSS	C-1	MCSS	F_C	PV_C
Residual	ESS	(R-1)(C-1)	MESS		
Total	SS	RC-1			

for respective row and column effects. They are

$$F_R = \frac{\text{MRSS}}{\text{MESS}} \quad \text{with d.f.} = [R - 1, (R-1)(C-1)]$$
$$F_C = \frac{\text{MCSS}}{\text{MESS}} \quad \text{with d.f.} = [C - 1, (R-1)(C-1)] \tag{8.18}$$

Our first task is to enter the raw data in such a way as to ultimately output these quantities. We use ST–06 again for data entry, but this time the two-way form listed on page 3–11 of ASM. If you will refer to that manual, several differences will be noted. First of all, you initialize with [E], not [E′] this time, and the very next step is to inform the calculator of the dimension of the problem by entering the number of rows with [B] and the number of columns with [C]. There is a restriction regarding the number of rows and columns, namely, $R + C$ must never be greater than 16. (Some printings of ASM say 15, but that is an error that has been corrected in later versions.) This restriction is minor, however, and will not be a consideration for the problems in this (and most) textbooks. You may not repartition the calculator, because of the number of steps in ZS–8. If necessary, use Step 7b whenever the data bank becomes full, that is, press [D], and continue entering data.

The next step is to start inputting the data by rows, and this time you do not—repeat *not*—press [B′] and [C′] after each row, as before. You just use Step 6 for each successive row, and press [B′] and [C′] *after* all the data are entered. When that step is completed, you must press [RST] to exit ST–06, lower all flags, and prepare for data compilation in ZS–8 to produce the AOV table entries. The necessary program has been coded there under AOV–2.

To construct the AOV table, refer to Step 2 of AOV–2 of the user instructions for ZS–8. Unlike the one-way case, we did not—and cannot, as a matter of fact—conveniently obtain either row means $\bar{x}_{i.}$ or column means $\bar{x}_{.j}$ as part of the data entry in ST–06. If we are ultimately going to examine contrasts, which we would naturally want to do if we reject either null hypothesis, then we had better compute and record these. Step 2 of AOV–2 has been conveniently arranged for doing this.

First, the row means are produced by pressing [B] and following with $(R - 1)$ successive [R/S] presses. Next, you generate the column means, starting at the left by pressing [B′] and following with $(C - 1)$ [R/S] presses. If you should lose your place for any reason, you may begin anew with either [B] or [B′]. They operate independently of each other and may be repeated at any time, provided the program has been completed once, and the appropriate flags are set before each repetition (see user instructions). Having completed that step, you may, then, as with the 1-way case, output the various sums of squares and d.f. terms row-by-row in the AOV table by pressing [C] and continuing with [R/S] presses. Again, you may repeat this display, provided it has been completed once. (You should be sure that flag 5 is set before any such repetition, however.)

EXAMPLE 8.4

An experiment was conducted to examine the influence of school location, as well as teachers, in a school district. Three schools were matched with five teachers, a unit of instruction given, and average scores on a standard scale with mean supposedly at 50 points recorded as follows:

	TEACHERS				
SCHOOLS	A	B	C	D	E
I	53	47	46	50	49
II	61	55	52	58	54
III	51	51	49	54	50

Examine the results for significant differences.

Solution Following the instructions for ST–06, initialize the program by pressing [E]. Next, enter 3, press [B], and 5 press [C]. Next come the data 53, 47, 46, 50, 49 of the first row entered successively by pressing [A]. You should see a running count as you go along, ending with 5 in the display. Without pausing, enter the data 61, 55, 52, 58, 54 in the same way, noting that the running count begins anew with 1 and ends in 5. Proceed similarly with the data of the third row, 51, 51, 49, 54, 50. Take it easy, and remember not to attempt to enter a value until the running count appears.

It is at this point that you press [B′] and [C′], yielding, for the example, 52 and 14.9333... . The first is the value of $\bar{x}$, of minor interest, and the second is an overall MSD, in which we have no interest at all, except that it happens to be the same as RCSS and is needed for the IDB. But, the latter is stored in R_{07}, in any case, and we will output what is needed when we construct the AOV table. You must press [RST], as per instruction 16, to exit ST–06 and lower all flags. At this point, we want to generate and record the various row and column means, starting with the former. So, press [B] and observe $\bar{x}_{1.} = 49$, in R_D. Record this, and proceed through [R/S] presses to find $\bar{x}_{2.} = 56$, and finally, $x_{3.} = 51$. That takes care of the 3 row means. Now, find column means by pressing [B′] to see $\bar{x}_{.1} = 55$. Successive presses of [R/S] produce $\bar{x}_{.2} = 51$, $\bar{x}_{.3} = 49$, $\bar{x}_{.4} = 54$, and $\bar{x}_{.5} = 51$. To construct the AOV table, press [C] to see RSS and successive [R/S] presses to find and verify the following AOV table.

AOV TABLE

Source	Sum of squares	d.f.	MSS	F	P
Rows	130	2	65	23.6	.0004
Column	72	4	18	6.5	.012
Residual	22	8	2.75		
Total	224	14			

At the 5% level of significance, both null hypotheses would be rejected, indicating that both teachers *and* schools have an effect on the scores. Which ones?

To answer the last question posed in the example, we turn to Scheffe confidence intervals. Procedures have been worked out to cover this 2-way case also. In fact, the formulas are quite similar to the 1-way case. For row means, the CI for any contrast

$$\sum_{i=1}^{R} c_i \mu_{i.}$$

is given by

$$\sum_{i=1}^{R} c_i \bar{x}_{i.} \pm \sqrt{(R-1)F_\alpha \text{ MESS} \sum_{i=1}^{R} \frac{c_i^2}{C}}$$

$$\text{d.f.} = [R-1, (R-1)(C-1)] \tag{8.19}$$

For columns, the CI for the contrast

$$\sum_{j=1}^{C} c_j \mu_{.j}$$

would be

$$\sum_{j=1}^{C} c_j \bar{x}_{.j} \pm \sqrt{(C-1)F_\alpha \text{MESS} \sum_{j=1}^{C} \frac{c_j^2}{R}}$$

$$\text{d.f.} = [C-1, (R-1)(C-1)] \tag{8.20}$$

These formulas are programmed for you in AOV–2 at Step 4. After initializing by pressing [E] this time, the contrast data are entered as pairs c_i and $\bar{x}_{i.}$ (or $\bar{x}_{.j}$), in that order. Then, simply look up F_α with the right d.f., enter it, and press [D] or [D′], as the case may be, and the interval (l, u) will be discovered in the usual manner.

EXAMPLE 8.5

Continuing with the data of Example 8.4, examine rows and columns for significant differences.

Solutions If data are no longer in your calculator, it is only necessary to store $R - 1 = 2$ in R_{58}, $C - 1 = 4$ in R_{59}, and MESS $= 2.75$ from the AOV table in R_{48}. Then, if you will initialize by pressing [E], you may find the CI for, say, $\mu_{1.} - \mu_{2.}$, as follows. Input 1, press [R/S], then $49 = \bar{x}_{1.}$ and [R/S] to see a count of 1 in R_D. Follow this with -1, [R/S], then $56 = \bar{x}_{2.}$ and [R/S] for a count of 2. Since we are dealing with row differences, d.f. $= (2,8)$ and $F_{.05} = 4.46$. Enter this, and press [D] to find $l = -10.13$, followed by [$x \leftrightarrows t$], to discover $u = -3.87$. Since this 95% CI for $\mu_{1.} - \mu_{2.}$ fails to include 0, conclude that there is a difference between Schools I and II. Now, from R_{49}, you will find that $e = 3.13$ was used in this calculation, so any row difference that exceeds this quantity in absolute value will detect a significant difference in population row means. Using this guide, you should be able to verify that Schools II and III also differ, but not I and III.

To compare teachers, first compute a 95% CI for, say, $\mu_{.1} - \mu_{.3}$. Press [E] again, and follow with the sequence 1, [R/S], 55, [R/S], -1, [R/S], 49, [R/S]. This time d.f. $= (4,8)$, and $F_{.05} = 3.84$. Hence $.69 < \mu_{.1} - \mu_{.3} < 11.31$ with confidence 95%. There is a significant difference between teachers A and C. With $e = 5.31$ from R_{49} used for comparison, you should be able to discover that they are the only pair for which such a difference can be detected.

Of course, it is possible to find CIs for other contrasts, as well, although we are typically interested in differences. Again, the latter may fail to show up, even when one or the other set of hypotheses is rejected by the *F*-test.

There are many more extensions of AOV to cover a surprisingly wide variety of applications. Each time, the theme is the same however. The total sum of squares is partitioned into parts corresponding to each of the effects being modeled. The remainder, the residual sum of squares, is a common denominator used to construct one-sided *F*-tests of hypotheses concerning whether or not the classification under study has any real effect on the observations. Naturally, each such complication adds to the housekeeping, though, in a natural way, and creates a new burden on computing capacity. We have limited our treatment to those standard models found in most elementary textbooks in keeping with the theme adopted throughout this book. With this much development, and practice with the problems that follow, you should have a solid basis for implementing and interpreting more sophisticated models, and you will often resort to a computer, then, for computation.

Problems

For each of the following problems, construct the two-way AOV table, and then determine significant differences ($\alpha = .05$) by finding e for each of the two classifications. Correct tables follow problems 21–30.

21. The following table represents crop yield in bushels corresponding to four different fertilizers used in four different counties.

FERT \ COUNTY	A	B	C	D	$\bar{x}_{i.}$
I	8	9	11	12	10
II	8	10	10	12	10
III	10	6	7	5	7
IV	9	6	7	6	7
$\bar{x}_{.j}$	8.75	7.75	8.75	8.75	

Ans: No significant differences among rows or columns.

22. A sample of 10 European families reported the following hourly wage rates by region and sex.

SEX \ REGION	NORTH	SOUTH	EAST	WEST	$\bar{x}_{i.}$
Male	4.80	4.20	4.30	4.70	4.5
Female	4.20	4.00	3.90	4.15	4.06
$\bar{x}_{.j}$	4.5	4.1	4.1	4.425	

Ans: Males differ from females; no differences among regions.

23. The manager of a production operation which uses four identical machines, each with 1 operator, has been concerned that the output varies among his operators. He conducts a test with each operator using each one of the machines with the following results.

MACHINE \ OPERATOR	1	2	3	4	$\bar{x}_{i.}$
1	21	22	21	23	21.75
2	22	24	22	22	22.50
3	23	25	26	25	24.75
4	21	24	22	23	22.50
$\bar{x}_{.j}$	21.75	23.75	22.75	23.25	

Ans: Significant differences between rows 1,3; 2,3; 3,4 but none among columns.

24. An experiment was designed to test four different jet fuels and three different fuel injection systems, with the following consumption results.

INTAKE SYSTEM \ FUEL TYPE	1	2	3	4	$\bar{x}_{i.}$
1	4.40	4.55	4.60	4.55	4.53
2	4.20	4.30	4.55	4.35	4.35
3	4.15	4.20	4.20	4.40	4.24
$\bar{x}_{.j}$	4.25	4.35	4.45	4.43	

Ans: Significant differences between rows 1, 3; no others.

25. The data tabled below are coded observations corresponding to yields for three different varieties of tomatoes, on which two competing brands of spray were used. Is there any significant difference in sprays?

SPRAY \ VARIETY	#1	#2	#3	$\bar{x}_{i.}$
A	7	9	5	7
B	6	5	8	6.33
$\bar{x}_{.j}$	6.5	7	6.5	

Ans: No significant differences for either sprays or varieties.

26. The pilot study of Problem 4 was later extended to compare departments by class levels, as well. The data reported are grade point averages.

DEPARTMENT \ LEVEL	FRESHMAN	SOPHOMORE	JUNIOR	SENIOR	$\bar{x}_{i.}$
Physics	1.72	1.82	2.04	2.49	2.02
Statistics	1.99	2.22	2.18	2.16	2.14
Chemistry	1.92	2.02	2.56	2.53	2.26
Mathematics	1.82	1.87	2.18	2.41	2.07
Biology	1.95	1.97	2.12	2.20	2.06
$\bar{x}_{.j}$	1.88	1.98	2.22	2.36	

Ans: There are no significant differences between departments. As for grade levels, freshmen differ from juniors and seniors, sophomores from seniors, while juniors differ only from freshmen.

27. Grade point averages for a required course given each of 3 quarters by 4 different professors were recorded 1 year as follows.

PROFESSOR \ QUARTER	FALL	WINTER	SPRING	$\bar{x}_{i.}$
A	2.00	1.95	1.83	1.93
B	1.98	2.11	1.94	2.01
C	2.02	1.97	1.98	1.99
D	2.07	2.09	2.08	2.08
$\bar{x}_{.j}$	2.02	2.03	1.96	

Ans: It appears not to matter which professor is selected nor which quarter the course is taken.

28. The following data are coded observations that represent output for different lathe operators in a machine shop.

OPERATOR \ LATHE	A	B	C	$\bar{x}_{i.}$
Roy	30	33	37	33.33
Frank	25	26	23	24.67
John	25	30	30	28.33
$\bar{x}_{.j}$	26.67	29.67	30	

Ans: Frank and Roy differ, but no other differences are detectable.

29. Two types of smog devices were tested on four brands of automobiles, with the coded results shown below.

DEVICE \ AUTOMOBILE	A	B	C	D	$\bar{x}_{i.}$
Type A	72	77	65	75	72.25
Type B	70	75	62	70	69.25
$\bar{x}_{.j}$	71.0	76.0	63.5	72.5	

Ans: The two devices differ significantly and automobiles A, B, and D differ from C.

30. Use the Data of Table 4.6 on page 4–26 of ASM, and test for differences among diets and among litters.

Ans: Rows 1,2 and 2,3 differ significantly. No significant differences among the columns.

AOV TABLES FOR PROBLEMS 21 THROUGH 30

21. AOV TABLE

Source	Sum of squares	d.f.	MSS	F	P	
Rows	36	3	12	3.09	.08	$e_R = 4.75$
Column	3	3	1	.26	.85	$e_C = 4.75$
Residual	35	9	3.89			
Total	74	15				

22. AOV TABLE

Source	Sum of squares	d.f.	MSS	F	P	
Rows	.3828	1	.3828	23.71	.02	$e_R = .29$
Column	.2684	3	.0895	5.54	.10	$e_C = .67$
Residual	.0484	3	.0161			
Total	.6997	7				

23. AOV TABLE

Source	Sum of squares	d.f.	MSS	F	P	
Rows	20.25	3	6.75	9.0	.005	$e_R = 2.08$
Column	8.75	3	2.92	3.9	.049	$e_C = 2.08$
Residual	6.75	9	.75			
Total	35.75	15				

24. AOV TABLE

Source	Sum of squares	d.f.	MSS	F	P	
Rows	.1679	2	.0840	10.33	.01	$e_R = .20$
Column	.0756	3	.0252	3.10	.11	$e_C = .28$
Residual	.0488	6	.0081			
Total	.2923	11				

25. AOV TABLE

Source	Sum of squares	d.f.	MSS	F	P	
Rows	.6667	1	.6667	.11	.77	$e_R = 8.72$
Column	.3333	2	.1667	.03	.97	$e_C = 15.3$
Residual	12.3333	2	6.1667			
Total	13.3333	5				

26. AOV TABLE

Source	Sum of squares	d.f.	MSS	F	P	
Rows	.14063	4	.03516	1.51	.26	$e_R = .39$
Column	.71266	3	.23755	10.21	.001	$e_C = .31$
Residual	.27917	12	.02326			
Total	1.132455	19				

27. AOV TABLE

Source	Sum of squares	d.f.	MSS	F	P	
Rows	.0359	3	.012	3.48	.09	$e_R = .18$
Column	.0120	2	.006	1.75	.25	$e_C = .13$
Residual	.0207	6	.003			
Total	.0686	11				

28. AOV TABLE

Source	Sum of squares	d.f.	MSS	F	P	
Rows	113.56	2	56.78	8.81	.03	$e_R = 7.72$
Column	20.22	2	10.11	1.57	.31	$e_C = 7.72$
Residual	25.78	4	6.44			
Total	159.56	8				

29. AOV TABLE

Source	Sum of squares	d.f.	MSS	F	P
Rows	18.0	1	18.0	18.0	.02
Column	166.5	3	55.5	55.5	.004
Residual	3.0	3	1.0		
Total	187.5	7			

$e_R = 2.25$

$e_C = 5.28$

30. AOV TABLE

Source	Sum of squares	d.f.	MSS	F	P
Rows	51.082	2	25.54	8.13	.004
Column	54.887	8	6.86	2.18	.09
Residual	50.258	16	3.14		
Total	156.227	26			

$e_R = 2.24$
$F_{.05} = 3.60$
$e_C = 6.60$
$F_{.05} = 2.60$

$\bar{x}_{1.} = 17.17$ $\bar{x}_{2.} = 19.44$ $\bar{x}_{3.} = 16.16$

$\bar{x}_{.1} = 15.9$ $\bar{x}_{.2} = 15.6$ $\bar{x}_{.3} = 17.8$ $\bar{x}_{.4} = 19.2$ $\bar{x}_{.5} = 17.8$

$\bar{x}_{.6} = 18.0$ $\bar{x}_{.7} = 19.7$ $\bar{x}_{.8} = 18.6$ $\bar{x}_{.9} = 15.7$

STEP	PROCEDURE	ENTER	PRESS	DISPLAY
AOV–1	*One-way Analysis of Variance*			
1	a. Enter data using ST–06 (one-way). *Note*: Record each row mean.			
	b. Exit ST–06		[RST]	
2	*Calculate AOV Table Entries.* *Note*: These steps may be repeated any time.		[A]	RSS
			[R/S]	$K - 1$
			[R/S]	MRSS
			[R/S]	ESS
			[R/S]	$N - K$
			[R/S]	MESS
			[R/S]	SS
			[R/S]	$N - 1$
			[R/S]	F
			[R/S]	P
3	*Confidence Intervals for Contrasts* (After Step 2)			
	a. Initialize.		[E′]	0
	b. Enter contrast data (Repeat for each i; ignore any $c_i = 0$.)	c_i	[R/S]	c_i
		$\bar{x}_i$	[R/S]	$\bar{x}_i$
		n_i	[R/S]	i
	c. Enter F-percentile with d.f. = $(K - 1, N - K)$, and calculate CI.	F_α	[A′]	l
			[$x \rightleftharpoons t$]	u
	Note 1: Steps 3abc may be repeated. *Note 2*: These steps are also valid, if the contents of R_{14} and R_{48} below are manually stored.			

REGISTER CONTENTS

00		10	Used	20	Used	30	F	40		50	$\Sigma c_i \bar{x}_i$
01	Used	11	Used	21	by	31	P	41		51	Last c_i
02	Used	12	SS	22	ST–22	32		42		52	Last $\bar{x}$
03		13	ESS	23		33		43		53	$\Sigma c_i^2/n_i$
04		14	K-1	24		34		44		54	Used
05		15		25		35		45		55	
06	Last $\bar{x}$	16		26		36		46		56	
07	Used	17		27		37		47		57	
08	RSS	18		28		38		48	MESS	58	
09	N	19		29		39		49	e	59	

STEP	PROCEDURE	ENTER	PRESS	DISPLAY
AOV–2	*Two-Way Analysis of Variance*			
1	a. Enter data using ST–06 (two-way) ($R + C \leq 16$).			
	b. Exit ST–06		[RST]	
2	*Calculate Row and Column Means*			
	a. Calculate row means (record).		[B]	$\bar{x}_{1.}$
			[R/S]	$\bar{x}_{2.}$
			.	.
			.	.
			.	.
	Note: To repeat, be certain that Flag 3 is set.		[R/S]	$\bar{x}_{R.}$
	b. Calculate column means (record).		[B′]	$\bar{x}_{.1}$
			[R/S]	$\bar{x}_{.2}$
			.	.
			.	.
			.	.
	Note: To repeat, be certain that Flag 4 is set.		[R/S]	$\bar{x}_{.C}$
3	*Calculate AOV Table Entries*		[C]	RSS
			[R/S]	$R - 1$
	Note: To repeat, be certain that Flag 5 is set.		[R/S]	MRSS
			[R/S]	CSS
			[R/S]	$C - 1$
			[R/S]	MCSS
			[R/S]	ESS
			[R/S]	$(R-1)(C-1)$
			[R/S]	MESS
			[R/S]	SS
			[R/S]	$RC - 1$
			[R/S]	F_R
			[R/S]	P_R
			[R/S]	F_C
			[R/S]	P_C

STEP	PROCEDURE	ENTER	PRESS	DISPLAY
4	*Confidence Intervals for Posterior Contrast (after Step 3)*			
	a. Initialize.		[E]	0.
	b. Enter contrast data.	c_i	[R/S]	c_i^2
	(Repeat for each i or j.)	$\bar{x}_{i.}$ (or $\bar{x}_{.j}$)	[R/S]	i (or j)
	c. CI for row contrast $\Sigma c_i \mu_{i.}$	F_α	[D]	l
	d.f. = $[R-1, (R-1)(C-1)]$		[x ⇆ t]	μ
	d. CI for column contrast $\Sigma c_i \mu_{.j}$	F_α	[D′]	l
	d.f. = $[C-1, (R-1)(C-1)]$	0	[x ⇆ t]	u
	Note 1: Steps 4abc or 4abd may be repeated.			
	Note 2: These steps are also valid, if the contents of registers R_{48}, R_{58}, R_{59} below are stored manually.			

REGISTER CONTENTS

00 Used	10 RSS		20		30	40	50 $\Sigma c\bar{x}$
01 R	11 $(R-1)(C-1)$		21		31	41	51 last c
02 C	12		22	⋮	32 $\bar{x}_{.1}$	42	52 last $\bar{x}$
03 RC	13		23	and	33 $\bar{x}_{.2}$	43	53 Σc^2
04 $x..$	14	Used	24	ST–22	34 $\bar{x}_{.3}$	44	54 Used
05 $x^2..$	15	by	25		35 ⋮	45 ⋮	55
06 $\bar{x}..$	16	ST–16	26		36	46	56
07 Used	17		27		37	47 $\bar{x}_{R.}$	57
08 SS	18		28		38	48 MESS	58 $R-1$
09 CSS	19		29		39	49 e	59 $C-1$

TABLE F. F-PERCENTILES

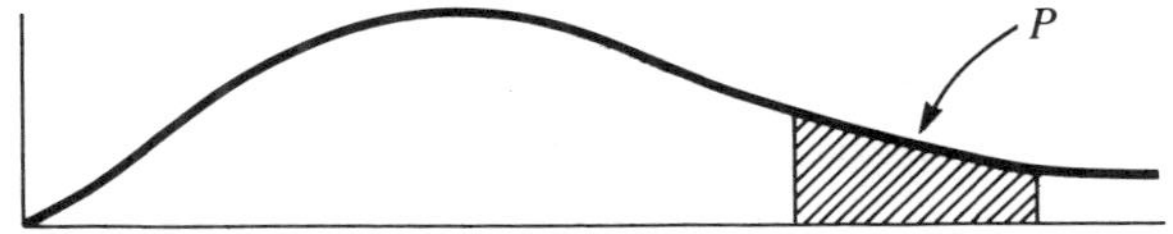

P	$\downarrow \nu_2$	$\nu_1 \rightarrow$ 1	2	3	4	5	6	7	8	9	10	12	15	20	30	50	100	200	> 200
.10	1	39.9	49.5	53.6	55.8	57.2	58.2	58.9	59.4	59.9	60.2	60.7	61.2	61.7	62.3	62.7	6.30	63.2	63.3
.05	1	161	200	216	225	230	234	237	239	241	242	244	246	248	250	252	253	254	254
.025	1	648	800	864	900	922	937	948	957	963	969	977	985	993	1000	1008	1013	1016	1020
.01	1	4050	5000	5400	5620	5760	5860	5930	5980	6020	6060	6110	6160	6210	6260	6303	6334	6350	6370
.005	1	16200	20000	21600	22500	23100	23400	23700	23900	24100	24200	24400	24600	24800	25000	25200	25300	25400	25500
.10	2	8.53	9.00	9.16	9.24	9.29	9.33	9.35	9.37	9.38	9.39	9.41	9.42	9.44	9.46	9.47	9.48	9.49	9.49
.05	2	18.5	19.0	19.2	19.2	19.3	19.3	19.4	19.4	19.4	19.4	19.4	19.4	19.5	19.5	19.5	19.5	19.5	19.5
.025	2	38.5	39.0	39.2	39.2	39.3	39.3	39.4	39.4	39.4	39.4	39.4	39.4	39.4	39.5	39.5	39.5	39.5	39.5
.01	2	98.5	99.0	99.2	99.2	99.3	99.3	99.4	99.4	99.4	99.4	99.4	99.4	99.4	99.5	99.5	99.5	99.5	99.5
.005	2	199	199	199	199	199	199	199	199	199	199	199	199	199	199	199	199	199	199
.10	3	5.54	5.46	5.39	5.34	5.31	5.28	5.27	5.25	5.24	5.23	5.22	5.20	5.118	5.17	5.15	5.14	5.14	5.13
.05	3	10.1	9.55	9.28	9.12	9.01	8.94	8.89	8.85	8.81	8.79	8.74	8.70	8.66	8.62	8.58	8.55	8.54	8.53
.025	3	17.4	16.0	15.4	15.1	14.9	14.7	14.6	14.5	14.5	14.4	14.3	14.3	14.2	14.1	14.0	14.0	13.9	13.9
.01	3	34.1	30.8	29.5	28.7	28.2	27.9	27.7	27.5	27.3	27.2	27.1	26.9	26.7	26.5	26.3	26.2	26.2	26.1
.005	3	55.6	49.8	47.5	46.2	45.4	44.8	44.4	44.1	43.9	43.7	43.4	43.1	42.8	42.5	42.2	42.0	41.9	41.8
.10	4	4.54	4.32	4.19	4.11	4.05	4.01	3.98	3.95	3.93	3.92	3.90	3.87	3.84	3.82	3.80	3.78	3.77	3.76
.05	4	7.71	6.94	6.59	6.39	6.26	6.16	6.09	6.04	6.00	5.96	5.91	5.86	5.80	5.75	5.70	5.66	5.65	5.63
.025	4	12.2	10.6	9.98	9.60	9.36	9.20	9.07	8.98	8.90	8.84	8.75	8.66	8.56	8.46	8.38	8.32	8.29	8.26
.01	4	21.2	18.0	16.7	16.0	15.5	15.2	15.0	14.8	14.7	14.5	14.4	14.2	14.0	13.8	13.7	13.6	13.5	13.5
.005	4	31.3	26.3	24.3	23.2	22.5	22.0	21.6	21.4	21.1	21.0	20.7	20.4	20.2	19.9	19.7	19.5	19.4	19.3
.10	5	4.06	3.78	3.62	3.52	3.45	3.40	3.37	3.34	3.32	3.30	3.27	3.24	3.21	3.17	3.15	3.13	3.12	3.11
.05	5	6.61	5.79	5.41	5.19	5.05	4.95	4.88	4.82	4.77	4.74	4.68	4.62	4.56	4.50	4.44	4.41	4.39	4.37
.025	5	10.0	8.43	7.76	7.39	7.15	6.98	6.85	6.76	6.68	6.62	6.52	6.43	6.33	6.23	6.14	6.08	6.05	6.02
.01	5	16.3	13.3	12.1	11.4	11.0	10.7	10.5	10.3	10.2	10.1	9.89	9.72	9.55	9.38	9.24	9.13	9.08	9.02
.005	5	22.8	18.3	16.5	15.6	14.9	14.5	14.2	14.0	13.8	13.6	13.4	13.1	12.9	12.7	12.4	12.3	12.2	12.1
.10	6	3.78	3.46	3.29	3.18	3.11	3.05	3.01	2.98	2.96	2.94	2.90	2.87	2.84	2.80	2.77	2.75	2.73	2.72
.05	6	5.99	5.14	4.76	4.53	4.39	4.28	4.21	4.15	4.10	4.06	4.00	3.94	3.87	3.81	3.75	3.71	3.69	3.67
.025	6	8.81	7.26	6.60	6.23	5.99	5.82	5.70	5.60	5.52	5.46	5.37	5.27	5.17	5.07	4.98	4.92	4.88	4.85
.01	6	13.7	10.9	9.78	9.15	8.75	8.47	8.26	8.10	7.98	7.87	7.72	7.56	7.40	7.23	7.09	6.99	6.93	6.88
.005	6	18.6	14.5	12.9	12.0	11.5	11.1	10.8	10.6	10.4	10.2	10.0	9.81	9.59	9.36	9.17	9.03	8.95	8.88

TABLE F. F-PERCENTILES

P	↓ ν_2	$\nu_1 \rightarrow$ 1	2	3	4	5	6	7	8	9	10	12	15	20	30	50	100	200	> 200
.10	7	3.59	3.26	3.07	2.96	2.88	2.83	2.78	2.75	2.72	2.70	2.67	2.63	2.59	2.56	2.52	2.50	2.48	2.47
.05	7	5.59	4.74	4.35	4.12	3.97	3.87	3.79	3.73	3.68	3.64	3.57	3.51	3.44	3.38	3.32	3.27	3.25	3.23
.025	7	8.07	6.54	5.89	5.52	5.29	5.12	4.99	4.90	4.82	4.76	4.67	4.57	4.47	4.36	4.28	4.21	4.18	4.14
.01	7	12.2	9.55	8.45	7.85	7.46	7.19	6.99	6.84	6.72	6.62	6.47	6.31	6.16	5.99	5.86	5.75	5.70	5.65
.005	7	16.2	12.4	10.9	10.1	9.52	9.16	8.89	8.68	8.51	8.38	8.18	7.97	7.75	7.53	7.35	7.22	7.15	7.08
.10	8	3.46	3.11	2.92	2.81	2.73	2.67	2.62	2.59	2.56	2.54	2.50	2.46	2.42	2.38	2.35	2.32	2.31	2.29
.05	8	5.32	4.46	4.07	3.84	3.69	3.58	3.50	3.44	3.39	3.35	3.28	3.22	3.15	3.08	3.02	2.97	2.95	2.93
.025	8	7.57	6.06	5.42	5.05	4.82	4.65	4.53	4.43	4.36	4.30	4.20	4.10	4.00	3.89	3.81	3.74	3.71	3.67
.01	8	11.3	8.65	7.59	7.01	6.63	6.37	6.18	6.03	5.91	5.81	5.67	5.52	5.36	5.20	5.07	4.96	4.91	4.86
.005	8	14.7	11.0	9.60	8.81	8.30	7.95	7.69	7.50	7.34	7.21	7.01	6.81	6.61	6.40	6.22	6.09	6.02	5.95
.10	9	3.36	3.01	2.81	2.69	2.61	2.55	2.51	2.47	2.44	2.42	2.38	2.34	2.30	2.25	2.22	2.19	2.17	2.16
.05	9	5.12	4.26	3.86	3.63	3.48	3.37	3.29	3.23	3.18	3.14	3.07	3.01	2.94	2.86	2.80	2.76	2.73	2.71
.025	9	7.21	5.71	5.08	4.72	4.48	4.32	4.20	4.10	4.03	3.96	3.87	3.77	3.67	3.56	3.47	3.40	3.37	3.33
.01	9	10.6	8.02	6.99	6.42	6.06	5.80	5.61	5.47	5.35	5.26	5.11	4.96	4.81	4.65	4.52	4.42	4.36	4.31
.005	9	13.6	10.1	8.72	7.96	7.47	7.13	6.88	6.69	6.54	6.42	6.23	6.03	5.83	5.62	5.45	5.32	5.26	5.19
.10	10	3.29	2.92	2.73	2.61	2.52	2.46	2.41	2.38	2.35	2.32	2.28	2.24	2.20	2.15	2.12	2.09	2.07	2.06
.05	10	4.96	4.10	3.71	3.48	3.33	3.22	3.14	3.07	3.02	2.98	2.91	2.84	2.77	2.70	2.64	2.59	2.56	2.54
.025	10	6.94	5.46	4.83	4.47	4.24	4.07	3.95	3.85	3.78	3.72	3.62	3.52	3.42	3.31	3.22	3.15	3.12	3.08
.01	10	10.0	7.56	6.55	5.99	5.64	5.39	5.20	5.06	4.94	4.85	4.71	4.56	4.41	4.25	4.12	4.01	3.96	3.91
.005	10	12.8	9.43	8.08	7.34	6.87	6.54	6.30	6.12	5.97	5.85	5.66	5.47	5.27	5.07	4.90	4.77	4.71	4.64
.10	12	3.18	2.81	2.61	2.48	2.39	2.33	2.28	2.24	2.21	2.19	2.15	2.10	2.06	2.01	1.97	1.94	1.92	1.90
.05	12	4.75	3.89	3.49	3.26	3.11	3.00	2.91	2.85	2.80	2.75	2.69	2.62	2.54	2.47	2.40	2.35	2.32	2.30
.025	12	6.55	5.10	4.47	4.12	3.89	3.73	3.61	3.51	3.44	3.37	3.28	3.18	3.07	2.96	2.87	2.80	2.76	2.72
.01	12	9.33	6.93	5.95	5.41	5.06	4.82	4.64	4.50	4.39	4.30	4.16	4.01	3.86	3.70	3.57	3.49	3.41	3.36
.005	12	11.8	8.51	7.23	6.52	6.07	5.76	5.52	5.35	5.20	5.09	4.91	4.72	4.53	4.33	4.17	4.04	3.97	3.90
.10	15	3.07	2.70	2.49	2.36	2.27	2.21	2.16	2.12	2.09	2.06	2.02	1.97	1.92	1.87	1.83	1.79	1.79	1.76
.05	15	4.54	3.68	3.29	3.06	2.90	2.79	2.71	2.64	2.59	2.54	2.48	2.40	2.33	2.25	2.18	2.12	2.10	2.07
.025	15	6.20	4.77	4.15	3.80	3.58	3.41	3.29	3.20	3.12	3.06	2.96	2.86	2.76	2.64	2.55	2.47	2.44	2.40
.01	15	8.68	6.36	5.42	4.89	4.56	4.32	4.14	4.00	3.89	3.80	3.67	3.52	3.37	3.21	3.08	2.98	2.92	2.87
.005	15	10.8	7.70	6.48	5.80	5.37	5.07	4.85	4.67	4.54	4.42	4.25	4.07	3.88	3.69	3.52	3.39	3.33	3.26

.10	20	2.97	2.59	2.38	2.25	2.16	2.09	2.04	2.00	1.96	1.94	1.89	1.84	1.79	1.74	1.69	1.65	1.63	1.61
.05		4.35	3.49	3.10	2.87	2.71	2.60	2.51	2.45	2.39	2.35	2.28	2.20	2.12	2.04	1.97	1.91	1.88	1.84
.025		5.87	4.46	3.86	3.51	3.29	3.13	3.01	2.91	2.84	2.77	2.68	2.57	2.46	2.35	2.25	2.17	2.13	2.09
.01		8.10	5.85	4.94	4.43	4.10	3.87	3.70	3.56	3.46	3.37	3.23	3.09	2.94	2.78	2.64	2.54	2.48	2.42
.005		9.94	6.99	5.82	5.17	4.76	4.47	4.26	4.09	3.96	3.85	3.68	3.50	3.32	3.12	2.96	2.83	2.76	2.69
.10	30	2.88	2.49	2.28	2.14	2.05	1.98	1.93	1.88	1.85	1.82	1.77	1.72	1.67	1.61	1.55	1.51	1.48	1.46
.05		4.17	3.32	2.92	2.69	2.53	2.42	2.33	2.27	2.21	2.16	2.09	2.01	1.93	1.84	1.76	1.70	1.66	1.62
.025		5.57	4.18	3.59	3.25	3.03	2.87	2.75	2.65	2.57	2.51	2.41	2.31	2.20	2.07	1.97	1.88	1.84	1.79
.01		7.56	5.39	4.51	4.02	3.70	3.47	3.30	3.17	3.07	2.98	2.84	2.70	2.55	2.39	2.25	2.13	2.07	2.01
.005		9.18	6.35	5.24	4.62	4.23	3.95	3.74	3.58	3.45	3.34	3.18	3.01	2.82	2.63	2.46	2.32	2.25	2.18
.10	50	2.81	2.41	2.20	2.06	1.97	1.90	1.84	1.80	1.76	1.73	1.68	1.63	1.57	1.50	1.44	1.39	1.36	1.33
.05		4.03	3.18	2.79	2.56	2.40	2.29	2.20	2.13	2.07	2.03	1.95	1.87	1.78	1.69	1.60	1.52	1.48	1.44
.025		5.34	3.97	3.39	3.05	2.83	2.67	2.55	2.46	2.38	2.32	2.22	2.11	1.99	1.87	1.75	1.65	1.60	1.54
.01		7.17	5.06	4.20	3.72	3.41	3.19	3.02	2.89	2.79	2.70	2.56	2.42	2.27	2.10	1.95	1.82	1.75	1.68
.005		8.63	5.90	4.83	4.23	3.85	3.58	3.38	3.22	3.09	2.99	2.82	2.65	2.47	2.27	2.10	1.95	1.87	1.78
.10	100	2.76	2.36	2.14	2.00	1.91	1.83	1.78	1.73	1.69	1.66	1.61	1.56	1.49	1.42	1.35	1.29	1.26	1.21
.05		3.94	3.09	2.70	2.46	2.31	2.19	2.10	2.03	1.97	1.93	1.85	1.77	1.68	1.59	1.48	1.39	1.34	1.28
.025		5.18	3.83	3.25	2.92	2.70	2.54	2.42	2.32	2.24	2.18	2.08	1.97	1.85	1.71	1.59	1.48	1.42	1.35
.01		6.90	4.82	3.98	3.51	3.21	2.99	2.82	2.69	2.59	2.50	2.38	2.22	2.07	1.89	1.74	1.60	1.52	1.43
.005		8.24	5.59	4.54	3.96	3.59	3.33	3.13	2.97	2.85	2.74	2.58	2.41	2.23	2.02	1.84	1.68	1.59	1.48
.10	200	2.73	2.33	2.11	1.97	1.88	1.80	1.75	1.70	1.66	1.63	1.58	1.52	1.46	1.38	1.31	1.24	1.20	1.14
.05		3.89	3.04	2.65	2.42	2.26	2.14	2.06	1.98	1.93	1.88	1.80	1.72	1.62	1.52	1.41	1.32	1.26	1.19
.025		5.18	3.76	3.18	2.85	2.63	2.47	2.35	2.26	2.18	2.11	2.01	1.90	1.78	1.64	1.51	1.39	1.32	1.23
.01		6.76	4.71	3.88	3.41	3.11	2.89	2.73	2.60	2.50	2.41	2.27	2.13	1.97	1.79	1.63	1.48	1.39	1.28
.005		8.06	5.44	4.41	3.84	3.47	3.21	3.01	2.86	2.73	2.63	2.47	2.30	2.11	1.91	1.71	1.54	1.44	1.31
.10	> 200	2.71	2.30	2.08	1.94	1.85	1.77	1.72	1.67	1.63	1.60	1.55	1.49	1.42	1.34	1.26	1.18	1.13	1.00
.05		3.84	3.00	2.60	2.37	2.21	2.10	2.01	1.94	1.88	1.83	1.75	1.67	1.57	1.46	1.35	1.24	1.17	1.00
.025		5.02	3.69	3.12	2.79	2.57	2.41	2.29	2.19	2.11	2.05	1.94	1.83	1.71	1.57	1.43	1.29	1.21	1.00
.01		6.63	4.61	3.78	3.32	3.02	2.80	2.64	2.51	2.41	2.32	2.18	2.04	1.88	1.70	1.52	1.35	1.25	1.00
.005		7.88	5.30	4.28	3.72	3.35	3.09	2.90	2.74	2.62	2.52	2.36	2.19	2.00	1.79	1.59	1.40	1.28	1.00

Analysis of Variance

1. Three types of solvents are tested on grease-soaked material, and the amount of grease removed in milligrams is noted for several specimens, with the following results.

Solvent A:	11	12	12	
Solvent B:	13	15		
Solvent C:	12	10	11	11

Test the hypothesis of no differences in solvents.

Solutions

ZS STEP	ENTER	PRESS	DISPLAY	COMMENTS
AOV–1		[Pgm]06	*xxx*	Call for ST–06 in the module.
1a		[E′]	0.	Initialize program ST–06.
	11	[A]	1.	Enter data for the first row.
	12	[A]	2.	Enter data for the first row.
	12	[A]	3.	Row 1 data entry concluded
		[B′]	11.67	First row mean $\bar{x}_1$ calculated—record!
		[C′]	.2222	MSD for row 1: ignore and go on to enter data for second row.
	13	[A]	1.	Running count begins anew
	15	[A]	2.	Row 2 data entry concluded
		[B′]	14.	Row mean $\bar{x}_2$ calculated—record!
		[C′]	1.	Row 2 MSD: ignore and proceed
	12	[A]	1.	to enter data for third row.
	10	[A]	2.	
	11	[A]	3.	
	11	[A]	3.	Row 3 data entry concluded
		[B′]	11.	Value of $\bar{x}_3$—record!
		[C′]	0.5	Value of MSD: ignore. Data
1b		[RST]	0.5	entry concluded; exit ST–03.
2		[A]	12.22 ...	Value of RSS displayed
		[R/S]	2.	d.f. for RSS
		[R/S]	6.11 ...	MRSS displayed
		[R/S]	4.66 ...	ESS displayed
		[R/S]	6.	d.f. for ESS displayed
		[R/S]	0.77 ...	MESS displayed
		[R/S]	16.88 ...	Total sum of squares, SS displayed
		[R/S]	8.	Coincidentally, d.f. or SS
		[R/S]	7.86	*F*-ratio = MRSS/MESS
		[R/S]	.02	*P*-value of the *F*-ratio

2. Find Scheffe 95% confidence intervals for the contrast $\mu_1 - .5\mu_2 - .5\mu_3$ in Problem 5, where MESS = 105.97, and $K = 3$, $N - k = 18$; $\bar{x}_1 = 103.71$, $\bar{x}_2 = 92$, $\bar{x}_3 = 89$, $n = 7$.

Solution

ZS STEP	ENTER	PRESS	DISPLAY	COMMENTS
	2	[STO]14	2.	$\nu_1 = K - 1 = 2$ stored in R_{14}
	105.97	[STO]48	105.97	MESS stored in R_{48}, as required
AOV–1				
3a		[E′]	0.	Initialize CI routine.
3b	1	[R/S]	1.	First member of triple c_1, $\bar{x}$, n, entered
	103.71	[R/S]	103.71	Second member of triple
	7	[R/S]	1.	Sample size n; count of 1 (triple) displayed
	−.5	[R/S]	−.5	Beginning entry of c_2, $\bar{x}_2$, n
	92	[R/S]	92.	
	7	[R/S]	2.	Running count of 2 displayed
	−.5	[R/S]	−.5	Entering final triple
	89	[R/S]	89.	
	7	[R/S]	3.	Entry completed
3c	3.49	[A′]	.62	$F_{.05}$ for d.f. = (2,20) entered
		[$x \rightleftarrows t$]	25.80	and confidence limits displayed. Conclude contrast significantly different from 0.

3. Example 8.4. Five teachers were matched with 3 schools to produce the following average scores on a standardized examination after a unit of instructions.

SCHOOLS \ TEACHERS	A	B	C	D	E	$x_{i.}$
I	53	47	46	50	49	49
II	61	55	52	58	54	56
III	51	51	49	54	50	51
$\bar{x}_{.j}$	55	51	49	54	51	

 Construct a 2-way AOV table, and find CI's for $\mu_{1.} - \mu_{2.}$ and $\mu_{.1} - \mu_{.3}$.

Solution

ZS STEP	ENTER	PRESS	DISPLAY	COMMENTS
AOV–2		[Pgm]06	*xxx*	Call for ST–06 from the module.
1a		[E]	0.	Initialize ST–6 two-way AOV.
	3	[B]	3.	Enter $R = 3$, the number of rows.
	5	[C]	5.	Enter $C = 5$, the number of columns.
	53	[A]	1.	Enter first data value from row 1.
	47	[A]	2.	Continue entering data from row 1.
	46	[A]	3.	.
	50	[A]	4.	:
	49	[A]	5.	until all the data from row 1 are entered.
	61	[A]	1.	Go on with first value from row 2
	55	[A]	2.	and continue
	52	[A]	3.	until all of
	58	[A]	4.	the data from row 2
	54	[A]	5.	have been entered.
	51	[A]	1.	Continue nonstop until data
	51	[A]	2.	from the third through
	49	[A]	3.	last rows
	54	[A]	4.	have finally
	50	[A]	5.	been entered.
		[B′]	52.	At this point, you press [B′] to find $\bar{\bar{x}}$.
		[C′]	14.93	Also press [C′] to find MSD (of no interest).
1b		[RST]	14.93	Exit ST–06 to prepare for ZS–8.
AOV–2		[B]	49.	Calculate and display $\bar{x}_{1.} = 49$.
2a		[R/S]	56.	Find second row mean $\bar{x}_{2.} = 56$.
		[R/S]	51.	Find row mean of $\bar{x}_{3.} = 51$.
2b		[B′]	55.	Calculate and display $\bar{x}_{.1} = 55$.
		[R/S]	51.	Find $\bar{x}_{.2} = 51$.
		[R/S]	49.	Find $\bar{x}_{.3} = 49$.

ZS STEP	ENTER	PRESS	DISPLAY	COMMENTS
		[R/S]	54.	Find $\bar{x}_{.4} = 54$, and finally,
		[R/S]	51.	$\bar{x}_{.5} = 51$.
3		[C]	130.	Data compiled and RSS displayed
		[R/S]	2.	Row d.f. $= R - 1 = 2$
		[R/S]	65.	MRSS computed and displayed
		[R/S]	72.	CSS displayed
		[R/S]	4.	Column d.f. $= C - 1 = 4$
		[R/S]	18.	MCSS $=$ CSS$/(C - 1)$ displayed
		[R/S]	22.	Residual or error ESS
		[R/S]	8.	Residual d.f. $= (R - 1)(C - 1)$
		[R/S]	2.75	MESS calculated, stored, displayed
		[R/S]	224.	Total SS
		[R/S]	14.	Total d.f. $= RC - 1$
		[R/S]	23.6	F-ratio for rows, $F_R = 23.6$
		[R/S]	.0004	P-value for F_R computed and displayed
		[R/S]	6.5	F-ratio for columns, $F_c = 6.5$
		[R/S]	.012	P-value for F_C
4a		[E]	0.	Initialize for Scheffe CI's.
4b	1	[R/S]	1.	Enter $c_1 = 1$ to find CI for $\mu_{1.} - \mu_{2.}$.
	49	[R/S]	1.	Enter row mean $\bar{x}_{1.}$.
	-1	[R/S]	1.	Enter $c_2 = -1$ for contrast $\mu_{1.} - \mu_{2.}$.
	56	[R/S]	2.	All non-zero c's now entered
4c	4.46	[D]	$-$ 10.13	$F_{.05} = 4.46$ for d.f. $= (2,8)$
		[$x \leftrightarrows t$]	$-$ 3.87	l is displayed, followed by u.
		[RCL]49	3.13	The value of e is retrieved from R_{49} for further comparisons.
4a		[E]	0.	Reinitialize CI program.
4b	1	[R/S]	1.	Enter c_1 for contrast $\mu_{.1} - \mu_{.3}$.
	55	[R/S]	1.	$\bar{x}_{.1}$, the first column mean in entered.
	-1	[R/S]	1.	Enter $c_2 = -1$
	49	[R/S]	2.	Enter third column mean $\bar{x}_{.3}$
4d	3.84	[D']	0.69	Enter $F_{.05}$ for d.f. $= (4,8)$, and calculate lower limit.
		[$x \leftrightarrows t$]	11.31	Upper limit retrieved from R_T
		[RCL]49	5.31	Value of e found in R_{49} for further comparisons.

9

Simple Linear Regression

Section 9.1: The Simple Linear Model

In statistical analysis the problem of trying to determine the relationship between one variable, such as weight of an automobile, and another, such as rate of fuel consumption (mileage) often arises. In actuality the "curve" relating the independent variable (weight) to the dependent variable (mileage) could be very complicated, especially as the independent variable ranges over a broad interval of values. You are probably aware that, even for a *given* automobile, mileage will vary from one tankful to another. We now discuss how you can "fit" a line to *x*-*y* data, such as a set of mileage data collected for several different automobile weights. (Later, we will discuss other forms of curves.) You will see how to use the regression features of the TI–59 to estimate the "best fitting line" for any given set of data.

The term "regression" is somewhat of a misnomer. It was coined by Sir Francis Galton, who was involved in a study of human inheritance during the late 1800's. In particular, Galton found that tall fathers tended to have rather tall sons, but *on the average*, the sons were not as tall as their fathers. Similarly, sons of short fathers tended not to be as short as their fathers. In other words, there seemed to be a tendency for offspring to "regress," or be drawn toward, the general population mean height. Galton coined the term *regression* to describe this phenomenon, and we have lived with the term ever since. The phenomenon Galton noticed is a tricky one, and we have to be very careful how we interpret it. Actually, there is a simple mathematical explanation for this apparently mysterious phenomenon, once the model is understood. We will return to this subject toward the end of the chapter.

It is useful to imagine the data plotted in the Cartesian plane, with the independent variable values associated with the x-axis and dependent variable values along the y-axis. We shall denote the i-th data pair in the form (x_i, y_i), where the index i is simply a bookkeeping device for our convenience. In our example, the data collected at the i-th mileage measurement would then be in the form (auto weight, mileage). If we plot the data pairs for, say, 10 different autos, we might get an array such as the one shown on page V–37 of the Personal Programming Manual. The line we wish to find is characterized by two parameters, the slope m and the intercept b, so the linear relationship we wish to estimate is of the form $y = mx + b$. This is accomplished by estimating the parameters m and b.* If we denote these estimates by $\hat{m}$ and $\hat{b}$, the estimated line is the form $\hat{y} = \hat{m}x + \hat{b}$, where $\hat{y}$ denotes the estimated "response" for any input value of the independent variable x.

Because of variations in the *response*, y, for each fixed value of the independent variable, x (also called the *carrier*), we generally find there is no single line that can exactly fit the n data pairs. We must agree on a criterion for measuring the goodness of fit of any line to a data set; once that is done, we seek estimates $\hat{m}$ and $\hat{b}$, giving the line that best fits the data according to our fit criterion.

The most commonly used criterion of fit is the "least squares" criterion. This is based on the sum of squared deviations between the observed responses and the responses that would have been predicted by the fitted line. Formally, the *least squares estimates* are those choices for $\hat{m}$ and $\hat{b}$ that minimize the *error sum of squares* (ESS) defined by

$$\text{ESS} = \sum_{i=1}^{n} (y_i - \hat{y}_i)^2 \tag{9.1}$$

Here,

$$\hat{y}_i = \hat{m}x_i + \hat{b}$$

and (x_i, y_i) $i = 1,2,...,n$, are the observed data pairs. The line defined in Equation 9.2 is sometimes called the "least squares line," or the "fitted regression line," for the observed data.

* You may encounter other choices of notation for slope and intercept such as β and α or β_1 and β_0, respectively. We defer to the notation of PPM to remain consistent with their descriptions.

Formulas for the least squares estimates $\hat{m}$ and $\hat{b}$ can be given in various forms. The most common form is

$$\hat{m} = \frac{\sum_{i=1}^{n} (x_i - \bar{x})(y_i - \bar{y})}{\sum_{i=1}^{n} (x_i - \bar{x})^2} \tag{9.3}$$

$$\hat{b} = \bar{y} - \hat{m}\bar{x} \tag{9.4}$$

and we will use this form in what follows. An alternate, equivalent form is given in PPM, page V–37. Yet another form is sometimes given, in which the x's and y's are coded by subtracting $\bar{x}$ from each x_i and $\bar{y}$ from each y_i. The resulting coded data is said to be *centered.* You can see how Equation 9.3 would superficially appear simpler, if centered data were being used, rather than raw data. The computations and results with all of these forms of least squares estimates are equivalent, requiring only careful bookkeeping to transform from one to the other.

As you expect, the TI–59 will perform the calculations in Equations 9.3 and 9.4, once you input the data. The data input routine and the computation routine for $\hat{m}$ and $\hat{b}$ are resident within the TI–59 itself, as functions and Op codes. That is why you find linear regression discussed in PPM, starting on page V–32 and extending through page V–40.

The data entry for linear regression uses the [Σ+] key, and compiles a data base in registers R_{01} through R_{06}. To initialize this data base, simply press [CMs]. Then enter each of the n data pairs following the sequence x_i, [$x \rightleftharpoons t$], y_i, [Σ+]. Be sure to allow enough time after pressing [Σ+] for the data base compilation. Completion of each entry is signaled by the appearance of i in R_D. A procedure for correcting an error in the input sequence is described on page V–33 of PPM; you will find it is wise to enter the data carefully, so that such correction is unnecessary. The contents of the data registers in the data base are listed on page V–33 of PPM. You will rarely, if ever, need to examine these quantities directly, but the regression-related programs access them.

EXAMPLE 9.1

The mileage (mpg) of 6 automobiles was measured in a total of 10 controlled trials. The resulting (weight, mileage) data follow.

Weight	2,830	2,700	3,492	2,590	4,100	3,050
mpg	27.2	25	15.1	31.3	12.5	25.9
	26.4		17.8		13.1	
	26.6					

Clear the data base registers, and enter the data for linear regression.

Solution Press [CMs], then enter the data in pairs, as follows.

$i = 1$:	2,830	[$x \rightleftarrows t$]	27.2	[$\Sigma+$]	(see "1" displayed)
$i = 2$:	2,830	[$x \rightleftarrows t$]	26.4	[$\Sigma+$]	(see "2" displayed)
$i = 3$:	2,830	[$x \rightleftarrows t$]	26.6	[$\Sigma+$]	(see "3" displayed)
$i = 4$:	2,700	[$x \rightleftarrows t$]	25	[$\Sigma+$]	(see "4" displayed)
$i = 5$:	3,492	[$x \rightleftarrows t$]	15.1	[$\Sigma+$]	(see "5" displayed)
$i = 6$:	3,492	[$x \rightleftarrows t$]	17.8	[$\Sigma+$]	(see "6" displayed)
$i = 7$:	2,590	[$x \rightleftarrows t$]	31.3	[$\Sigma+$]	(see "7" displayed)
$i = 8$:	4,100	[$x \rightleftarrows t$]	12.5	[$\Sigma+$]	(see "8" displayed)
$i = 9$:	4,100	[$x \rightleftarrows t$]	13.1	[$\Sigma+$]	(see "9" displayed)
$i = 10$:	3,050	[$x \rightleftarrows t$]	25.9	[$\Sigma+$]	(see "10" displayed)

Notice the repeated entry of weight corresponding to *each* mpg value.

If you wish, you may recall the contents of R_{01} through R_{06} to verify that

$\Sigma y_i = 220.9$	$\in R_{01}$	$\Sigma x_i = 32{,}014$	$\in R_{04}$
$\Sigma y_i^2 = 5{,}292.57$	$\in R_{02}$	$\Sigma x_i^2 = 105{,}335{,}428$	$\in R_{05}$
$n = 10$	$\in R_{03}$	$\Sigma x_i y_i = 674{,}374.8$	$\in R_{06}$

(Do not turn your calculator off; we will access this data base in the following example.)

When the data have been entered, the sample mean $\bar{y}$, of the y_i's, is computed by pressing [$\bar{x}$]; following this by [$x \rightleftarrows t$], displays $\bar{x}$. Similarly [INV][$\bar{x}$] computes the sample standard deviation of the y_i's. The least squares estimates of slope and intercept are calculated with [Op] 12, as discussed on page V–36 of PPM. The use of [Op] 11 to calculate MSD, discussed in PPM on pages V–33 and V–34, has already been discussed in Chapter 2.

EXAMPLE 9.2

For the weight-mileage data of Example 9.1, calculate the sample mean and standard deviation of the y_i's, and calculate the least squares estimates of slope and intercept for the fitted regression line.

Solution Press [$\bar{x}$] to display $\bar{y}$ = 22.09, the average of the mileages in the 10 trials. Press [$x \rightleftarrows t$] to display $\bar{x}$ = 3201.4. Press [INV][$\bar{x}$] to display s_y = 6.77. Press [Op] 12 to display $\hat{b}$ = 59.0, and [$x \rightleftarrows t$] to display $\hat{m}$ = −.0115. (Do not destroy the data base yet!)

You might find it instructive to plot the 10 data pairs, together with a graph of the fitted line $\hat{y} = -.0115\,x + 59$ in Example 9.2. You might observe that the negative coefficient is expected in this model, since we expect mileage, generally, to decrease with increasing vehicle weight.

When the data have been entered, you can predict the response for any value x of the carrier by finding the point $\hat{y} = \hat{m}x + \hat{b}$ on the line. This is accomplished by entering the desired carrier in R_D and pressing [Op] 14. Similarly, if you wanted to find the carrier value which gives a specified estimated response $\hat{y}$, enter $\hat{y}$ in R_D, and press [Op] 15. Using this feature, you could calculate ESS, the fit criterion given in Equation 9.1, by successively entering each x_i, using [Op] 14 to get $\hat{y}_i$. Then, calculate and accumulate the sum of $(y_i - \hat{y}_i)^2$. You will see in the next section why you need ESS, and we will use ZS–9 to do these and other calculations.

EXAMPLE 9.3

If a vehicle weighs 2,900 lb, what mileage would you estimate it might get? What weight of vehicle would give an estimated 17 mpg?

Solution With the data base from Example 9.2, enter 2,900, and press [Op] 14 to display $\hat{y}$ = 25.57. You can easily verify that $25.57 = \hat{m}(2{,}900) + \hat{b} = (-.0115)\,(2{,}900) + 59$. Next, enter 17 in R_D, and press [Op] 15 to display x = 3642.83. Again, you can easily verify that $17 = (-.0115)\,(3642.83) + 59$, to within rounding error.

We end this section with a few hints on use of the regression features of your TI–59.

1. For data entry where there are several y_i's at the same carrier value, as in our mileage example, you should resist the temptation to short-cut the carrier value entry for *each* y_i, since R_T is incremented during data base compilation.
2. If the data are all at the same carrier value, then all the x_i's are equal to $\bar{x}$, and the denominator in Equation 9.3 will be 0. You can expect a division by 0 error, resulting in a flashing display if you attempt Ops 12, 14, or 15 with such a data base. If you think about how a plot of such data would look, you can easily see why no sensible slope or intercept estimates can be made in such a case.
3. If the fitted line has $\hat{m} = 0$, you will generally run into division by 0 problems with [Op] 15. Again, with a horizontal line, there will be

no x value which gives a given $\hat{y}$, except for the case $\hat{y} = \hat{b}$.

4. You should be made aware of the problem of attempting to extrapolate your linear regression predictions too far from the carrier values used in the experiment. The reason is that the linear model is a reasonable approximation only over a restricted range of carrier values. For example, if you "predict" the mileage of a vehicle weighing 0 lb, (enter 0, press [Op] 14, with the data base of Example 9.2), you will see $\hat{y} = 59.00$ (which is, of course, $\hat{b}$) displayed. But what interpretation might such a mileage prediction have? Clearly, extrapolating to carrier values too far from $\bar{x}$ gives "predictions," in which we have little or no confidence, or, in some cases, no meaning.

Problems

Note: Problems 1–8 concern the following situation. The fat content of hamburger samples, in percent by weight, has been determined in a cook-down test with 6 samples purchased at various supermarkets. The object of the experiment was to determine a relationship between price and quality (as measured by fat content) of hamburger. The results follow.

x = Price ($/pound)*	1.69	1.05	.97	1.43	1.55	1.39
y = Fat content (% by weight)	19	40	46	26	30	41

1. Calculate the mean price.

Ans: $\bar{x} = \$1.35/\text{pound}$

2. Calculate the sample *variance* of fat content.

Ans: $s_y^2 = 106.7$

3. Estimate the slope and intercept of a fitted regression of fat content on price. (That is, consider price to be the carrier, fat content to be the response.)

Ans: $\hat{b} = 76.4 \quad \hat{m} = -31.7$

4. Give a formula for estimating fat content, as a function of price of hamburger.

Ans: $\hat{y} = -31.7\,x + 76.4$, where x is price.

5. What fat content would you predict for hamburger costing $1.39 per pound?

Ans: 32.3%

6. Why is the prediction in Problem 5 different from the fat content (41%) actually observed in the experiment?

Ans: $\hat{y} = 32.3$ is on the fitted line; the fitted line generally does not contain the observed values.

* Coded to 1975 dollars.

7. What fat content would you predict for hamburger costing \$2.50 per pound?

   Ans: This is a dangerous extrapolation, and such predictions should be avoided. If you persist, you get $\hat{y} = -2.9\%$!

8. For what cost would the predicted fat content be 35%? 5%?

   Ans: \$1.30/pound; \$2.25/pound (dangerous extrapolation) Problems 9–15 refer to the following data.

   The salaries of a random sample of teachers in a certain school district, together with their lengths of employment, are shown below.

y SALARY (PER MONTH)	x LENGTH OF EMPLOYMENT (MONTHS)
1,519	330
1,636	569
1,558	375
1,408	113
1,442	215
1,502	343
1,408	252
1,490	348
1,432	352
1,399	256
1,336	87
1,544	337

Estimate the regression of salary on length of employment.

Ans: $\hat{y} = .584x + 1{,}298.7$, where the carrier ($x$) is length of employment, and response (y) is salary.

9. Predict the salary of a teacher with length of employment 200 months.

   Ans: \$1,415.53 per month

10. For the data, calculate Σy_i^2 and $\bar{x}$.

    Ans: 26,108,274 (recall R_{02}); 298.1 months.

11. Where does the fitted line cross the x-axis? the y-axis?

    Ans: $-2{,}223$ (use 0 [Op] 15); \$1,298.7 ($= \hat{b}$)

12. What is the "error," $y_3 - \hat{y}_3$, for the 3rd data pair?

    Ans: \$40 = 1,558 − 1,518

13. What is the estimated length of employment of a teacher earning \$1500 per month?

    Ans: 345 months

14. If the salaries had been coded by subtracting 1,000 from each value, what would be the prediction equation?

    Ans: $\hat{y}' = .584x + 298.7$ (translates line down 1,000 units; does not change the slope). You may verify this by actual computation.

15. What are the coded answers to Problems 9, 11, and 12?

Ans: \$415.53 for 9, −511.5 and \$298.7 for 11; 12 is the same.

Section 9.2: Inference about Regression Parameters

In the preceding section, you saw how to make least squares calculations. We now wish to construct a random model with an error term (which is a random variable), to account for the fact that we generally have no control over the observed y-values. Even if we knew exactly the slope and intercept of the straight line relationship between the carrier and response, the y's would not fall on the line; the distance above or below this line would be due to chance "errors." But in practice, we do not even know the equation of the line. We assume only that there is some underlying linear relationship. (Even that "knowledge" is a strong assumption, which may only hold, if at all, over a restricted range of carrier values.) The model we have in mind accounts for random "error" in the response by an added error term, E. Thus, for the i-th measurement with the carrier at level x_i, we will observe $y_i = mx_i + b + E_i$. We assume the E_i's are independent and have mean 0. The *mean* response for the carrier held fixed at any x is thus $mx + b$, the linear regression we have been discussing.

One thing needs to be emphasized: the E's are random variables in this model, and the x's *are not.* Of course, if the E's are random, then so are the Y's, whose values are to be observed in the experiment. The x's are values we fix deliberately (or observe without error), and they enter into the mean of the Y's explicitly in the linear regression function. As in AOV, we assume the E_i's all have common variance σ^2. Then, for each i, with fixed x_i,

$$
\begin{aligned}
Y_i &= mx_i + b + E_i \\
E(Y_i) &= mx_i + b \qquad (9.5) \\
V(Y_i) &= \sigma^2
\end{aligned}
$$

So you see, in the simple linear regression model, the Y's are independent and have a common variance, but differ in their means precisely because of the presence of x_i. In all but trivial cases, the x's will be selected to be different, so there is never a question in this case of the means being tested for equality, as in AOV. We do not know the values of m and b and, of course σ^2, and that is where statistical estimation comes in. Theorists have shown that the least squares formulas produce estimators that have optimal properties, particularly when the E's, hence the Y's, are assumed to be normally distributed—the only case we will consider. In summary then, we assume $Y_1, Y_2, \ldots, Y_n$ are independent normal random variables with means $mx_i + b$ and common variance σ^2. We shall refer to this as the "simple linear regression model."

Viewed in advance, the least squares equations define *estimators*, since then the Y_i's are random. The estimator for slope

$$\hat{M} = \frac{\Sigma(x_i - \bar{x})(Y_i - \bar{Y})}{\Sigma(x_i - \bar{x})^2}$$

is a normal random variable with mean and variance

$$E(\hat{M}) = m \qquad V(\hat{M}) = \frac{\sigma^2}{\Sigma(x_i - \bar{x})^2} \tag{9.6}$$

The estimator $\hat{B}$ for the intercept is also normal, with mean and variance

$$E(\hat{B}) = b \qquad V(\hat{B}) = \frac{\sigma^2\,\Sigma x_i^2}{n\Sigma(x_i - \bar{x})^2} \tag{9.7}$$

The estimator for $\sigma^2 = V(E_i)$ is essentially the sample variance of the observed errors, $Y_1 - \hat{Y}_1, Y_2 - \hat{Y}_2, \ldots, Y_n - \hat{Y}_n$, except that the error sum of squares ESS $= (Y_i - \hat{Y}_i)^2$ is divided by d.f. $= n - 2$, rather than $n - 1$. Hence, with $S^2 = \text{ESS}/(n - 2)$, we have the estimator

$$S^2 = \frac{\Sigma(Y_i - \hat{M}x_i - \hat{B})^2}{(n - 2)}$$

whose mean is σ^2. The random variable $(n - 2)\,S^2/\sigma^2$ has a Chi-square distribution, with $n - 2$ degrees of freedom.

With the above theoretical facts, we can find confidence intervals and test hypotheses about the parameters m and b. The appropriate formulas are programmed in ZS–9 and are based on the following facts.

$$\frac{\hat{M} - m}{\sqrt{\dfrac{S^2}{\Sigma(x_i - \bar{x})^2}}} \quad \text{has a } t\text{-distribution with } n - 2 \text{ d.f.}$$

$$\frac{\hat{B} - b}{\sqrt{\dfrac{S^2\,\Sigma x_i^2}{n\Sigma(x_i - \bar{x})^2}}} \quad \text{has a } t\text{-distribution with } n - 2 \text{ d.f.}$$

For the record, a $100\,(1 - \alpha)\%$ CI for m is given by

$$\hat{m} \pm t_{\alpha/2}\, s \sqrt{\frac{1}{\Sigma(x_i - \bar{x})^2}}$$

and the value of a test statistic for a null hypothesis about m, with hypothetical value m_o, is

$$ts = \frac{(\hat{m} - m_o)\sqrt{\Sigma(x_i - \bar{x})^2}}{s}$$

When $m = m_o$, the statistic TS has a t-distribution with $n - 2$ degrees of freedom. You will probably have little need to refer to these expressions or the ones we state below, since the computations are automatic in ZS–9.

Similarly, again for the record, a 100 $(1 - \alpha)$ % CI for b is given by

$$\hat{b} \pm t_{\alpha/2}\, s\sqrt{\frac{\Sigma x_i^2}{n\Sigma(x_i - \bar{x})^2}}$$

and the value of a test statistic for a null hypothesis about b, with hypothetical value b_o, is

$$\text{ts} = \frac{(\hat{b} - b_o)\sqrt{n\Sigma(x_i - \bar{x})^2}}{s\sqrt{\Sigma x_i^2}}$$

When $b = b_o$, the statistic TS has a t-distribution with $n - 2$ degrees of freedom. In all these cases, the percentile, $t_{\alpha/2}$, is from a t-distribution with d.f. $= n - 2$.

In your reading of statistics texts, you may encounter the remark that the "least squares" estimators, $\hat{M}$ and $\hat{B}$, satisfy an optimality property known as the *Gauss-Markov* property. To understand this property, you need to know that $\hat{M}$ and $\hat{B}$ are both linear functions of the Y_i's, considering the x_i's to be known constants. We mentioned before that $E(\hat{M}) = m$ and $E(\hat{B}) = b$, that is, the estimators are unbiased. The Gauss-Markov property of $\hat{M}$ and $\hat{B}$ is the property that, among *all* linear functions of the Y_i's which are unbiased estimators for m or b, and hence might have been used to estimate m or b, the least squares estimators $\hat{M}$ and $\hat{B}$ have the smallest possible variance. Such estimators are sometimes said to be *best* (for minimum variance) *linear unbiased estimators.*

Now let us discuss how ZS–9 can be used to calculate confidence intervals or significance levels. Load that program and consult the user instructions. You begin by entering your data pairs and compiling a data base, as instructed under code DE, Steps 1 and 2. At Step 3 the data are compiled. Since the confidence intervals for m and b both involve the same t-distribution, if you wish to calculate confidence intervals for m or b (or both), you may enter the required percentile, $t_{\alpha/2}$, at this point or manually store it in R_{31} at any time. Confidence intervals are then calculated and displayed in the usual way by simply pressing [A] or [B] as directed under codes SLOPE and INT in ZS–9. These steps may be repeated at will with any desired percentile in R_{31}.

For testing hypotheses about m or b, you enter an H_1-code and hypothesized value, just as you have been doing in other programs. The significance level, P, of the test is then calculated, as indicated in the user instructions under the same codes as CIs. We will illustrate the use of these instructions by example.

EXAMPLE 9.4

The mileage data in Example 9.1 are reproduced here.

Weight	2,830	2,700	3,492	2,590	4,100	3,050
mpg	27.2	25	15.1	31.3	12.5	25.9
	26.4		17.8		13.1	
	26.6					

Assume the simple linear regression model for mileage, regressed on weight.

a. Find a 95% CI for m.
b. Find a 95% CI for b.
c. Conduct a significance test of $H_o{:}m = 0$ vs. $H_1{:}m \neq 0$.
d. Test $H_o{:}b \geq 65$ vs. $H_1{:}b < 65$, with a test of size $\alpha = .10$.

Solutions Follow the data entry procedure, DE in ZS–9 through data compilation at Step 3. Enter $t_{.025} = 2.306$ at Step 4, and press [R/S] to prepare for 95% CIs. Op codes all function.

a. Press [A] to display $l = -.014$; [$x \rightleftarrows t$] to display $u = -.009$.
b. Press [B] and see $l = 49.8$; [$x \rightleftarrows t$] to see $u = 68.2$.
c. Enter H_1-code $= 0$ and press [A′], enter $m_o = 0$ and press [R/S] to display the significance level $P = 1.4 \times 10^{-5}$ (reject H_0).
d. Enter H_1-code $= -1$ and press [B′], enter $b_o = 65$, and press [R/S] to see $P = .086$. Reject H_0.

You are invited to practice these routines with the problems that follow. You may want to consult the Register Contents on page 271 for various individual quantities of interest in a given problem.

Problems

16. The weights of fathers and their oldest sons when fully grown are given for a large southern Texas city.

Father:	130	148	155	170	140
Son:	135	152	160	155	140

a. Find the equation of the least squares regression line, regressing son's weight (y) on father's weight (x).

Ans: $\hat{y} = .58x + 61.9$

b. What is a 95% CI for the slope?

Ans: $-.11 < m < 1.27$

c. Compute an estimate of σ.

Ans: $s = 6.58$ (R_{33})

d. If a father weighs 150 pounds, what would you estimate his son's weight to be when fully grown?

Ans: 149.2

e. If a son weighs 150 pounds when fully grown, what would you estimate his father's weight to be?

Ans: 151.35

f. Test $H_o: b = 0$ vs. $H_1: b \neq 0$ with size $\alpha = .10$.

Ans: $P = .15$. Do not reject H_o.

17. The public relations officer of a large resort hotel has noticed a relationship between flyers mailed to persons on her mailing list and registrations at the resort. Her records indicate the following relationship.

Flyers mailed:	10,000	15,000	20,000	30,000
Registrations:	400	500	620	780

a. Estimate the regression of registrations on flyers.

Ans: $\hat{y} = .019x + 217$

b. If the resort mails out 25,000 flyers, how many registrations should they expect?

Ans: 694

c. Construct a 95% CI for the intercept.

Ans: $114.1 < b < 320.2$

d. Test the hypothesis $H_o{:}m \leq 0$ vs. $H_1{:}m > 0$ (use $\alpha = .05$).

Ans: $P = .002$ Reject H_o.

e. Estimate the standard deviation of $\hat{M}$.

Ans: .001 (R_{37})

18. The Recruiting Command for one of the armed services is trying to develop a method for predicting performance of inservice recruiters. At the beginning of a 6-month training course, the recruiters were given an aptitude test to measure sales ability. Records were then kept on the job performance for one month, with the following results.

Aptitude score:	50	40	60	75	55	80
Soldiers recruited:	15	10	21	27	17	28

a. Construct the regression line relating performance to the aptitude score as the carrier.

Ans: $\hat{y} = .46x - 7.99$

b. If a recruiter scored 65 on the aptitude test, how many soldiers would you expect him to recruit?

Ans: 22

c. What is a 95% CI for m? for b?

Ans: $.39 < m < .53$; $-12 < b < -4$

19. Assume the following values for 2 variables satisfying the regression model:

x:	7	9	6	3	10
y:	13	17	12	7	21

a. Estimate the regression line of y on x.

Ans: $\hat{y} = 1.9x + .7$

b. Estimate y when $x = 8$.

Ans: 15.9

c. Estimate x when $y = 10$.

Ans: 4.9

d. Construct a 95% CI for m.

Ans: $1.3 < m < 2.5$

e. Test $H_o{:}b \leq 0$ vs. $H_1{:}b > 0$ at level $\alpha = .05$.

Ans: $P = .34$. Do not reject H_o.

f. Estimate σ^2.

Ans: $s^2 = 1.23$

20. Suppose a random sample of 5 families had the following annual incomes and expenditures for automobiles.

Family	Income	Auto Expenditure
1	20,000	9,000
2	40,000	15,000
3	12,000	4,000
4	25,000	8,800
5	30,000	10,500

a. Estimate the regression of Auto Expenditures on Income.

Ans: $\hat{y} = .36x + 200$

b. Construct a 95% CI for m.

Ans: $.21 < m < .52$

c. Construct a 95% CI for b.

Ans: $-4{,}110 < b < 4{,}509$

d. If a family's income was $15,000, what would they most likely spend for an automobile?

Ans: $5,668

e. What is the sample standard deviation of Auto Expenditures?

Ans: $s_y = \$3{,}944$

21. To encourage participation, a bowling alley gave free instruction to all who desired it. From a sample of 20 of the participants, long-term average scores before instruction and observed scores after the instruction period were obtained. Based on the results given below, would you regard the instruction as beneficial?

PARTICIPANT	BEFORE SCORE	AFTER SCORE
1	90	95
2	105	112
3	78	90
4	125	126
5	150	148
6	85	90
7	100	102
8	102	115
9	94	99
10	129	125
11	152	153
12	80	90
13	76	80
14	140	145
15	95	110
16	104	112
17	89	100
18	110	115
19	115	122
20	99	108

Ans: Regress the after score on the before score to obtain $\hat{y} = .87$

$x + 20$. For before scores in the range 80 to 150, you can use [Op] 14 to see that the predicted after scores are larger than the corresponding before scores. This is evidence that the instruction was beneficial. However, with $\hat{m} = .87$, you can see that for before scores larger than 150 or so, the predicted after score is not larger than the corresponding before score. Perhaps the instruction is most beneficial for those who need it (before scores less than 150).

Section 9.3: Prediction and Discrimination

We have discussed how you can use ZS–9 to make point and interval estimates for the parameters of the simple linear regression model. You also know how to use [Op] 14 and [Op] 15 to make point estimates of the expected value of Y for a given x and for the x which produces a given $\hat{y}$. We shall now extend the latter 2-point estimation procedures to include interval estimates.

Let us begin with the "forward" problem of making interval estimates for μ_y and Y, corresponding to a given value of x, say x_0. There are two senses in which we can use intervals to estimate the response corresponding to x_0. One sense is concerned with the *mean* response $\mu_o = E(Y)$ at x_0, that is, a confidence interval estimate of $mx_0 + b$. Clearly, the point estimator for this value is $\hat{M}x_0 + \hat{B}$, and, since $\hat{M}$ and $\hat{B}$ are normal, so is this estimator. Consequently,

$$\frac{\hat{M}x_0 + \hat{B} - (mx_0 - b)}{\sqrt{\text{Estimated variance of } \hat{M}x_0 + \hat{B}}} \tag{9.8}$$

has a t-distribution with $n - 2$ degrees of freedom, and this can be exploited to give CIs for $mx_0 + b$.† Program ZS–9 contains a routine for carrying out the computations under code μ at x_0. We refer to this as a CI for $mx_0 + b$. It is found by entering x_0 and pressing [C].

The second sense of estimating the response with an interval concerns giving an interval which will contain the actual response, say, Y_0, with probability $1 - \alpha$. We refer to such an interval as a $100(1 - \alpha)\%$ *prediction* interval for $Y_0 = mx_0 + b + E$, when x is fixed at value x_0. We must, of course, still estimate m and b as before, but predicting the response also includes compensating for the random error term E. Consequently, the prediction interval for Y_0 is based on the t-distribution, that of

†It turns out that the variance of $\hat{M}x_0 + \hat{B}$ is estimated by

$$\frac{s^2}{n}\left[1 + \frac{n(\bar{x} - x_o)^2}{\Sigma(x_i - \bar{x})^2}\right]$$

$$\frac{(Y_0 - \hat{Y}_0)}{\sqrt{\text{estimated variance of } (Y_0 - \hat{Y}_0)}} \tag{9.9}$$

Here, Y_0 is the (future) response that will be observed with carrier value x_0, and $\hat{Y}_0 = \hat{M}x_0 + \hat{B}$ is the point estimator of the mean of Y_0.† Program ZS–9 contains a *prediction interval* routine under code Y at x_0 for carrying out the computations for this interval. It is found by entering x_0 and pressing [C′]. Since the prediction interval must contend with the additional error term E, you should expect that, for a given data set and x_0, the prediction interval for Y_0 will be wider than is the CI for the mean of Y_0.

EXAMPLE 9.5

Crop yield of dry land wheat in southwestern Colorado for 10 years, and the annual rainfall during those years, were as follows.

Rainfall:	12.8	20.0	14.9	15.0	11.3	12.9	16.1	20.0	13.1	15.6
Yield:	10	25	14	16	7	9	19	20	8	17

a. Compute a 95% CI for the expected yield, when rainfall is 16.
b. Find a 95% PI for the yield that will be observed, if rainfall is 16.

Solution Enter the data, according to the DE procedures for ZS–9, including entry of $t_{.025} = 2.306$ (d.f. $= 8$). Enter $x_0 = 16$, press [C] to obtain the CI (14.4, 17.7). Again enter the carrier value $x_0 = 16$, and press [C′] to obtain the PI (10.9, 21.3). Notice the PI is wider than the CI. Retain the data base for the next example.

Finally, consider the problem of obtaining a CI for the "inverse" problem, in which we wish to estimate the value of a carrier, x, which gave rise to an observed response, y. This has come to be known as *discrimination.* An example of this application is as follows: suppose an experiment was carried out in which the responses $y_1, y_2, \ldots, y_n$ of subjects to drug doses at respective fixed levels $x_1, x_2, \ldots, x_n$ were observed. Suppose the regression of response on dosage has been calculated. Now, at a hospital, an emergency case has been admitted in which the response, say, y^*, is observed, and it is desired to estimate the dosage, x^*, responsible. You have used [Op] 15 before to calculate point estimates of x^*. But now we want to extend this to confidence intervals for x^*. The theory that yields a CI for x^* is more complicated than that used for prediction of Y_0, although it is similar. The interval, based once more on a t-distribution with $n - 2$ d.f., is given by

† For the record, the estimated variance of $Y_0 - \hat{Y}_0$ is

$$\frac{s^2}{n}\left[1 + n + \frac{n(\bar{x} - x_o)^2}{\Sigma(x_i - \bar{x})^2}\right]$$

$$\bar{x} + \frac{m(y^* - \bar{y})}{d} \pm \frac{s}{d} \cdot t_{\alpha/2}\sqrt{(1 + \frac{1}{n})d + \frac{(y^* - \bar{y})^2}{\Sigma(x_i - \bar{x})^2}} \tag{9.10}$$

where

$$d = \hat{m}^2 - \frac{(s \cdot t_{\alpha/2})^2}{\Sigma(x_i - \bar{x})^2}$$

We call this a *discrimination* interval for x^*. Program ZS–9 (code DISC) can be used to carry out the computations. You need only input y^* and press [D′] to find the confidence limits.

EXAMPLE 9.6

For the wheat yield data of Example 9.5, suppose someone reports the current dry land wheat yield is 20. Give a 95% discrimination interval for this year's rainfall.

Solution Enter $y^* = 20$, and press [D′] to obtain the interval (15.3, 21.3).

The following remarks concerning the prediction and discrimination procedures discussed above should be noted:

1. In the discrimination procedure, for data sets in which $\hat{m}$ is close to 0, the factor d may be sufficiently negative, that the expression under the square root sign in Equation 9.10 is negative. In such cases, the CI for x^* does not exist,† and the display of your calculator will flash to signal this fact. (See problem 29.) This problem is analogous to the one experienced by [Op] 15 when $\hat{m}$ is 0.

2. As mentioned before, when x_0 is selected outside the range of the data, there is a problem with extrapolation, and it is fraught with danger. Of course, the calculator will calculate an estimate for you, regardless of where you choose x_0. Fortunately, in the case of PIs and CIs, there is a built in safeguard. As you can see by examining Equations 9.8 and 9.9, the expressions involving the variance have $(\bar{x} - x_0)^2$ in their numerators, which causes the resulting CIs to have shortest width when x_0 is chosen near the "center" of the data, $\bar{x}$. As you chose x_0 values farther away from $\bar{x}$, the confidence and prediction intervals will get wider. A similar comment holds for the location of y^* relative to $\bar{y}$ in the discrimination interval. This effect is particularly evident when $\hat{m}$ is near 0 (for discrimination intervals) or $|\hat{m}|$ is very large (for CIs or PIs).

† Put another way, the CI is the entire real line of values.

22. A test was performed to determine the relationship between the chemical content of a particular constituent (y) in solution and the crystallization temperature (x). The data follow.

Degrees centigrade x:	-1.7	-0.4	0.2	1.1	2.3	3.1	4.2	5.3
Grams per liter y:	1.1	2.3	3.2	4.3	5.4	6.6	7.8	8.8

a. Estimate the relationship by the method of least squares.

Ans: $\hat{y} = 1.13x + 2.95$

b. Test the hypothesis that $m = 1$ with a significance test.

Ans: $P = .0015$. Reject $H_0{:}m = 1$.

c. Determine 95% CIs for each of m and b.

Ans: $1.07 < m < 1.19$; $2.79 < b < 3.11$

d. Estimate the expected value of Y_0, given $x_0 = 5$, by a 95% CI.

Ans: $8.37 < \mu_{.5} < 8.82$, that is, $8.37 < (.5)\, m + b < 8.82$

e. Predict the amount of the constituent in a future trial if crystallization temperature is 6 degrees Centigrade.

Ans: $\hat{y}_0 = 9.72$

f. Predict the amount in (e) with a 95% PI.

Ans: $9.27 < Y_0 < 10.17$

g. Compute a point and 95% interval estimate of the temperature at which a new reading of $y^* = 3.5$ was observed.

Ans: $.489$; $.141 < x^* < .831$

23. The elongation of a piece of boiler plate (y) is related to the amount of force applied in tons psi (x). The data follow.

Force x:	1.33	2.68	3.57	4.46	5.35	6.24	7.14	8.93	9.82	10.70
Elongation Y:	27	50	67	83	101	117	134	150	188	206

a. Estimate the regression of elongation on force.

Ans: $\hat{y} = 18.5x + .698$

b. Test the hypothesis that $m = 20$ at the 5% level.

Ans: $P = .069$. Do not reject $H_o{:}m = 20$.

c. If the elastic limit is reached when $x = 16$, predict the elongation for this limit by a 95% interval.

Ans: $275 < Y_0 < 320$

d. If elongation is 175, what is the most likely amount of applied force?

Ans: 9.4

e. Calculate a 95% discrimination interval for x^*, if $y^* = 100$.

Ans: $4.49 < x^* < 6.22$

24. Under a certain set of conditions, 5 temperatures were read on a pyrometer and compared with true temperatures. (The data are in °C.)

Pyrometer readings x:	950	975	1,000	1,025	1,050
True temperatures y:	1,082	1,118	1,156	1,193	1,231

a. Estimate the relationship by the method of least squares.

Ans: $\hat{y} = 1.492x - 336$

b. Predict the true temperature for a pyrometer reading of 1,100°C by a 99% interval.

Ans: $1{,}299 < Y_o < 1{,}311$

c. Test the hypothesis $m = 1.3$ at the 5% level.

Ans: $P = .0001$. Reject $H_o{:}m = 1.3$.

d. Estimate the pyrometer reading which would give a temperature of 1,200 with a 99% interval.

Ans: $1{,}027 < x^* < 1{,}032$

25. A piece of metal ½ inch in diameter and 4 in. in length is subjected to a test to determine how much strain is set up by specified values of twisting movement. The data follow.

Twisting movement (pounds/inch):	100	300	500	700	1,000	1,200	1,360
Strain (torsion units):	111	329	551	769	1,103	1,322	1,497

a. Estimate the regression of strain on movement by the method of least squares.

Ans: $\hat{y} = 1.1x - .22$

b. Determine a 95% CI for the expected value of strain if the elastic limit of 1,400 pounds/inch has been reached.

Ans: $1{,}539 < 1{,}400m + b < 1{,}544$

c. Test the hypothesis that $m = 1$ at the 5% level.

Ans: $P \doteq 0$. Reject $H_o{:}m = 1$.

d. Predict by a 95% interval the strain at the elastic limit (see part b).

Ans: $1{,}537 < Y_o < 1{,}546$

26. The gauging laboratory of a manufacturing firm obtained the following data (in inches) by measuring a group of standard gauge blocks with a micrometer used by the inspection department.

STANDARD x	MICROMETER READING y
0.5000	0.501
0.5500	0.552
0.6000	0.597
0.6500	0.646
0.7000	0.703
0.7500	0.754
0.8000	0.798
0.8500	0.853
0.9000	0.902
1.0000	1.006

a. Compute estimates of m and b, where $E(Y) = mx + b$.

Ans: $\hat{m} = 1.01 \quad \hat{b} = -.006$

b. Determine a 90% CI for m.

Ans: $.998 < m < 1.021$

c. Test the hypothesis that $m = 1$ at level $\alpha = .05$.

Ans: $P = .15$. Do not reject the hypothesis.

d. Determine a 90% CI for the expected value of Y_0, given that $x_0 = .8000$.

Ans: $.800 < .8m + b < .804$

e. Guess the value of x that corresponds to a micrometer reading of .750, and give a 90% discrimination interval.

Ans: $\hat{x}^* = .7486$; $.743 < x^* < .754$

27. In the construction of mutual characteristic curves for the Type 6J5 triode, the plate voltage is held constant at 100 volts. Then the grid voltage is varied and the plate current observed. The data follow.

Grid Voltage, x (volts):	−4.00	−3.50	−3.00	−2.50	−2.00	−1.50	−1.00	−0.50	0.00
Plate Current, y (milliamperes):	1.32	1.73	2.71	3.42	4.59	5.41	6.33	9.29	12.00

a. Fit a straight line to the data.

Ans: $\hat{y} = 2.49x + 10.18$

b. Determine a CI for the slope.

Ans: 95% CI: $1.82 < m < 3.15$

c. Predict the plate current by an interval, when the grid voltage is −2.21 volts.

Ans: $1.96 < Y_0 < 7.40$ (95%)

d. What is the estimated standard deviation of $\hat{B}$?

Ans: .67 (R_{36})

e. Calculate a 95% CI for $b - 2.21m$.

Ans: $3.81 < b - 2.21m < 5.55$. (Compare with part c.)

28. The manager of Freshaire Hotel has become concerned with the number of complaints received from guests about poor room service. The hotel has 50 rooms on each of floors 2 through 7. For the past 2 months, records have been maintained as follows.

floor number x:	7	6	5	4	3	2
complaints y:	50	40	45	36	32	25

The manager suspects that the extra time required to reach the upper floors may be causing more complaints from guests on those floors. Is there a relationship between floor number and number of complaints?

Ans: Regress number of complaints on floor number, and test $H_0{:}m \leq 0$ vs. $H_1{:}m > 0$. Since $P = .003$, there is evidence of a positive relationship.

29. The growth (in inches) of a certain hybrid corn is related to time (in weeks).

Time x:	1	2	3	4	5
Growth y:	.5	.8	.9	2.0	1.8

a. Regress growth on time, and predict the actual growth for the sixth week.

Ans: $\hat{y} = .38x + .06$; $\hat{y} = 2.34$ when $x = 6$

b. Find a 99% discrimination interval for time corresponding to a growth of 2.5 in.

Ans: The entire real line, as indicated by the flashing display. In R_{42}, you will find that $d = -.19$ for this case.

c. Find a 99% CI for the true slope m.

Ans: (−.2, .96)

30. In a study of weight gain (y) versus excess calorie intake (x), some data were recorded as follows.

x (in 100 cal):	0	1	2	3	4	5	6	7	8	9	10
y (in lb):	−0.2	0.1	0.8	0.4	0.1	1.1	0.9	0.2	1.6	1.9	2.0

a. Determine the regression of y on x.

Ans: $\hat{y} = .19x - .145$

b. Estimate the mean weight gain for the average calorie excess, both by a point and a confidence interval estimate.

Ans: $\bar{x} = 5$ and $.497 < 5m + b < 1.12$. Confidence 95% around a point estimate of $\hat{\mu}_5 = .81$.

c. Test the hypothesis that there is no relationship between weight gain and calorie excess.

Ans: H_o:$m = 0$ yields $P = .0018$, so that H_o should be rejected at almost any standard level.

d. If a person gains 1 lb, what are the plausible levels of calorie excess? Some people claim they gain weight with no calorie increase. Is this plausible?

Ans: With confidence 95%, the excess calories range from −28 to 1,301, too wide to be of much use. Yes, 0 is in the CI.

e. Test the hypothesis that the regression line passes through the origin.

Ans: For H_0:$b = 0$ against H_1:$b \neq 0$, $P = .59$, tending to support H_0.

f. Compare the 95% CI for mean weight gain corresponding to a 500 cal increase to the PI for actual weight gain.

Ans: For mean, CI is (.5, 1.12), a span of .6 lb, while PI is (−.27, 1.89), a span of 2.2 lb.

Section 9.4: Correlation

In Chapter 6, you were introduced briefly to the concept of the correlation, ρ, between two random variables and the use of [Op] 13 to compute its point estimate, r, from paired data. The model for simple linear regression involves pairs (x, Y), where x is a fixed (nonrandom) carrier value, and Y is a random response with that x. In many situations, it may be reasonable to view x as the observed value of a random variable X. In this case, (X, Y) has a bivariate distribution, and the correlation between X and Y may be meaningfully discussed.

Several facts should be observed concerning r and ρ. First, both are *symmetric*, in the sense that if you interchange x and y (or X and Y), the value of r (or ρ) remains the same. Thus, whatever r or ρ means about X and Y, in that order, it means the same thing about Y and X, in *that* order. Second, r is positive or negative, according to whether y generally increases or decreases with x. Third, r, like ρ, will always be between -1 and 1. The closer r is to its extremes -1 or $+1$, the closer the (x, y) data must be to falling on a line. If the data were actually on a line, the correlation would be -1 or $+1$.

It seems obvious that r, then, would be used in any problem to estimate ρ. Of course, it would hardly ever be the case that r, subject to random fluctuations, would exactly equal ρ. Nor would we expect to see values of r exactly 0 or 1 or -1 or any other specified number. So even, for example, if we *knew* $\rho = 0$ and drew a sample from that bivariate population, chances are pretty good that the computed r will not be 0, although we might expect it to fall around 0.

What then does r tell us about ρ and the underlying population? A quick answer is this—unless X and Y have a bivariate normal distribution, hardly anything! Why bother with it then? Because, as a matter of fact, the joint distribution of many of the phenomena we try to model in real life (like mathematics ability versus verbal ability, for example) can be safely assumed to have such a distribution. It turns out that, in this case, the conditional distribution of Y, given $X = x$, is normal, with a mean which is a linear function of x, say, $mx + b$, and a constant variance. The latter (conditional) distribution is exactly the model we have used in simple linear regression. So, you see, there is a connection between regression and bivariate normal distributions through the conditional distribution. You should then suspect there is a connection between ρ and the regression parameters; indeed, it turns out that $m = \rho\sigma_Y/\sigma_X$!

Since ρ measures "degree of linear association" between X and Y, it is of interest to develop interval estimates for ρ and tests of hypotheses about ρ. But we confine our attention here to testing the hypothesis that $\rho = 0$. This is of special interest in the current setting of normal distribution assumptions, for, in that case (and only then), X and Y are independent. But it turns out that, when $\rho = 0$, the statistic

$$t = \frac{\sqrt{n-2}\,R}{\sqrt{1-R^2}}$$

has a t-distribution with $n - 2$ degrees of freedom. We can use this fact to conduct a test of $\rho = 0$. This procedure is available in Program ZS–9, under code CORR. All you have to do is enter the data and initialize as you did for regression tests. Then enter the usual H_1-code and press [E]. At the conclusion, the significance level P will be displayed to allow you to test H_0 with any size you desire.

EXAMPLE 9.7

Suppose a sample of 8 freshmen yields the following pairs of calculus (x) and chemistry (y) scores.

Student number:	1	2	3	4	5	6	7	8
Calculus score:	36	81	50	50	73	66	56	69
Chemistry score:	34	66	62	39	48	44	46	62

a. Calculate $\hat{\rho}$.

b. Conduct a significance test of $H_o{:}p = 0$ vs. $H_1{:}p \neq 0$.

Solution Clear the data base, (press [CMs]), enter the data using DE, Steps 1, 2, 3, and press [Op] 13 to see $r = .63$. Then enter H_1-code $= 0$ and press [E] to obtain $P = .09$. Thus, $H_0{:}\rho = 0$ would be rejected at any level above 9%.

This will cover all tests for $\rho = 0$ for most of the cases you will encounter, since usually you will have to start with raw data. Occasionally, however, you may be given only the summary data n and r. In that case, you may simply store $\nu = n - 2$ in R_{15}, r in R_{44}, and proceed as above.

For example, with $r = .6$ for 10 students, this procedure leads to $P = .067$ for $H_1{:}\rho \neq 0$; with $r = -.2$ for 40 students and $H_1{:}\rho < 0$, $P = .11$.

We already mentioned the fact that when X and Y have a bivariate normal distribution, the conditional mean of Y given $X = x_i$ is a linear function of x_i, in which the slope is given by $m = \rho\sigma_Y/\sigma_X$. That means $m = 0$ if and only if $\rho = 0$, and so you see that there must be an intimate relationship between correlation and regression.

But that only applies to this particular set of assumptions! You have seen regression models arising in other ways, for example, when x is a physical variable, like time, or a fixed level of concentration. In cases like this, it may not even make sense to talk about the correlation between x and Y. Of course, the calculator will calculate an r anyway, for it knows nothing about the model you are assuming. The point is, there may be no meaning to ρ when x is not random. In both models, the x's are fixed, as we have repeatedly emphasized. The difference—and this is important—is that in the bivariate normal model, the x's are fixed values of a *random variable* (like a prescribed set of IQ levels, for example). Then, and only then, can we relate the slope of the regression line to the correlation between X and Y. This makes regression models more general than correlation models, but when the bivariate normal assumptions are made in a model, there is a close connection.

First, if the data are really regression data, that is, the pairs (x_i, y_i) were obtained by *fixing* $x_1,\ldots,x_n$, and letting y_i be a random observation at x_i, then we know the true regression line has the property that

$$m = \frac{\rho\sigma_Y}{\sigma_X} \tag{9.11}$$

In addition to all the other considerations regarding m and b that we have discussed previously, we see that, additionally, we may test $H_0{:}\rho =$

0 by simply applying the t-test for H_0:$m = 0$, using SLOPE in ZS–9. That is because $\rho = 0$ if, and only if, $m = 0$.

There is also a relationship among the estimates analogous to Equation 9.11, namely,

$$\hat{m} = r\frac{s_y}{s_x} = r\sqrt{\frac{\Sigma(y_i - \bar{y})^2}{\Sigma(x_i - \bar{x})^2}} \tag{9.12}$$

From this, it may be shown that

$$r^2 = \frac{\Sigma\,(y_i - \bar{y})^2 - \Sigma\,(y_i - \hat{y}_i)^2}{\Sigma\,(y_i - \hat{y})^2} = \frac{\Sigma\,(\hat{y}_i - \bar{y})^2}{\Sigma\,(y_i - \bar{y})^2} \tag{9.13}$$

You see that the quantity r^2, called the *coefficient of determination* in the regression, is the proportion of the "total variation" of y, $\Sigma(y_i - \bar{y})^2$, explained by fitting the regression line. The coefficient of determination is often used as a measure of how useful the regression line is for predicting—the closer to 1, the better.

If you study Equation 9.13, you see that, if you predict y_i from $\hat{y}_i$ (the point on the regression line), instead of $\bar{y}$ (which is what you would use without any regression model), then the numerator is the reduction in sum of squares that results, and, algebraically, that amounts exactly to $\Sigma(\hat{y}_i - \bar{y})^2$. But this is r^2 times the total sum of squares of the y's about $\bar{y}$, which is the total variation without any regression. That is why we often refer to r^2 as the proportion of total variation "explained" by regression. Of course, what remains, $\Sigma(y_i - \hat{y}_i)^2$, is unexplained by regression, and that would be $1 - r^2$. For instance, in Example 9.7, if we regress chemistry scores on calculus scores, then about 40% ($.63^2$) of the total variation in chemistry scores can be explained by the regression of chemistry scores on calculus scores, indicating only modest success in predicting chemistry scores from calculus scores.

Finally, as we promised earlier, we will discuss the so-called regression "fallacy" that we brought up in the beginning of this chapter. Now that we understand the model, we can talk sensibly about why Galton's regression phenomenon is not as mysterious as it looks. First, we can imagine a regression scatter of father-son heights in Galton's study as something like that given in Figure 9.1. The dotted barrier, let us say, marks off what might reasonably be called "tall" fathers, and the ellipse represents a contour on the bivariate normal distribution of (fathers' heights, sons' heights). You can imagine the ellipse to represent an outline of a plot of Galton's data. Now you can see that area B in the figure is larger than area A, so that within the region of sons of tall fathers (A plus B), the son's mean height will fall below the diagonal line. This implies:

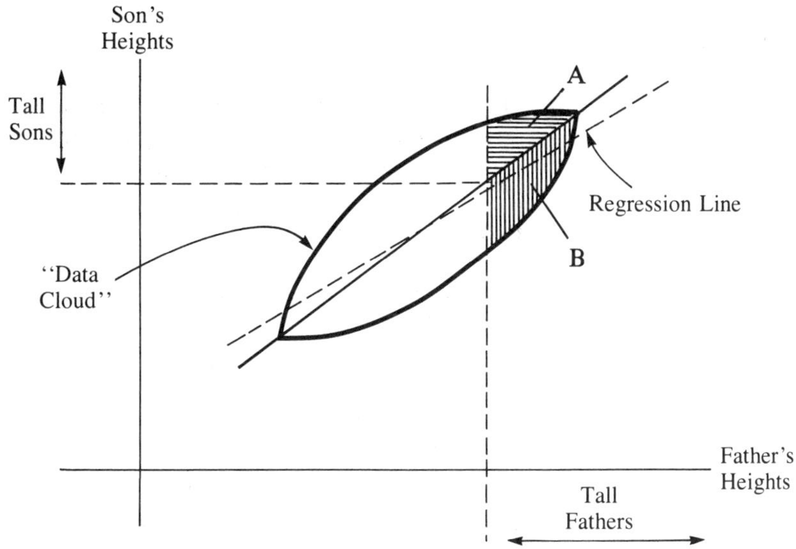

Figure 9.1. Father-son heights

a. The mean height of sons of tall fathers will be below the mean height of those fathers.
b. The regression line of sons' heights on fathers' heights (dotted line in Figure 9.1) will have slope less than 1.
c. If you were to regress fathers' heights on sons' heights, you would find *that* slope also less than 1, for the same reasons.

EXAMPLE 9.8

The following data pairs represent hypothetical data in Galton's experiment. If you plot the pairs, you will find that they are symmetric with respect to the line $y = x$.

a. Find the regression of y on x.
b. Find the regression of x on y, and compare with (a).
c. Find the coefficient of determination.
d. Discuss $\hat{m}$ in terms of the "regression fallacy."

x:	55	58	60	65	66	68	70	72
y:	58	55	65	60	68	66	72	70

Solution

a. $\hat{y} = .83x + 10.64$.
b. $\hat{x} = .83y + 10.64$; The estimates of m and b are the same as in (a).
c. $r^2 = .70$, in either case.
d. The estimated slope is less than 1, indicating "sons of tall fathers have mean heights tending to be not so tall." But the data are symmetric!

31. Managers of the Postal Inspection Service are interested in the relationship between salary level and years of education of their employees. A random sample of 10 employees yielded the following data.

EDUCATION (years)	SALARY (in thousands of dollars)
12.5	20.2
14.0	23.7
12.0	19.8
16.0	21.5
11.0	21.8
16.0	35.7
18.0	39.2
11.2	22.6
15.0	25.9
12.7	17.3

a. Test the null hypothesis $\rho = 0$ at the 5% level of significance.

Ans: $P = .009$; H_0 should be rejected.

b. Determine the regression of salary on education, and find the coefficient of determination.

Ans: $\hat{y} = 2.34x - 7.55$; $r^2 = .593$

c. Test the null hypothesis $m = 0$ and compare with (a).

Ans: $P = .009$ [Same as (a), because $\rho = 0$, if, and only if, $m = 0$.]

d. Determine the regression of education on salary, and find the coefficient of determination.

Ans: $\hat{y} = .254x + 7.55$; $r^2 = .593$ necessarily the same as (b).

e. Test the null hypothesis $m = 0$ for (d), and compare with (a) and (c).

Ans: $P = .009$. Necessarily the same, for $\rho = 0$, if, and only if, m (this time $\rho\, \sigma_x / \sigma_y$) is 0.

32. A study was carried out to compare IQ scores with vocabulary scores, as measured by a Vocabulary Test form. The results, based on a random selection of 20 students, follow.

IQ	VOCABULARY
89	125
71	107
90	121
80	105
88	108

88	107
87	104
76	88
82	93
81	90
84	93
88	93
84	86
110	109
102	100
82	78
84	78
85	77
96	85
100	100

a. Regress vocabulary on IQ.

Ans: $\hat{y} = .369x + 65.1$

b. Find the coefficient of determination.

Ans: .06, a very low value, indicating that IQ is a poor carrier for determining or predicting vocabulary.

c. Test $H_0{:}\rho = 0$.

Ans: $P = .296$, so that H_0 would be retained.

d. Find the sample variance of IQ scores.

Ans: $s_x^2 = 82.1$

e. Find a 99% discrimination interval for the IQ of a student having vocabulary 105.

Ans: The flashing display indicates the discrimination interval is the entire real line, that is, you cannot discriminate IQ values at all for a vocabulary of 105.

33. A local law enforcement officer believes there is a direct relationship between maximum temperature on a given day and the number of major crimes in his community. He bases this on observing the following data, which he pulled out of the files.

Maximum temperature:	12	89	40	52	75	60	50	20	32	90
Major crimes:	2	16	6	8	14	12	7	3	5	16

a. What is the quality of temperature as a predictor of major crimes?

Ans: $r^2 = .97$, so a regression of number of crimes on temperature would "account for" a large part (97%) of the variation in the number of crimes.

b. Test $H_0:\rho = 0$.

Ans: $P = 1.9 \times 10^{-7}$, so that H_0 should be rejected.

c. How many major crimes would you predict, if the temperature were to get up to 100?

Ans: This is dangerous extrapolation, but $\hat{y} = 18.1$.

d. Compute a 90% CI for m.

Ans: $.169 < m < .213$

34. For each of the following QPR ratings (x), a student was drawn at random and given a test which resulted in a score (y).

x:	2.0	2.1	2.2	2.3	2.4	2.5	2.6	2.7	2.8	2.9	3.0
y:	59	60	64	67	68	69	71	69	77	81	79
x:	3.1	3.2	3.3	3.4	3.5	3.6	3.7	3.8	3.9	4.0	
y:	80	86	82	88	89	93	96	95	99	96	

In the past, the equation $y = 20 + 20x$ was used to predict Y from x, thereby using a linear prediction model.

a. Determine the regression of y on x from these data.

Ans: $\hat{y} = 20.08x + 19.19$

b. Find 95% CIs for the true values m and b in the linear model. On this basis, are the values $m = 20$ and $b = 20$, currently being used, reasonable?

Ans: $18.5 < m < 21.7$; $14.3 < b < 24.1$. The values in use seem reasonable.

c. Would the assumption $b = 0$ be reasonable? Discuss why any such conclusion drawn from these data might be dangerous?

Ans: This concerns a value when $x = 0$, which is serious extrapolation; in any case, 0 is not in the CI for b.

d. After the experiment, a paper with a score of 90, but no record of a name or QPR, had been turned in. What is the most likely QPR value?

Ans: 3.53

e. One of the faculty members claimed that all of these results were nonsense, because in his view QPR had little to do with what was measured by y. Do the data support him? What is the value of r^2 for the data?

Ans: $r^2 = .97$, so there appears to be a relationship. Testing $\rho = 0$, $P \doteq 0$ further refutes the faculty member.

f. Of all the students past, present, and to come, whose QPR is 3.0, how would they rate *on the average* on this test?

Ans: 79.4 or, better, $78.5 < \mu_3 < 80.4$ with confidence 95%.

g. Later, one student with a QPR of 3.2 scored 90 on this test. Is this within a range of score points you would have predicted for such a QPR?

Ans: Not for a 95% prediction interval, which is $78.9 < Y_0 < 88.0$.

h. What range of values would you predict for the average QPR?

Ans: The average QPR is $\bar{x} = 3$. For $x_0 = 3$, the 95% prediction interval is $74.9 < Y_0 < 84.0$.

35. A comparison of motor skills versus ability in mathematics was carried out by administering appropriate tests to 20 subjects. A correlation of $r = .379$ was found. On this basis, it was concluded that $\rho = 0$ was entirely feasible. Would you agree?

Ans: A test of $H_0{:}\rho = 0$ produces a *P*-value of .099. It is feasible. H_0 would not be rejected for $\alpha < 10\%$.

36. The correlation between two achievement tests in entirely unrelated subjects given to a class of 36 students was found to be .89. Is this compatible with the assumption that the subjects are unrelated?

Ans: No. $P = 6 \times 10^{-12}$, with a large test statistic of $t = 11.38$. Not compatible with the assumption.

Section 9.5: Curve Fitting

We have mentioned that the assumption of a straight line model in regression may be questionable in some situations. If you have a set of (x, y) data, and you have an idea of the *form* of the curve relating y and x, a least squares fit of the curve can be obtained, in many cases, by *transforming* the data.

For example, in a population growth situation, we expect the population size, y, to grow exponentially with time, x. Thus, given a set of measurements of the population size at various times, we may try to fit the curve, $y = be^{mx}$, by estimating the parameters, b and m. If we take the natural logarithm of both sides of the equation, we obtain $lny = lnb + mx = mx + b^*$, say, where $b^* = lnb$. Now, an ordinary least squares fit of the *line* $y^* = mx + b^*$ to the transformed data (x_1, lny_1), $(x_2, lny_2), \ldots, (x_n, lny_n)$ can be obtained by entering these data with [Σ+] and using [Op] 12, as before. It is then common practice to "transform back" to the original units, once $\hat{b}^*$ and $\hat{m}$ are obtained, so that the prediction equation is of the form $\hat{y} = \hat{b}e^{\hat{m}x}$, where $\hat{b} = e^{\hat{b}^*}$.

Since the distribution properties of transformed errors can be very complicated, we shall restrict ourselves to the problem of estimating parameters through the transformation process described above. No attempt to test hypotheses or derive CIs will be made. *You should not attempt such inferences with transformed data.* The statistics module contains a program, ST–12, which makes it easy to carry out the least squares process with transformed data. With ST–12, you have "automatic" handling of four types of transformations, shown in Table 9.1 below, called "Preprogrammed Curves" in ASM (pages 5–8 through 5–13).

TABLE 9.1. PREPROGRAMMED CURVES WITH ST–12.

CURVE	MODEL	TRANSFORMED DATA-LINEAR	LINEAR PARAMETERS
exponential	$y = be^{mx}$	$y^* = lny \quad x^* = x$	$m^* = m,\ b^* = lnb$
power	$y = bx^m$	$y^* = lny \quad x^* = lnx$	$m^* = m,\ b^* = lnb$
logarithmic	$y = mlnx + b$	$y^* = y \quad x^* = lnx$	$m^* = m,\ b^* = b$
linear†	$y = mx + b$	none	

†The linear curve is the standard case covered by methods discussed for simple linear regression.

You can easily obtain the prediction model for the preprogrammed curves, following the user instructions on pages 5–10 and 5–11 of ASM. If you follow those instructions, then [C] at Step 6 outputs the original parameter estimates, $\hat{b}$ and $\hat{m}$, for you. That is, you do not have to convert manually from linear estimates $\hat{m}^*$ and $\hat{b}^*$ back to m and b.

EXAMPLE 9.9

Suppose the following data are believed to be fit by a power model, $y = bx^m$, where the values of the parameters b and m are unknown.

x:	1	2	3	4	5
y:	3.5	11.1	29.0	50	70

a. Transform the data by logarithms, and use ordinary linear regression to estimate $b^* = lnb$ and $m^* = m$.

b. Write the prediction equation for y in terms of $\hat{m}$ and $\hat{b}$.

c. Enter the data in ST–12 to estimate the parameters b and m directly.

d. Predict y, if $x = 2.5$.

Solution a. According to Table 9.1, we must transform the data (by taking the logarithms of each x and each y) before using [Σ+] entry and [Op] 12 solution. Using the [*lnx*] key, you can accomplish these transformations as you enter the data. Accordingly, clear memories with [CMs], enter 1 in R_D, and press [*lnx*][$x \rightleftarrows t$]. Then enter 3.5 in R_D, and press [lnx][$\Sigma+$]. This procedure has entered the first transformed data pair (ln 1, ln 3.5) = (0, 1.253) in the [$\Sigma+$] data entry sequence. Similarly enter the remaining four transformed data pairs, and press [Op] 12 to display $\hat{b}^* = 1.208$; press [$x \rightleftarrows t$] to display $\hat{m}^* = 1.911$.

b. To recover original parameters, $\hat{m} = \hat{m}^* = 1.91$ is already in R_D. Press [$x \rightleftarrows t$] to put $\hat{b}^*$ back in R_D, then [INV][lnx] produces $e^{\hat{b}^*} = \hat{b} = 3.35$. Thus, $\hat{y} = 3.35x^{1.91}$ is the prediction equation.

c. Follow Steps 1–6b of the user instructions for ST–12 on page 5–10 of ASM, choosing the Power option. Incidentally, at Steps 5a and 5b, it is x_i and y_i that you enter, *not* x_i^2 and y_i^2; the "exponent" refers to Note 2 on page 5–11. Following this procedure, you find $\hat{b} = 3.347$ by pressing [C] and $\hat{m} = 1.911$ is in R_T.

d. As directed in ASM, $\hat{y} = 19.27$ is found by entering 2.5 and pressing [D].

It is important to notice once again that, when you use the preprogrammed curves in ST–12, the "original" estimates are displayed at Step 6 (in our example, $\hat{b}$ is displayed at 6a), whereas when you transform by hand and use [$\Sigma+$] data entry with [Op] 12, the "transformed" estimates are displayed ($lnb = \hat{b}^*$ in our example). If you want to fit a curve with a form other than one of the preprogrammed curves, provision is made in ST–12 for entry of user-defined transforms. At Step 10, you first press [CP] to clear the calculator of all program steps. Naturally, this will erase any magnetic card programs that have been entered, so you may not use this routine with ZS–9, for example. Then you press [LRN] to enter the learn mode, and enter algorithms for the functions to perform the desired transforms. The transform $f(x)$ for x is set up as subroutine A′, and the transform $g(y)$ for y as subroutine B′. In this case, the estimates given at Steps 16a and 16c are "transformed" estimates; you must manually retransform these, if you want to write down the prediction equation in original form.

EXAMPLE 9.10

Suppose a theoretical argument shows that x and y are related by a curve of the form

$$y = \tan^{-1}\left(b + \frac{1}{a\sqrt{x}}\right) \qquad y \text{ in degrees} \tag{9.14}$$

The following data were collected.

x:	1.5	2.1	4.6	.5	1.8
y:	64.0	71.0	67.5	73.0	70.0

a. Obtain a least squares fit of the appropriate transformed data.
b. Write the prediction equation (transform the estimates to original form).
c. Predict y, if $x = .9$.

Solutions

a. Taking the tangent of both sides of Equation 9.16, you get

$$\tan y = b + m \frac{1}{\sqrt{x}} \tag{9.15}$$

where $m = 1/a$. The transformations are thus $f(x) = 1/\sqrt{x}$ and $g(y) = \tan y$. First, place the calculator in degree mode by pressing [Deg] to accommodate y. Then, enter these transformations at Step 10 of ST–12 as follows:

[CP][LRN][L*bl*][A'][$\sqrt{x}$][1/*x*][INV][SBR]
[L*bl*][B'][tan][INV][SBR][LRN]

Now select [Pgm] 12, initialize with [E'] and then press [D']. Now enter the data following Step 15, entering x_i and y_i, as given in the table. At the conclusion of data entry, press [C] to display the "transformed" intercept, $\hat{b} = 2$, and [$x \rightleftharpoons t$] to display the transformed slope, $\hat{m} = .83$.

b. The parameter b was unchanged by the transformation, so $\hat{b} = 2.00$ in original form; $a = 1/m$, so the procedure used at Step 16d of ST–12 requires the retransformed estimate $\hat{a} = 1/\hat{m} = 1.21$. Thus, the prediction equation is

$$\hat{y} = \tan^{-1}(2.00 + \frac{1}{1.21\sqrt{x}})$$

c. Using Steps 17a and 17b, enter .9, and press [RST][A'][Op] 14, which displays $g(\hat{y}) = 2.865$. Retransform to y by pressing [INV] [tan] to display $\hat{y} = 70.8$.

For more complicated transforms, such as Example 9.10, you would anticipate more complicated procedures. But the resulting freedom to define almost arbitrary transformations is worth the extra detail required. Usually, you will encounter, at most, the preprogrammed cases in standard elementary textbooks. Practice with the following problems.

Problems

37. The following data are supposed to follow an exponential curve, $y = be^{mx}$.

x:	−2	−1	0	1	2
y:	.42	1.0	2.8	8.4	25

a. Estimate the prediction equation.

Ans: $\hat{y} = 3.01\ e^{1.03x}$

b. Predict y if $x = 1.5$.

Ans: $\hat{y} = 14.1$

c. Estimate x if $y = 20$.

Ans: $\hat{x} = 1.84$

38. The following data are supposed to follow a power curve, $y = bx^m$

x:	2	1	3	5	7
y:	1.3	1.9	.6	.3	.2

a. Estimate the prediction equation.

Ans: $\hat{y} = 2.27\ x^{-1.22}$

b. Predict y, if $x = 1$.

Ans: 2.27

c. Predict y, if $x = 2.5$.

Ans: $\hat{y} = .7$

d. Estimate x, if $y = .3$.

Ans: 5

39. The following data are supposed to follow a logarithmic curve, $y = m \ln x + b$.

x:	.4	1	1.5	4	10
y:	3.08	.81	.33	−1.5	−3.9

a. Estimate the prediction equation.

Ans: $\hat{y} = -2.09\ \ln x + 1.09$

b. Predict y, if $x = 6.1$.

Ans: -2.69

c. Estimate x, if $y = 0$.

Ans: $\hat{x} = 1.7$

40. The following data represent observations of hand strength (y) and age (x) of a random sample of prisoners in a large federal prison.

x:	21	62	36	24	33	48	38
y:	2.9	1.1	1.8	2.7	3.1	1.0	1.2

a. Fit a linear curve, and calculate r^2. Predict y, if $x = 40$.

Ans: $\hat{y} = -.052x + 3.92$; $r^2 = .646$; $\hat{y} = 1.84$. (Note that $r < 0$.)

b. Fit an exponential curve, and calculate r^2. Predict y, if $x = 40$.

Ans: $\hat{y} = 5.25e^{-.029x}$; $r^2 = .686$; $\hat{y} = 1.66$

c. Fit a power curve, and calculate r^2. Predict y, if $x = 40$.

Ans: $\hat{y} = 91.4x^{-1.1}$; $r^2 = .71$; $\hat{y} = 1.55$

d. Fit a logarithmic curve and calculate r^2. Predict y, if $x = 40$.

Ans: $\hat{y} = -2 \ln x + 9.13$; $r^2 = .679$; $\hat{y} = 1.72$

e. Fit the transformation, $y^2 = b + m\sqrt[3]{x}$ $(y > 0)$, calculate r^2, and predict y, if $x = 40$.

Ans: $y = \sqrt{28 - 7.1\sqrt[3]{x}}$; $r^2 = .623$; $\hat{y} = 1.94$

f. Which of the curves best fits the data?

Ans: In terms of r^2, none of the curves does a spectacular job of fitting the transformed data. The best judgment may be obtained by plotting the observed points and fitted curves all on the same graph. You will probably agree that the linear curve appears to do about as well as any of the remaining curves for this data set.

ZS–9: User Instructions

STEP	PROCEDURE	ENTER	PRESS	DISPLAY
DE	*Data Entry*			
1	Clear registers.		[CMs]	*xxx*
2	Enter (x_i,y_i) pairs.	x_i	[D]	*xxx*
	(Repeat $i = 1, 2,...,n$)	y_i	[R/S]	i
3	Compile data.		[E′]	$n - 2$
4	Enter percentile for CIs; d.f. $= n - 2$ (may also store manually in R_{31} at any time).	$t_{\alpha/2}$	[R/S]	$t_{\alpha/2}$
SLOPE				
1	*CI for m*		[A]	l
			[$x \rightleftarrows t$]	u
2	*Test* $H_0{:}m = m_0$.			
	a. Enter H_1-code.*	H_1-code	[A′]	H_1-code
	b. Enter hypothesized value.	m_0	[R/S]	P
INT				
1	*CI for b*		[B]	l
			[$x \rightleftarrows t$]	u
2	*Test* $H_0{:}b = b_0$.			
	a. Enter H_1-code.	H_1-code	[B′]	H_1-code
	b. Enter hypothesized value.	b_0	[R/S]	P
μ at x_o	*CI for* $mx_0 + b$	x_0	[C]	l
			[$x \rightleftarrows t$]	u
Y at x_o	*PI for* $Y_0 = mx_0 + b + e$	x_0	[C′]	l
			[$x \rightleftarrows t$]	u
DISC	*CI for* x^*, *when* y^* *is observed*	y^*	[D′]	l
			[$x \rightleftarrows t$]	u
CORR	*Test* $H_0{:}\rho = 0$.			
	Enter H_1-code.	H_1-code	[E]	P
	Note: Valid whenever $n-2 \in R_{15}$ and $r \in R_{44}$			

$$\text{* Note: } H_1\text{-code} \begin{cases} -1 \text{ for } H_1{:}\theta < \theta_0 \\ 0 \text{ for } H_1{:}\theta \neq \theta \\ 1 \text{ for } H_1{:}\theta > \theta_0 \end{cases}$$

TABLE T. STUDENT −t PERCENTILES

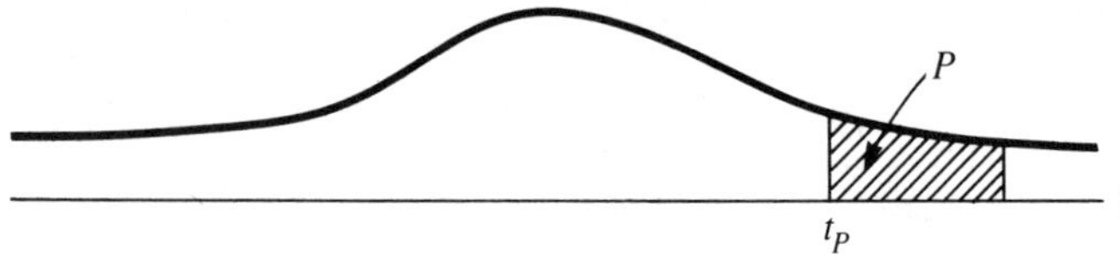

	P				
ν	.10	.05	.025	.01	.005
1	3.078	6.314	12.71	31.82	63.66
2	1.886	2.920	4.303	6.965	9.925
3	1.638	2.353	3.182	4.541	5.841
4	1.533	2.132	2.776	3.747	4.604
5	1.476	2.015	2.571	3.365	4.032
6	1.440	1.943	2.447	3.143	3.707
7	1.415	1.895	2.365	2.998	3.499
8	1.397	1.860	2.306	2.896	3.355
9	1.383	1.833	2.262	2.821	3.250
10	1.372	1.812	2.228	2.764	3.169
11	1.363	1.796	2.201	2.718	3.106
12	1.356	1.782	2.179	2.681	3.054
13	1.350	1.771	2.160	2.650	3.012
14	1.345	1.761	2.145	2.624	2.977
15	1.341	1.753	2.131	2.602	2.947
16	1.337	1.746	2.120	2.583	2.921
17	1.333	1.740	2.110	2.567	2.898
18	1.330	1.734	2.101	2.552	2.878
19	1.328	1.729	2.093	2.539	2.861
20	1.325	1.725	2.086	2.528	2.845
21	1.323	1.721	2.080	2.518	2.831
22	1.321	1.717	2.074	2.508	2.819
23	1.320	1.714	2.069	2.500	2.807
24	1.318	1.711	2.064	2.492	2.797
25	1.316	1.708	2.060	2.485	2.787
26	1.315	1.706	2.056	2.479	2.779
27	1.314	1.703	2.052	2.473	2.771
28	1.313	1.701	2.048	2.467	2.763
29	1.311	1.699	2.045	2.462	2.756
30	1.310	1.697	2.042	2.457	2.750
> 30	1.282	1.645	1.960	2.326	2.576

REGISTER CONTENTS				
00 Used	10	20 ↓	30 ts	40 $\hat{\theta}$
01 Σy_i	11	21	31 $t_{\alpha/2}$	41 Used
02 Σy_i^2	12	22	32	42 d
03 n	13	23 ST–19	33 s	43 $y^* - \bar{y}$
04 Σx_i	14 Used	24 ST–21	34	44 r
05 Σx_i^2	15 $\nu = n - 2$	25	35	45 $\hat{m}$
06 $\Sigma x_i y_i$	16 ↑ Used	26 ↓	36 $s_{\hat{B}}$	46 $\hat{b}$
07 $\Sigma(y_i - \bar{y})^2$	17 by	27	37 $s_{\hat{M}}$	47
08 $\Sigma(x_i - \bar{x})^2$	18	28 H_1-code	38 $s_{\hat{M}x_0 + \hat{B}}$	48
09	19	29 $P(ts)$	39 $e(\hat{\theta} \pm e)$	49

EXAMPLE ZS–9

The resistance of a length of wire is thought to be a linear function of the temperature of the wire. For a given temperature, errors in readings of resistance are normally distributed with mean 0 and variance σ^2. The following readings were made at the temperatures indicated.

Temperature:	0	10	20	30	40	50
Resistance:	22.6	25.1	29.0	29.9	33.4	34.8

a. Estimate the regression of resistance on temperature.
b. Estimate the resistance, if temperature is 25.
c. Estimate the temperature, if resistance is 30.
d. Find a 95% CI for the slope, m.
e. Find a 95% CI for the intercept b.
f. Find a 95% CI for the expected resistance when temperature is 25.
g. Find a 95% PI for the measured response when temperature is 25.
h. Find a 95% discrimination interval for the temperature at which a resistance of 30 is observed.
i. Test the hypotheses $H_0{:}m = 0$ vs. $H_1{:}m \neq 0$.
j. Test the hypotheses $H_0{:}b \leq 20$ vs. $H_1{:}b > 20$.
k. Calculate the coefficient of determination.
l. Test $H_0{:}\rho = 0$ vs. $H_1{:}\rho \neq 0$.

Solution

ZS STEP	ENTER	PRESS	DISPLAY	COMMETS
DE				
1		[CMs]	*xxx*	Clear data registers; ignore R_D.
2	0	[D]	*xxx*	Enter x_1; ignore R_D.
	22.6	[R/S]	1.	Enter y_1, update data base; (x,y) count displayed.
	10	[D]	1.	Enter x_2; ignore R_D.
	25.1	[R/S]	2.	Enter y_2, update data base.
	20	[D]	11.	Enter x_3; ignore R_D.
	29	[R/S]	3.	(x,y)–count displayed
	30	[D]	21.	Ignore R_D.
	29.9	[R/S]	4.	(x,y)–count displayed
	40	[D]	31.	Ignore R_D.
	33.4	[R/S]	5.	(x,y)–count displayed
	50	[D]	41.	Ignore R_D.
	34.8	[R/S]	6.	Value of $n = 6$ concludes data entry
3		[E′]	4.	Compile data; display d.f. = 4.
4	2.776	[R/S]	2.776	Enter $t_{.025}$ for d.f. = 4.
OP CODES		[Op]12	22.93	$\hat{b}$
		[$x \rightleftarrows t$]	.248	$\hat{m}$; $\hat{y} = .248x + 22.93$ answers (a).
	25	[Op]14	29.13	$\hat{y}$ for $x = 25$ answers (b).
	30	[Op]15	28.49	$\hat{x}$ for $y = 30$ answers (c).
SLOPE				
1		[A]	.198	l
		[$x \rightleftarrows t$]	.298	u, so CI is $.198 < m < .298$, answers (d).
INT				
1		[B]	21.4	l
		[$x \rightleftarrows t$]	24.4	u, so CI is $21.4 < b < 24.4$, answers (e).
μ at x_o	25	[C]	28.28	l
		[$x \rightleftarrows t$]	29.98	u, so CI is $28.28 < 25m + b < 29.98$, answers (f).
Y at x_o	25	[C′]	26.89	l
		[$x \rightleftarrows t$]	31.38	u, so PI is $26.89 < Y_0 < 31.38$, answers (g).
DISC	30	[D′]	19.37	l
		[$x \rightleftarrows t$]	37.91	u, so CI is $19.37 < x^* < 37.91$, answers (h).

Solution

ZS STEP	ENTER	PRESS	DISPLAY	COMMETS
SLOPE				
2a	0	[A′]	0.	Enter H_1-code for $H_1{:}m \neq 0$.
2b	0	[R/S]	.00016	Significance of test; reject H_0; answers (i).
INT				
2a	1	[B′]	1	Enter H-code for $H_1{:}b > 20$.
2b	20	[R/S]	.0028	Significance of test; reject H_0; answers (j).
OP CODE		[Op]13[x^2]	.98	$r^2 = .98$ answers (k).
CORR	0	[E]	.00016	Significance of test of $H_0{:}\rho = 0$ [must agree with (i)].

10 Multiple Regression

If simple regression is really as important as we say it is, you can bet that multiple regression is even more important. It represents a powerful extension of simple regression that allows us to consider the effect on y of more than one carrier. In a way, it opens up too much, for where do we stop adding variables? In a lot of the phenomena that we try to model, we can almost always think up yet another variable that y might depend upon. In the end, you have to decide on a reasonable number of the most important carriers that might be used to predict y. That is not an easy task, and we avoid it here completely. Our *theory* will not tell us if we have a good model.

In this brief chapter, we are only going to study the simplest extension, namely, to two carriers, with four parameters to be estimated (three coefficients, plus a variance). This has several advantages. For one thing, you can still draw pictures, only you will need to mimic 3-dimensional space to do it in. Secondly, the algebra is still (just barely) manageable, so that we can see the analogues of the simple linear case in the formulas we shall be using. Once you add just one more carrier (to 3), both of those conveniences vanish, and the only way to handle the results decently is with matrix algebra. Curiously enough, that not only makes things easier, it immediately opens up the door for admitting as many carriers as you like—up to the practicality of being able to invert matrices (your TI–59 will invert up to 9 × 9 matrices with the master module).

Following the notation in ASM, we will examine the model

$$Z = a_0 + a_1x + a_2y + E \tag{10.1}$$

where E is a normal error term having mean 0 and variance σ^2, x and y are fixed values of the two carriers, and a_0, a_1, and a_2 are parameters we wish to estimate. We shall continue to use the least squares criterion to fit a plane this time to the n trivariate data points, (x_1, y_1, z_1), (x_2, y_2, z_2),..., (x_n, y_n, z_n), in 3-dimensional space. The coefficients can be estimated using program ST–18. First, the data triples are entered using ST–05. Then the user instructions for ST–18 on page 5–14 of ASM can be followed. Since we are interested in confidence intervals and tests of hypotheses as well, we have provided program ZS–10, which accesses ST–18 for estimates and stores required summaries, as noted in the register contents. Consequently, you will *not* need to refer to ST–18 directly.

Analogous to the correlation coefficient between two variables, there is, with three variables, as in the current model, the possibility of three such correlations. In particular, there is the notion of the correlation between Z and $\hat{Z}$, where $\hat{Z}$ is the prediction resulting from estimating the coefficients. Such a correlation, denoted R, is called the (sample) multiple correlation. It measures how Z is related to the regressors x and y all at once. In fact, it may be shown that R^2 is again the ratio of the variation, explained by regressing Z on x and y, namely, $\Sigma(\hat{Z}_i - \bar{Z})^2$ to the total variation $\Sigma(Z_i - \bar{Z})^2$. As such, it is called the coefficient of determination, and, like r^2, is often reported as a percentage. It is used as a measure of how well the regression plane fits the data. Often, it is used to decide when a regressor is worthwhile adding or deleting from a list of possibilities. You will always find R^2 in R_{44} after data have been processed with ZS–10.

If you refer to the user instructions for ZS–10, you see how to process the data for multiple regression under the data entry routine coded DE. The process that requires the most care is Step 1, where you input the trivariate data using program ST–05. The order in which you use that program is essential and is shown on page 3–8 of ASM. You input the value of the first regressor variable, x, with label A of ST–05 (after [Pgm]05 and the usual initialization with [E′] in that program), the value of the second regressor, y, with label B, and, finally, the observed z with label C, to complete one triple. Repeat that procedure for each triple,† and then exit the program using Step 2 in DE (press [RST]). Step 3 then merely requires that you press [E′] in the card program ZS–10, and all the necessary data are compiled, with a stop displaying the degrees of freedom d.f. = $n - 3$. This is to allow

†Avoid repartitioning and reset the pointer with [D] after entry of 9 triples.

you to enter $t_{\alpha/2}$ in anticipation of calculating CI. Of course, if you do not plan to calculate CIs, you may ignore this step. If you change your mind later on, you can manually retrieve $n - 3$ from R_{15} and store $t_{\alpha/2}$ in R_{31}, if you like. Also, $\hat{a}_0$, $\hat{a}_1$, and $\hat{a}_2$ are stored, respectively, in R_{41}, R_{42}, and R_{43}. These may be displayed by pressing [D] any time after Step 3 of DE.

Once you complete the data compilation, you may proceed to find CIs and run tests much the same as you did in ZS–9. The register contents will preserve all the necessary outputs needed for most questions asked about this regression. Of course, the original data are destroyed, so please exercise caution in entering the data with ST–05.

EXAMPLE 10.1

The following data (x, y, z) were gathered under a regression model, where each z is the response to the carrier values x and y.

x	1	2	3	4	5	6
y	−1	3	2	5	9	1
z	.2	5.5	5.0	8.6	15.1	6.6

a. Determine the regression of Z on x and y.
b. Find a 90% CI for a_0.
c. Find a 90% CI for a_1.
d. Find a 90% CI for a_2.
e. Test $H_0{:}a_0 = 0$ vs. $H_1{:}a_0 \neq 0$.
f. Test $H_0{:}a_1 \geq 1$ vs. $H_1{:}a_1 < 1$.
g. Test $H_0{:}a_2 \leq 1$ vs. $H_1{:}a_2 > 1$.
h. Predict z for $x = 3.5$ and $y = 6$.
i. Find the coefficient of determination.

Solution Enter the data using ST–05, then press [RST] and then [E′] to see d.f. = 3 displayed. Enter $t_{.05} = 2.353$, and press [R/S]. You are now prepared for the questions posed.

a. Press [D] to see $\hat{a}_0 = .318$, [R/S] to find $\hat{a}_1 = .835$, and [R/S] again to find $\hat{a}_2 = 1.135$. Thus, $\hat{z} = .318 + .835x + 1.135y$.
b. Press [A] to see $l = -.787$ and [$x \rightleftarrows t$] to see $u = 1.42$.
c. Press [B] to see $l = .513$ and [$x \rightleftarrows t$] to see $u = 1.16$.
d. Press [C] to see $l = .962$ and [$x \rightleftarrows t$] to see $u = 1.31$.
e. Enter 0, and press [A′]. Then, enter $a'_0 = 0$, and press [R/S] to see $P = .547$.

f. Enter −1, press [B′]. Enter $a'_1 = 1$, and press [R/S] to see $P =$ .157.

g. Enter 1, press [C′]. Enter $a'_2 = 1$, and press [R/S] to see $P = .082$.

h. Enter 3.5, press [E]. Then enter 6, and press [R/S] to see $\hat{z} = 10.05$.

i. Press [RCL]44 to find $R^2 = .99$.

It is very common to find the estimated coefficients in a multiple regression being interpreted individually, and you should be aware that there is a problem. An individual coefficient is sometimes referred to as the "marginal" change in the response, z, associated with a unit increase in the associated carrier, "everything else being equal." But, as authorities* on the subject point out, this interpretation is fraught with woes and pitfalls. We paraphrase Mosteller and Tukey in the following comments on the subject of interpreting individual coefficients in multiple regression for your information.

Historically, some of the problem *may* (we are not really sure) have arisen from the success of interpreting fitted coefficients in physical science experiments. But those cases usually invoke very tightly controlled experiments with a small number of independent physical variables that are well understood with a long history of experimentation, so that the only error involved is essentially one of measurement. However, hardly any of those conditions are usually satisfied in the circumstances under which we generally are trying to apply statistical analysis. Since any regression coefficient depends in a critical way (often) on what *other* variables are used in the model, we cannot hope to buy the quantitative effect of an increase of one unit on that one variable so cheaply.

To put it in simple terms, if we wish to know what the effect on z is of increasing one of the carriers 1 unit, while all of the other carriers are held fixed, we will have to conduct an experiment in which z's are observed under circumstances where only the carrier in question can change, and there is no chance for the other carriers to change. Of course, ordinarily it is quite out of the question to carry out such an experiment. And, even if we could, it might not produce what we want. After all, we should ask, "Why are we doing regression in the first place?" Certainly one answer is to be able to predict z from the variables that we have used as carriers. If this task is reasonably accomplished with the coefficients as estimated from the data, then a main goal is accomplished. It may be asking too much to be able further to interpret these estimates in a more severe sense.

You should be able to perform the analyses required for the exercises that follow with ease. The general layout is quite similar to the model example.

* See F. Mosteller, and J. W. Tukey, *Data Analysis and Regression* (Reading, Mass.: Addison-Wesley, 1977), for example.

Problems

The following data relate to Problems 1–4. A company sells its dog food through supermarkets. The sales manager is seeking a way of predicting sales when the company enters a new market. He decides to investigate the feasibility of doing this on the basis of the following data.

AREA	DOG LICENSES (HUNDREDS)	NUMBER OF SUPERMARKETS	SALES (IN THOUSANDS)
1	15	25	$76
2	10	29	38
3	7	15	25
4	21	25	98
5	14	11	93
6	8	13	54
7	14	22	78
8	24	14	85
9	9	13	65
10	23	11	88

1. Find the regression of sales on dog licenses and supermarkets. (*Note*: You must reset the raw data pointer by pressing [D] in ST–05 before entering the 10th ordered triple.)

   Ans: $\hat{z}_0 = 32.5 + 3.03x - .360y$, where $x =$ licenses and $y =$ supermarkets.
2. Predict sales for a community having 750 dog licenses and 5 supermarkets.

   Ans: For $x_0 = 7.5$ and $y_0 = 5$, we have $\hat{z} =$ \$53,400.
3. Do sales really depend upon the number of supermarkets?

   Ans: Testing $H_0{:}a_2 = 0$ vs. $H_1{:}a_2 \neq 0$, $P = .68$. Probably not.
4. Find a 90% CI for the intercept, a_0.

   Ans: $-6.6 < a_0 < 71.7$

The next data set refers to Problems 5–8. A record was kept of beer sales according to temperature and traffic count for each of several days at a corner convenience store.

TEMPERATURE	TRAFFIC COUNT	NUMBER OF 6-PACKS SOLD
78	720	61
80	623	83
40	301	15
70	850	90
68	352	45
45	424	38

92	605	103
42	457	58
75	783	79
50	311	27

5. Find the regression of sales on temperature and traffic.

   Ans: $\hat{z} = -27.7 + .785x + .069y$, where x = temperature and y = traffic.

6. Find a 95% CI for the coefficient of traffic counts.

   Ans: $-.008 < a_2 < .146$

7. Test whether or not sales depend on temperature.

   Ans: Testing $H_0{:}a_1 = 0$ vs. $H_1{:}a_1 \neq 0$, $P = .06$.

8. Predict sales when temperature is 85, and the traffic count is 400.

   Ans: $\hat{z} = 66.6$

The next data set refers to Problems 9–17. The Fortune 500 consists of the 500 largest industrial corporations ranked by sales. We would like to be able to predict the net income of a corporation based upon sales and assets. A random sample of 27 corporations was selected from the Fortune 500 for the year 1977, with the results shown in the table on p. 280.

9. Assuming net income is linearly related to sales and assets, use the least squares method to estimate the regression coefficients.

   Ans: $\hat{a}_0 = -17.156$; $\hat{a}_1 = .02487$; $\hat{a}_2 = .04526$

10. Predict the net income for a corporation that has sales of $1 billion and has assets of $1 billion.

    Ans: $52.97 million

11. Determine whether there is a significant linear relationship between net income and each of the 2 independent variables (sales and assets) at the .05 level of significance.

    Ans: Testing $H_0{:}a_1 = 0$ gives $P = .015$, so reject H_0. Testing $H_0{:}a_2 = 0$ gives $P = .00097$, so reject H_0.

12. Find a 95% CI for the intercept.

    Ans: $-34.2 < a_0 < -.15$

13. Find a 95% CI for the coefficient of sales.

    Ans: $.00519 < a_1 < .04454$

14. Find a 95% CI for the coefficient of assets.

    Ans: $.02039 < a_2 < .07012$

15. What is the estimated standard deviation of $\hat{A}_1$?

    Ans: $s_{\hat{A}_1} = .0095$ (recall from R_{37})

16. What is the average of the incomes of the 27 corporations?

    Ans: $\bar{z} = 100.76$ million (recall from R_{08})

17. Predict the income of a Fortune 500 corporation having sales $200 million and assets $200 million.

 Ans: $\hat{z} = \$-3.13$ million (this is useless extrapolation)

CORPORATION	SALES (millions)	ASSETS (millions)	NET INCOME (millions)
General Electric	$17,518.60	$13,696.8	$1,088.20
United Technologies	5,550.67	2,979.3	195.98
Cities Service	4,388.20	3,739.6	210.20
Gulf & Western, Inc.	3,643.00	4,159.1	150.33
General Mills	2,909.40	1,447.3	117.03
Dresser Industries	2,538.80	2,173.3	185.10
Singer	2,294.30	1,461.9	94.20
Iowa Beef Processors	2,923.77	277.5	29.97
Interco	1,566.40	778.0	76.76
SCM	1,377.64	767.9	37.41
Cummins Engine	1,263.81	818.4	67.02
Oscar Mayer	1,188.43	377.7	35.02
Crown Cork & Seal	1,049.14	631.1	53.80
Libbey–Owens–Ford	978.69	652.3	58.94
Akzona	808.85	674.9	7.52
A. O. Smith	727.28	355.6	18.68
Joy Manufacturing	677.69	551.2	48.22
General Cable	620.51	496.6	20.38
Adolph Coors	593.12	691.6	67.70
Cahot	574.84	579.1	39.90
Macmillan	512.73	525.0	19.40
Bell & Howell	491.57	366.8	11.80
Eagle–Picher Inc.	474.04	282.7	26.00
Ward Foods	443.67	125.2	7.74
Hanes	414.16	229.0	20.82
Inland Container	396.73	308.3	21.92
Koehring	361.61	277.2	10.53

* Source: Based on information from *Fortune,* May 8, 1978, pages 238–265.

The next data set is for use with Problems 18–22. The Fortune Second 500 consists of the industrial corporations that are ranked 501st to 1,000th in sales. As in the preceding Problems for the Fortune 500, we would like to predict the net income of a corporation based upon sales and assets. A random sample of 12 corporations was selected from the Fortune Second 500 for the year 1977, with the results shown in the following table:

CORPORATION	SALES (millions)	ASSETS (millions)	NET INCOME (millions)
Royal Crown Companies	$349.62	$168.6	$18.67
Wyman–Gordon	291.50	195.9	20.97
Rorer Group	243.80	169.2	18.14
Prentice–Hall	230.64	177.9	23.72
Capital Industries—EMI	209.77	137.1	16.16
King–Seeley Thermos	209.31	134.6	19.16
Pettibone	206.68	180.7	8.79
Cox Broadcasting	186.43	284.5	25.46
Storage Technology	162.27	159.7	11.43
First Mississippi	161.49	210.7	14.44
Overhead Door	157.63	79.9	9.64
American Sterilizer	155.71	104.5	2.88

* Source: Based on information from *Fortune,* June 19, 1978, pages 170–178.

18. Predict the income of a Second 500 corporation having sales of $200 million and assets of $200 million, and compare with the prediction in Problem 17.

    Ans: $\hat{z} = 17.58$ million

19. Find estimates of a_0, a_1, and a_2.

    Ans: $\hat{a}_0 = -5.79$, $\hat{a}_1 = .0443$, $\hat{a}_2 = .0726$

20. Find 90% CIs for a_0, a_1, and a_2.

    Ans: $-17.8 < a_0 < 6.22$; $-.0014 < a_1 < .090$; $.021 < a_2 < .124$

21. Test $H_0{:}a_1 \geq 0$ vs. $H_1{:}a_1 < 0$.

    Ans: $P = .945$

22. Estimate the variance of the error term.

    Ans: $s^2 = 23.03$

The next data set is for Problems 23–29. The personnel department of a large industrial corporation is studying the relationship between weekly salary, length of employment, and age of its managerial employees. A random sample of 12 employees yielded the following results.

LENGTH OF EMPLOYMENT (months)	AGE (years)	SALARY
330	46	$519
569	65	636
375	57	558
113	47	408
215	41	442
343	59	502
252	45	408
348	57	490
352	55	432
256	61	399
87	28	336
337	51	544

24. Find the regression of salary on length of employment (x) and age (y).

    Ans: $\hat{z} = 332.66 + .6505x - 1.054y$

25. Predict the weekly salary for an employee who has been employed for 180 months and is 47 years old.

    Ans: \$400.23

26. Determine whether there is any significant dependence of salary on either length of employment or age.

    Ans: For $H_0{:}a_1 = 0$, $P = .0028$. For $H_0{:}a_2 = 0$, $P = .609$. The dependence on age is significant but not salary.

27. Compute R^2, and interpret its meaning.

    Ans: $R^2 = .795$. About 80% of the variation in salary is explained by the regression.

28. Find a 90% CI for the coefficient of age.

    Ans: $-4.70 < a_2 < 2.59$

29. Estimate the standard deviations of $\hat{A}_0$, $\hat{A}_1$, and $\hat{A}_2$ and the variance of Z.

    Ans: Respectively, 71.6, .160, 1.99 (registers R_{36}, R_{37}, R_{38}). $\hat{\sigma}^2 = s^2 = 1763$

30. A foreign automobile company wishes to examine the relationship (assumed linear) between its month sales (Z) and both advertising expenditures (X) and base price (Y). Ten cities are selected, and the collected data are given below.

ADVERTISING	BASE PRICE (in hundreds)	SALES
25	\$60	22
26	44	30
28	54	24
30	48	27
32	64	30
34	70	30
35	60	32
36	60	35
38	72	34
38	68	36

a. Determine the regression of Z on x and y.

b. What would you guess for monthly sales for cars with a base price of \$6,000 in a city where \$3,000 will be spent on advertising?

c. Do sales really depend on base price or could that variable just as well be dropped?

Ans:

a. $\hat{z} = 6 + 1.1x - .2y$

b. $\hat{z} = 28$

c. For a test of H_0:$a_2 = 0$, $P = .14$. Should drop base price.

ZS–10: User Instructions

STEP	PROCEDURE	ENTER	PRESS	DISPLAY
DE	*Data Entry*			
1	Enter data with program ST–05 using A for x, B for y, and C for z.			
2	Exit ST–05		[RST]	
3	Compile data.		[E′]	$n - 3$
4	Enter t-percentiles for CIs, d.f. = $n - 3$.	$t_{\alpha/2}$	[R/S]	$t_{\alpha/2}$
CIs				
1	*CI for* a_0			
	Complete DE.		[A]	l
			[$x \leftrightarrows t$]	u
2	*CI for* a_1			
	Complete DE.		[B]	l
			[$x \leftrightarrows t$]	u
3	*CI for* a_2			
	Complete DE.		[C]	l
			[$x \leftrightarrows t$]	u
Tests				
1	*Test* H_0:$a_0 = a'_0$			
	a. Complete DE			
	b. Enter H_1-code.*	H_1-code	[A′]	H_1-code
	c. Enter hypothesized value.	a'_0	[R/S]	P
2	*Test* H_0:$a_1 = a'_1$			
	a. Complete DE.			
	b. Enter H_1-code.	H_1-code	[B′]	H_1-code
	c. Enter hypothesized value.	a'_1	[R/S]	P
3	*Test* H_0:$a_2 = a'_2$			
	a. Complete DE.			
	b. Enter H_1-code.	H_1-code	[C′]	H_1-code
	c. Enter focal value.	a'_2	[R/S]	P

$$* H_1\text{-code} = \begin{cases} -1 \text{ if } H_1: 0 < 0_1 \\ 0 \text{ if } H_1: 0 \neq 0_0 \\ 1 \text{ if } H_1: 0 > 0_0 \end{cases}$$

STEP	PROCEDURE	ENTER	PRESS	DISPLAY
PMTRS	Display coefficients (after DE).		[D] [R/S] [R/S]	$\hat{a}_0$ $\hat{a}_1$ $\hat{a}_2$
PREDICT	Calculate $\hat{z} = \hat{a}_0 + \hat{a}_1 x_0 + \hat{a}_2 y_0$.	x_0 y_0	[E] [R/S]	0 $\hat{z}$

REGISTER CONTENTS				
00	10 Used	20 ↓ Used by PGM 21	30 ts	40 $\hat{\theta}$
01 Σy	11 Σxz	21	31 $t_{\alpha/2}$	41 $\hat{a}_0$
02 Σy^2	12 Σyz	22 PGM	32	42 $\hat{a}_1$
03 n	13 Σz^2	23 21	33 s	43 $\hat{a}_2$
04 Σx	14	24	34	44 R^2
05 Σx^2	15 $n - 3$	25	35	45
06 Σxy	16 ↑ Used by	26	36 $s_{\hat{A}_0}$	46
07 Σz	17	27 Used	37 $s_{\hat{A}_1}$	47
08 $\bar{z}$	18	28 H_1-code	38 $s_{\hat{A}_0}$	48
09 last x	19	29 $P(ts)$	39 $e(\hat{\theta} \pm e)$	49

EXAMPLE ZS–10

The data below represent characteristics of a sample of automobiles.

Weight:	3,810	4,220	2,900	3,290	3,400	3,920	4,350
Horsepower:	255	180	96	120	100	140	150
Cost:	7,999	9,221	8,222	9,010	10,099	11,019	11,219

a. Regress cost on weight and horsepower.
b. Predict the cost of an automobile weighing 5,000 pounds and having 160 hp?
c. Determine the significance of horsepower for predicting cost.
d. Find a 95% CI for the coefficient of weight.
e. Estimate σ.
f. Find the coefficient of determination.

Solution

ZS STEP	ENTER	PRESS	DISPLAY	COMMENTS
DE				
1		[Pgm]05 [E′]	0.	Initialize ST–05.
	3,810	[A]	1.	Enter x_1.
	255	[B]	1.	Enter y_1.
	7,999	[C]	1.	Enter z_1, completing 1 triple.
	4,220	[A]	2.	Enter x_2.
	180	[B]	2.	Enter y_2.
	9,221	[C]	2.	Enter z_2, completing 2 triples.
	2,900	[A]	3.	Enter x_3.
	96	[B]	3.	Enter y_3.
	8,222	[C]	3.	Enter z_3, completing 3 triples.
	3,290	[A]	4.	Enter x_4.
	120	[B]	4.	Enter y_4.
	9,010	[C]	4.	Enter z_4, completing 4 triples.
	3,400	[A]	5.	Enter x_5.
	100	[B]	5.	Enter y_5.
	10,099	[C]	5.	Enter z_5, completing 5 triples.
	3,920	[A]	6.	Enter x_6.
	140	[B]	6.	Enter y_6.
	11,019	[C]	6.	Enter z_6, completing 6 triples.
	4,350	[A]	7.	Enter x_y.
	150	[B]	7.	Enter y_7.
	11,219	[C]	7.	Enter z_7, completing last triple.
2		[RST]	7.	Exit ST–05.
3		[E′]	4.	Process trivariate data, displaying $\nu = 4$.
4	2.776	[R/S]	2.776	Enter $t_{.025}$.
PMTRS		[D]	3361.72	Display $\hat{a}_0$.
		[R/S]	2.466	Display $\hat{a}_1$.
		[R/S]	– 19.769	Recall and display $\hat{a}_2$.

(*Note*: The regression equation is $\hat{z} = 3361.72 + 2.466x - 19.769y$.)

Solution

ZS STEP	ENTER	PRESS	DISPLAY	COMMENTS
PREDICT	5,000	[E]	0	Prepare to predict z.
	160	[R/S]	12,527.	Predicted cost
TESTS				
3b	0	[C′]	0.	Testing $H_0{:}a_2 = 0$ vs. $H_1{:}a_2 \neq 0$
3c	0	[R/S]	.03	Significance level for a_2 ($ts = -3.2$)
CIs				
2		[B]	.68	Lower 95% limit for a_1
		[$x \rightleftarrows t$]	4.25	Upper 95% limit for a_1
		[RCL]33	691.7	$\hat{\sigma} = s$
		[RCL]44	.80	Recall and display R^2 = .80.

Appendix

In this appendix, you will find a list of all of the programs used for solving problems in this book. They are key-coded according to the chapter to which they apply. They have been arranged in eight groups and may, therefore, be recorded on as many magnetic cards. This arrangement makes it very convenient and natural for solving certain groups or types of problems as they arise in the text. But that is a matter of convenience, and you may wish to rearrange the programs a different way. If you have no desire to make such modifications, then you should read the guidelines provided in Chapter 1, proceed to record the programs *exactly* as they are written on magnetic cards for a permanent record, and then you are ready to apply these to the various problems starting in Chapter 2.

If you should want to modify the program arrangement in any way, there are some guidelines you should bear in mind. First of all, two or more subroutines having the same name may not be used on the same magnetic card, for, obviously, then the calculator would have no way of knowing which one was being called for. Consequently, for each of the programs listed, we have displayed the names and locations of each label used for a subroutine. This makes it easy for you to locate and rename subroutines that conflict. We avoided direct addressing as much as possible for this very reason.

Programs

ZS–2: LABELS AND LOCATIONS

```
001  13  C
029  14  D
050  18  C'
069  52  EE
163  19  D'
174  15  E
190  23  LNX
215  10  E'
228  12  B
241  17  B'
```

```
000  76  LBL
001  13  C
002  42  STO
003  00  00
004  75  -
005  43  RCL
006  17  17
007  95  =
008  55  ÷
009  43  RCL
010  18  18
011  95  =
012  42  STO
013  20  20
014  36  PGM
015  19  19
016  12  B
017  42  STO
018  07  07
019  43  RCL
020  20  20
021  36  PGM
022  19  19
023  13  C
024  32  X:T
025  43  RCL
026  07  07
027  92  RTN
028  76  LBL
029  14  D
030  13  C
031  42  STO
032  08  08
033  32  X:T
034  91  R/S
035  13  C
036  22  INV
037  44  SUM
038  08  08
039  43  RCL
040  08  08
041  85  +
042  01  1
043  95  =
044  32  X:T
045  43  RCL
046  08  08
047  94  +/-
048  92  RTN
049  76  LBL
050  18  C'
051  32  X:T
052  93  .
053  05  5
054  77  GE
055  52  EE
056  32  X:T
057  94  +/-
058  85  +
059  01  1
060  95  =
061  32  X:T
062  71  SBR
063  52  EE
064  94  +/-
065  42  STO
066  19  19
067  92  RTN
068  76  LBL
069  52  EE
070  32  X:T
071  33  X²
072  23  LNX
073  94  +/-
074  34  ΓX
075  42  STO
076  00  00
077  53  (
078  53  (
079  53  (
080  02  2
081  93  .
082  05  5
083  01  1
084  05  5
085  05  5
086  01  1
087  07  7
088  85  +
089  93  .
090  08  8
091  00  0
092  02  2
093  08  8
094  05  5
095  03  3
096  65  ×
097  43  RCL
098  00  00
099  85  +
100  93  .
101  00  0
102  01  1
103  00  0
104  03  3
105  02  2
106  08  8
107  65  ×
108  43  RCL
109  00  00
110  33  X²
111  54  )
112  55  ÷
113  53  (
114  01  1
115  85  +
116  01  1
117  93  .
118  04  4
119  03  3
120  02  2
121  07  7
122  08  8
123  08  8
124  65  ×
125  43  RCL
126  00  00
127  85  +
128  93  .
129  01  1
130  08  8
131  09  9
132  02  2
133  06  6
134  09  9
135  65  ×
136  43  RCL
137  00  00
138  33  X²
139  85  +
140  93  .
141  00  0
142  00  0
143  01  1
144  03  3
145  00  0
146  08  8
147  65  ×
148  43  RCL
149  00  00
150  45  YX
151  03  3
```

```
152  54  )
153  54  )
154  94  +/-
155  85  +
156  43  RCL
157  00  00
158  54  )
159  42  STO
160  19  19
161  92  RTN
162  76  LBL
163  19  D'
164  18  C'
165  65  ×
166  43  RCL
167  18  18
168  85  +
169  43  RCL
170  17  17
171  95  =
172  92  RTN
173  76  LBL
174  15  E
175  42  STO
176  00  00
177  87  IFF
178  01  01
179  23  LNX
180  43  RCL
181  16  16
182  35  1/X
183  42  STO
184  17  17
185  42  STO
186  18  18
187  86  STF
188  01  01
189  76  LBL
190  23  LNX
191  43  RCL
192  00  00
193  65  ×
194  43  RCL
195  16  16
196  95  =
197  94  +/-
198  22  INV
199  23  LNX
200  94  +/-
201  85  +
202  01  1
203  95  =
204  42  STO
205  07  07
206  94  +/-
207  85  +
208  01  1
209  95  =
210  32  X:T
211  43  RCL
212  07  07
213  92  RTN
214  76  LBL
215  10  E'
216  94  +/-
217  85  +
218  01  1
219  95  =
220  23  LNX
221  55  ÷
222  43  RCL
223  16  16
224  95  =
225  94  +/-
226  92  RTN
227  76  LBL
228  12  B
229  36  PGM
230  02  02
231  71  SBR
232  88  DMS
233  10  E'
234  42  STO
235  00  00
236  78  Σ+
237  43  RCL
238  00  00
239  92  RTN
240  76  LBL
241  17  B'
242  36  PGM
243  02  02
244  71  SBR
245  88  DMS
246  19  D'
247  42  STO
248  00  00
249  78  Σ+
250  43  RCL
251  00  00
252  92  RTN
```

ZS-3: LABELS AND LOCATIONS

```
001  13  C
010  12  B
028  44  SUM
053  17  B'
075  42  STO
107  18  C'
117  14  D
126  25  CLR
136  15  E
146  23  LNX
163  11  A
170  16  A'
```

```
000  76  LBL
001  13  C
002  43  RCL
003  13  13
004  75  -
005  43  RCL
006  12  12
007  95  =
008  92  RTN
009  76  LBL
010  12  B
011  79  X̄
012  32  X:T
013  94  +/-
014  42  STO
015  25  25
016  43  RCL
017  29  29
018  42  STO
019  02  02
020  03  3
021  01  1
022  42  STO
023  20  20
024  00  0
025  42  STO
026  19  19
027  76  LBL
028  44  SUM
029  53  (
030  73  RC*
031  20  20
032  85  +
033  43  RCL
034  25  25
035  54  )
036  50  I×I
037  44  SUM
038  19  19
039  01  1
040  44  SUM
041  20  20
042  97  DSZ
043  02  02
044  44  SUM
045  43  RCL
046  19  19
047  55  ÷
048  43  RCL
049  03  03
050  95  =
051  92  RTN
052  76  LBL
053  17  B'
054  79  X̄
055  32  X:T
056  94  +/-
057  42  STO
058  25  25
059  43  RCL
060  29  29
061  42  STO
062  02  02
063  03  3
064  02  2
065  42  STO
066  20  20
067  03  3
068  03  3
069  42  STO
070  21  21
071  00  0
```

```
072  42 STO
073  19  19
074  76 LBL
075  42 STO
076  53  (
077  53  (
078  73 RC*
079  20  20
080  85  +
081  43 RCL
082  25  25
083  54  )
084  50 I×I
085  65  ×
086  73 RC*
087  21  21
088  54  )
089  44 SUM
090  19  19
091  02  2
092  44 SUM
093  20  20
094  44 SUM
095  21  21
096  97 DSZ
097  02  02
098  42 STO
099  43 RCL
100  19  19
101  55  ÷
102  43 RCL
103  03  03
104  95  =
105  92 RTN
106  76 LBL
107  18 C'
108  85  +
109  43 RCL
110  13  13
111  75  -
112  43 RCL
113  12  12
114  95  =
115  92 RTN
116  76 LBL
117  14  D
118  87 IFF
119  01  01
120  25 CLR
121  03  3
122  00  0
123  42 STO
124  00  00
125  76 LBL
126  25 CLR
127  01  1
128  44 SUM
129  00  00
130  73 RC*
131  00  00
132  86 STF
133  01  01
134  92 RTN
135  76 LBL
136  15  E
137  42 STO
138  01  01
139  03  3
140  01  1
141  42 STO
142  02  02
143  42 STO
144  30  30
145  76 LBL
146  23 LNX
147  73 RC*
148  02  02
149  16 A'
150  36 PGM
151  03  03
152  11  A
153  01  1
154  44 SUM
155  02  02
156  97 DSZ
157  01  01
158  23 LNX
159  43 RCL
160  03  03
161  92 RTN
162  76 LBL
163  11  A
164  16 A'
165  36 PGM
166  03  03
167  11  A
168  92 RTN
169  76 LBL
170  16 A'
```

ZS-4: LABELS AND LOCATIONS

```
001  19 D'
015  42 STO
025  16 A'
029  11  A
072  48 EXC
094  85  +
101  37 P/R
120  57 ENG
132  65  ×
140  75  -
147  80 GRD
163  18 C'
167  13  C
181  30 TAN
207  60 DEG
235  59 INT
271  10 E'
275  15  E
290  17 B'
294  12  B
306  39 COS
322  38 SIN
343  52 EE
434  14  D
```

```
000  76 LBL
001  19 D'
002  36 PGM
003  09  09
004  11  A
005  42 STO
006  37  37
007  36 PGM
008  09  09
009  16 A'
010  42 STO
011  38  38
012  00  0
013  92 RTN
014  76 LBL
015  42 STO
016  42 STO
017  36  36
018  91 R/S
019  42 STO
020  34  34
021  86 STF
022  04  04
023  92 RTN
024  76 LBL
025  16 A'
026  71 SBR
027  42 STO
028  76 LBL
029  11  A
030  43 RCL
031  37  37
032  42 STO
033  40  40
034  43 RCL
035  38  38
036  55  ÷
037  43 RCL
038  03  03
039  34 √X
040  95  =
041  42 STO
042  32  32
043  35 1/X
044  65  ×
045  53  (
046  43 RCL
047  40  40
048  75  -
049  43 RCL
050  34  34
051  54  )
052  95  =
053  42 STO
054  30  30
055  87 IFF
056  05  05
057  39 COS
058  43 RCL
059  03  03
060  75  -
061  01  1
062  95  =
063  42 STO
064  15  15
065  87 IFF
066  04  04
067  80 GRD
068  91 R/S
069  42 STO
070  31  31
071  76 LBL
072  48 EXC
073  65  ×
074  43 RCL
075  32  32
```

```
076  95  =
077  42  STO
078  39  39
079  85  +
080  43  RCL
081  40  40
082  95  =
083  32  X:T
084  43  RCL
085  39  39
086  75  -
087  43  RCL
088  40  40
089  95  =
090  94  +/-
091  24  CE
092  92  RTN
093  76  LBL
094  85  +
095  43  RCL
096  30  30
097  36  PGM
098  19  19
099  12  B
100  76  LBL
101  37  P/R
102  42  STO
103  42  42
104  22  INV
105  86  STF
106  04  04
107  29  CP
108  43  RCL
109  36  36
110  22  INV
111  77  GE
112  57  ENG
113  22  INV
114  67  EQ
115  65  ×
116  43  RCL
117  42  42
118  85  +
119  76  LBL
120  57  ENG
121  43  RCL
122  42  42
123  95  =
124  32  X:T
125  01  1
126  22  INV
127  77  GE
128  75  -
129  32  X:T
130  92  RTN
131  76  LBL
132  65  ×
133  01  1
134  75  -
135  43  RCL
```

```
136  42  42
137  95  =
138  92  RTN
139  76  LBL
140  75  -
141  02  2
142  75  -
143  32  X:T
144  95  =
145  92  RTN
146  76  LBL
147  80  GRD
148  04  4
149  01  1
150  32  X:T
151  43  RCL
152  15  15
153  77  GE
154  85  +
155  43  RCL
156  30  30
157  36  PGM
158  21  21
159  15  E
160  61  GTO
161  37  P/R
162  76  LBL
163  18  C'
164  71  SBR
165  42  STO
166  76  LBL
167  13  C
168  43  RCL
169  38  38
170  33  X²
171  42  STO
172  40  40
173  43  RCL
174  03  03
175  75  -
176  01  1
177  95  =
178  42  STO
179  15  15
180  76  LBL
181  30  TAN
182  87  IFF
183  04  04
184  60  DEG
185  91  R/S
186  43  RCL
187  15  15
188  65  ×
189  43  RCL
190  40  40
191  95  =
192  42  STO
193  00  00
194  55  ÷
195  43  RCL
```

```
196  41  41
197  95  =
198  32  X:T
199  43  RCL
200  00  00
201  55  ÷
202  43  RCL
203  31  31
204  95  =
205  92  RTN
206  76  LBL
207  60  DEG
208  65  ×
209  43  RCL
210  40  40
211  55  ÷
212  43  RCL
213  34  34
214  95  =
215  42  STO
216  30  30
217  03  3
218  01  1
219  32  X:T
220  43  RCL
221  15  15
222  77  GE
223  59  INT
224  36  PGM
225  21  21
226  11  A
227  43  RCL
228  30  30
229  36  PGM
230  21  21
231  13  C
232  61  GTO
233  37  P/R
234  76  LBL
235  59  INT
236  65  ×
237  09  9
238  55  ÷
239  02  2
240  95  =
241  35  1/X
242  42  STO
243  29  29
244  75  -
245  01  1
246  85  +
247  53  (
248  53  (
249  43  RCL
250  30  30
251  55  ÷
252  43  RCL
253  15  15
254  54  )
255  45  Y^X
```

```
256  03  3
257  35  1/X
258  54  )
259  95  =
260  55  ÷
261  43  RCL
262  29  29
263  34  √X
264  95  =
265  36  PGM
266  19  19
267  12  B
268  61  GTO
269  37  P/R
270  76  LBL
271  10  E'
272  71  SBR
273  42  STO
274  76  LBL
275  15  E
276  43  RCL
277  37  37
278  42  STO
279  40  40
280  43  RCL
281  03  03
282  65  ×
283  02  2
284  95  =
285  42  STO
286  15  15
287  61  GTO
288  30  TAN
289  76  LBL
290  17  B'
291  71  SBR
292  42  STO
293  76  LBL
294  12  B
295  42  STO
296  00  00
297  86  STF
298  05  05
299  43  RCL
300  48  48
301  42  STO
302  38  38
303  61  GTO
304  11  A
305  76  LBL
306  39  COS
307  22  INV
308  86  STF
309  05  05
310  87  IFF
311  04  04
312  85  +
313  43  RCL
314  00  00
315  71  SBR
```

```
316  38 SIN
317  42 STO
318  31  31
319  61 GTO
320  48 EXC
321  76 LBL
322  38 SIN
323  42 STO
324  10  10
325  32 X:T
326  93  .
327  05  5
328  32 X:T
329  22 INV
330  77  GE
331  52 EE
332  43 RCL
333  10  10
334  94 +/-
335  85  +
336  01  1
337  95  =
338  71 SBR
339  52 EE
340  94 +/-
341  92 RTN
342  76 LBL
343  52 EE
344  33 X²
345  23 LNX
346  94 +/-
347  34 √X
348  42 STO
349  10  10
350  53  (
351  53  (
352  53  (
```

```
353  02  2
354  93  .
355  05  5
356  01  1
357  05  5
358  05  5
359  01  1
360  07  7
361  85  +
362  93  .
363  08  8
364  00  0
365  02  2
366  08  8
367  05  5
368  03  3
369  65  ×
370  43 RCL
371  10  10
372  85  +
373  93  .
374  00  0
375  01  1
376  00  0
377  03  3
378  02  2
379  08  8
380  65  ×
381  43 RCL
382  10  10
383  33 X²
384  54  )
385  55  ÷
386  53  (
387  01  1
388  85  +
389  01  1
```

```
390  93  .
391  04  4
392  03  3
393  02  2
394  07  7
395  08  8
396  08  8
397  65  ×
398  43 RCL
399  10  10
400  85  +
401  93  .
402  01  1
403  08  8
404  09  9
405  02  2
406  06  6
407  09  9
408  65  ×
409  43 RCL
410  10  10
411  33 X²
412  85  +
413  93  .
414  00  0
415  00  0
416  01  1
417  03  3
418  00  0
419  08  8
420  65  ×
421  43 RCL
422  10  10
423  45 YX
424  03  3
425  54  )
426  54  )
```

```
427  94 +/-
428  85  +
429  43 RCL
430  10  10
431  54  )
432  92 RTN
433  76 LBL
434  14  D
435  71 SBR
436  38 SIN
437  65  ×
438  53  (
439  43 RCL
440  15  15
441  65  ×
442  09  9
443  55  ÷
444  02  2
445  54  )
446  35 1/X
447  42 STO
448  29  29
449  34 √X
450  75  -
451  43 RCL
452  29  29
453  85  +
454  01  1
455  95  =
456  45 YX
457  03  3
458  65  ×
459  43 RCL
460  15  15
461  95  =
462  61 GTO
463  30 TAN
```

ZS–6: LABELS AND LOCATIONS

```
001  19 D'       237  65  ×
038  55  ÷       245  75  -
048  42 STO      252  17 B'
058  43 RCL      256  12  B
075  48 EXC      276  80 GRD
098  88 DMS      292  18 C'
109  16 A'       296  13  C
113  11  A       341  70 RAD
184  89  π       369  14  D
196  85  +       402  58 FIX
206  37 P/R      427  49 PRD
225  57 ENG      440  15  E
```

```
000  76 LBL      006  43 RCL      012  11  A       018  42 STO
001  19 D'       007  23  23      013  43 RCL      019  37  37
002  00  0       008  42 STO      014  15  15      020  22 INV
003  36 PGM      009  33  33      015  42 STO      021  79  x̄
004  13  13      010  36 PGM      016  13  13      022  24 CE
005  16 A'       011  13  13      017  79  x̄       023  42 STO
```

```
024  38  38
025  71  SBR
026  55  ÷
027  79  x̄
028  32  X:T
029  42  STO
030  47  47
031  22  INV
032  79  x̄
033  32  X:T
034  24  CE
035  42  STO
036  48  48
037  76  LBL
038  55  ÷
039  43  RCL
040  03  03
041  48  EXC
042  13  13
043  42  STO
044  03  03
045  00  0
046  92  RTN
047  76  LBL
048  42  STO
049  42  STO
050  36  36
051  91  R/S
052  42  STO
053  34  34
054  86  STF
055  04  04
056  92  RTN
057  76  LBL
058  43  RCL
059  42  STO
060  32  32
061  43  RCL
062  40  40
063  75  -
064  43  RCL
065  34  34
066  95  =
067  55  ÷
068  43  RCL
069  32  32
070  95  =
071  42  STO
072  30  30
073  92  RTN
074  76  LBL
075  48  EXC
076  43  RCL
077  32  32
078  65  ×
079  43  RCL
080  31  31
081  95  =
082  42  STO
083  39  39
084  85  +
085  43  RCL
086  40  40
087  95  =
088  32  X:T
089  43  RCL
090  39  39
091  94  +/-
092  85  +
093  43  RCL
094  40  40
095  95  =
096  92  RTN
097  76  LBL
098  88  DMS
099  43  RCL
100  47  47
101  75  -
102  43  RCL
103  37  37
104  95  =
105  42  STO
106  40  40
107  92  RTN
108  76  LBL
109  16  A'
110  71  SBR
111  42  STO
112  76  LBL
113  11  A
114  71  SBR
115  88  DMS
116  43  RCL
117  48  48
118  33  X²
119  55  ÷
120  43  RCL
121  13  13
122  95  =
123  42  STO
124  07  07
125  85  +
126  53  (
127  43  RCL
128  38  38
129  33  X²
130  55  ÷
131  43  RCL
132  03  03
133  54  )
134  42  STO
135  08  08
136  95  =
137  34  √X
138  71  SBR
139  43  RCL
140  87  IFF
141  05  05
142  85  +
143  43  RCL
144  07  07
145  55  ÷
146  53  (
147  24  CE
148  85  +
149  43  RCL
150  08  08
151  54  )
152  95  =
153  42  STO
154  09  09
155  94  +/-
156  85  +
157  01  1
158  95  =
159  33  X²
160  55  ÷
161  53  (
162  43  RCL
163  03  03
164  75  -
165  01  1
166  54  )
167  85  +
168  53  (
169  43  RCL
170  09  09
171  33  X²
172  55  ÷
173  53  (
174  43  RCL
175  13  13
176  75  -
177  01  1
178  54  )
179  54  )
180  95  =
181  35  1/X
182  59  INT
183  76  LBL
184  89  π
185  42  STO
186  15  15
187  87  IFF
188  04  04
189  80  GRD
190  91  R/S
191  42  STO
192  31  31
193  61  GTO
194  48  EXC
195  76  LBL
196  85  +
197  22  INV
198  86  STF
199  05  05
200  43  RCL
201  30  30
202  36  PGM
203  19  19
204  12  B
205  76  LBL
206  37  P/R
207  42  STO
208  42  42
209  22  INV
210  86  STF
211  04  04
212  29  CP
213  43  RCL
214  36  36
215  22  INV
216  77  GE
217  57  ENG
218  22  INV
219  67  EQ
220  65  ×
221  43  RCL
222  42  42
223  85  +
224  76  LBL
225  57  ENG
226  43  RCL
227  42  42
228  95  =
229  32  X:T
230  01  1
231  22  INV
232  77  GE
233  75  -
234  32  X:T
235  92  RTN
236  76  LBL
237  65  ×
238  01  1
239  75  -
240  43  RCL
241  42  42
242  95  =
243  92  RTN
244  76  LBL
245  75  -
246  02  2
247  75  -
248  32  X:T
249  95  =
250  92  RTN
251  76  LBL
252  17  B'
253  71  SBR
254  42  STO
255  76  LBL
256  12  B
257  71  SBR
258  88  DMS
259  43  RCL
260  27  27
261  55  ÷
262  43  RCL
263  03  03
```

```
264   34  √X
265   95   =
266   71  SBR
267   43  RCL
268   43  RCL
269   03   03
270   75   -
271   01   1
272   95   =
273   61  GTO
274   89   π
275   76  LBL
276   80  GRD
277   04   4
278   01   1
279   32  X:T
280   43  RCL
281   15   15
282   77   GE
283   85   +
284   43  RCL
285   30   30
286   36  PGM
287   21   21
288   15   E
289   61  GTO
290   37  P/R
291   76  LBL
292   18  C'
293   71  SBR
294   42  STO
295   76  LBL
296   13   C
297   71  SBR
298   88  DMS
299   29  CP
300   43  RCL
301   33   33
302   22  INV
303   67   EQ
304   70  RAD
305   53   (
306   43  RCL
307   13   13
308   75   -
309   01   1
310   54   )
311   42  STO
312   15   15
313   65   ×
314   43  RCL
315   48   48
316   33  X²
317   85   +
318   53   (
319   53   (
320   43  RCL
321   03   03
322   75   -
323   01   1
324   54   )
325   44  SUM
326   15   15
327   65   ×
328   43  RCL
329   38   38
330   33  X²
331   54   )
332   95   =
333   55   ÷
334   43  RCL
335   15   15
336   95   =
337   34  √X
338   42  STO
339   33   33
340   76  LBL
341   70  RAD
342   43  RCL
343   33   33
344   65   ×
345   53   (
346   43  RCL
347   03   03
348   35  1/X
349   85   +
350   43  RCL
351   13   13
352   35  1/X
353   54   )
354   34  √X
355   95   =
356   71  SBR
357   43  RCL
358   43  RCL
359   03   03
360   85   +
361   43  RCL
362   13   13
363   75   -
364   02   2
365   95   =
366   61  GTO
367   89   π
368   76  LBL
369   14   D
370   42  STO
371   36   36
372   43  RCL
373   48   48
374   55   ÷
375   43  RCL
376   38   38
377   95   =
378   33  X²
379   42  STO
380   30   30
381   43  RCL
382   13   13
383   75   -
384   01   1
385   95   =
386   36  PGM
387   22   22
388   11   A
389   43  RCL
390   15   15
391   87  IFF
392   04   04
393   03   03
394   96   96
395   91  R/S
396   43  RCL
397   03   03
398   75   -
399   01   1
400   95   =
401   76  LBL
402   58  FIX
403   36  PGM
404   22   22
405   12   B
406   43  RCL
407   16   16
408   87  IFF
409   04   04
410   49  PRD
411   91  R/S
412   43  RCL
413   41   41
414   65   ×
415   43  RCL
416   30   30
417   95   =
418   32  X:T
419   43  RCL
420   30   30
421   55   ÷
422   43  RCL
423   31   31
424   95   =
425   92  RTN
426   76  LBL
427   49  PRD
428   43  RCL
429   30   30
430   36  PGM
431   22   22
432   13   C
433   75   -
434   01   1
435   95   =
436   94  +/-
437   61  GTO
438   37  P/R
439   76  LBL
440   15   E
441   42  STO
442   36   36
443   43  RCL
444   47   47
445   55   ÷
446   43  RCL
447   37   37
448   95   =
449   42  STO
450   30   30
451   43  RCL
452   13   13
453   65   ×
454   02   2
455   95   =
456   36  PGM
457   22   22
458   11   A
459   43  RCL
460   15   15
461   87  IFF
462   04   04
463   04   04
464   66   66
465   91  R/S
466   43  RCL
467   03   03
468   65   ×
469   02   2
470   95   =
471   61  GTO
472   58  FIX
```

ZS–7: LABELS AND LOCATIONS

Location	Code	Key
001	38	SIN
092	48	EXC
115	13	C
142	33	X²
151	34	ГX
156	12	B
237	11	A
249	37	P/R
265	57	ENG
277	65	×
285	75	-
292	17	B'
335	30	TAN
339	18	C'
359	35	1/X
369	28	LOG
384	43	RCL
404	39	COS
416	44	SUM
425	16	A'

Step	Code	Key
000	76	LBL
001	38	SIN
002	33	X²
003	23	LNX
004	94	+/-
005	34	ГX
006	42	STO
007	10	10
008	53	(
009	53	(
010	53	(
011	02	2
012	93	.
013	05	5
014	01	1
015	05	5
016	05	5
017	01	1
018	07	7
019	85	+
020	93	.
021	08	8
022	00	0
023	02	2
024	08	8
025	05	5
026	03	3
027	65	×
028	43	RCL
029	10	10
030	85	+
031	93	.
032	00	0
033	01	1
034	00	0
035	03	3
036	02	2
037	08	8
038	65	×
039	43	RCL
040	10	10
041	33	X²
042	54	)
043	55	÷
044	53	(
045	01	1
046	85	+
047	01	1
048	93	.
049	04	4
050	03	3
051	02	2
052	07	7
053	08	8
054	08	8
055	65	×
056	43	RCL
057	10	10
058	85	+
059	93	.
060	01	1
061	08	8
062	09	9
063	02	2
064	06	6
065	09	9
066	65	×
067	43	RCL
068	10	10
069	33	X²
070	85	+
071	93	.
072	00	0
073	00	0
074	01	1
075	03	3
076	00	0
077	08	8
078	65	×
079	43	RCL
080	10	10
081	45	Yˣ
082	03	3
083	54	)
084	54	)
085	94	+/-
086	85	+
087	43	RCL
088	10	10
089	54	)
090	92	RTN
091	76	LBL
092	48	EXC
093	42	STO
094	32	32
095	65	×
096	43	RCL
097	31	31
098	95	=
099	42	STO
100	39	39
101	85	+
102	43	RCL
103	40	40
104	95	=
105	32	X⇄T
106	43	RCL
107	39	39
108	94	+/-
109	85	+
110	43	RCL
111	40	40
112	95	=
113	92	RTN
114	76	LBL
115	13	C
116	71	SBR
117	38	SIN
118	42	STO
119	31	31
120	43	RCL
121	03	03
122	32	X⇄T
123	43	RCL
124	40	40
125	71	SBR
126	28	LOG
127	71	SBR
128	48	EXC
129	32	X⇄T
130	42	STO
131	04	04
132	32	X⇄T
133	42	STO
134	05	05
135	29	CP
136	77	GE
137	33	X²
138	00	0
139	42	STO
140	05	05
141	76	LBL
142	33	X²
143	43	RCL

```
144  04  04
145  32  X↕T
146  01  1
147  77  GE
148  34  √X
149  32  X↕T
150  76  LBL
151  34  √X
152  43  RCL
153  05  05
154  92  RTN
155  76  LBL
156  12  B
157  43  RCL
158  40  40
159  65  ×
160  43  RCL
161  03  03
162  85  +
163  93  .
164  05  5
165  95  =
166  59  INT
167  42  STO
168  00  00
169  43  RCL
170  03  03
171  75  -
172  71  SBR
173  44  SUM
174  02  2
175  95  =
176  91  R/S
177  43  RCL
178  00  00
179  65  ×
180  02  2
181  95  =
182  91  R/S
183  71  SBR
184  44  SUM
185  02  2
186  95  =
187  91  R/S
188  43  RCL
189  03  03
190  75  -
191  43  RCL
192  00  00
193  95  =
194  65  ×
195  02  2
196  95  =
197  91  R/S
198  71  SBR
199  44  SUM
200  43  RCL
201  41  41
202  55  ÷
203  53  (
204  24  CE
205  85  +
206  53  (
207  43  RCL
208  03  03
209  75  -
210  43  RCL
211  00  00
212  54  )
213  54  )
214  95  =
215  32  X↕T
216  43  RCL
217  00  00
218  55  ÷
219  53  (
220  24  CE
221  75  -
222  53  (
223  24  CE
224  75  -
225  43  RCL
226  03  03
227  75  -
228  01  1
229  54  )
230  65  ×
231  43  RCL
232  31  31
233  54  )
234  95  =
235  92  RTN
236  76  LBL
237  11  A
238  71  SBR
239  38  SIN
240  42  STO
241  31  31
242  71  SBR
243  39  COS
244  71  SBR
245  43  RCL
246  61  GTO
247  48  EXC
248  76  LBL
249  37  P/R
250  42  STO
251  42  42
252  29  CP
253  43  RCL
254  36  36
255  22  INV
256  77  GE
257  57  ENG
258  22  INV
259  67  EQ
260  65  ×
261  43  RCL
262  42  42
263  85  +
264  76  LBL
265  57  ENG
266  43  RCL
267  42  42
268  95  =
269  32  X↕T
270  01  1
271  22  INV
272  77  GE
273  75  -
274  32  X↕T
275  91  R/S
276  76  LBL
277  65  ×
278  01  1
279  75  -
280  43  RCL
281  42  42
282  54  )
283  91  R/S
284  76  LBL
285  75  -
286  02  2
287  75  -
288  32  X↕T
289  54  )
290  91  R/S
291  76  LBL
292  17  B'
293  42  STO
294  36  36
295  91  R/S
296  42  STO
297  34  34
298  36  PGM
299  20  20
300  12  B
301  43  RCL
302  03  03
303  36  PGM
304  20  20
305  11  A
306  65  ×
307  43  RCL
308  40  40
309  95  =
310  42  STO
311  30  30
312  32  X↕T
313  43  RCL
314  36  36
315  32  X↕T
316  22  INV
317  77  GE
318  30  TAN
319  01  1
320  22  INV
321  67  EQ
322  03  03
323  26  26
324  94  +/-
325  85  +
326  43  RCL
327  30  30
328  95  =
329  36  PGM
330  20  20
331  14  D
332  61  GTO
333  37  P/R
334  76  LBL
335  30  TAN
336  01  1
337  91  R/S
338  76  LBL
339  18  C'
340  42  STO
341  34  34
342  43  RCL
343  40  40
344  75  -
345  43  RCL
346  34  34
347  95  =
348  55  ÷
349  53  (
350  53  (
351  43  RCL
352  03  03
353  32  X↕T
354  43  RCL
355  34  34
356  71  SBR
357  28  LOG
358  76  LBL
359  35  1/X
360  95  =
361  42  STO
362  30  30
363  36  PGM
364  19  19
365  12  B
366  61  GTO
367  37  P/R
368  76  LBL
369  28  LOG
370  65  ×
371  53  (
372  24  CE
373  94  +/-
374  85  +
375  01  1
376  54  )
377  54  )
378  55  ÷
379  32  X↕T
380  54  )
381  34  √X
382  92  RTN
383  76  LBL
```

```
384  43 RCL
385  43 RCL
386  13  13
387  32 X:T
388  43 RCL
389  47  47
390  71 SBR
391  28 LOG
392  33 X²
393  85  +
394  53  (
395  43 RCL
396  03  03
397  32 X:T
398  43 RCL
399  37  37
400  71 SBR
401  28 LOG
402  92 RTN
403  76 LBL
404  39 COS
405  53  (
406  43 RCL
407  47  47
408  75  -
409  43 RCL
410  37  37
411  54  )
412  42 STO
413  40  40
414  92 RTN
415  76 LBL
416  44 SUM
417  43 RCL
418  00  00
419  85  +
420  01  1
421  95  =
422  65  ×
423  92 RTN
424  76 LBL
425  16 A'
426  42 STO
427  36  36
428  43 RCL
429  13  13
430  35 1/X
431  85  +
432  43 RCL
433  03  03
434  35 1/X
435  95  =
436  42 STO
437  33  33
438  43 RCL
439  13  13
440  42 STO
441  00  00
442  65  ×
443  43 RCL
444  47  47
445  85  +
446  43 RCL
447  03  03
448  44 SUM
449  00  00
450  65  ×
451  43 RCL
452  37  37
453  95  =
454  55  ÷
455  43 RCL
456  00  00
457  65  ×
458  53  (
459  24 CE
460  75  -
461  01  1
462  54  )
463  95  =
464  94 +/-
465  49 PRD
466  33  33
467  71 SBR
468  39 COS
469  55  ÷
470  43 RCL
471  33  33
472  34 √X
473  42 STO
474  32  32
475  95  =
476  61 GTO
477  35 1/X
```

ZS–8: LABELS AND LOCATIONS

```
001  17 B'
016  70 RAD
034  38 SIN
040  80 GRD
050  12  B
071  60 DEG
089  39 COS
099  49 PRD
109  13  C
158  48 EXC
204  57 ENG
240  15  E
249  34 √X
273  14  D
292  52 EE
309  19 D'
330  11  A
387  10 E'
396  33 X²
428  16 A'
455  58 FIX
```

```
000  76 LBL
001  17 B'
002  87 IFF
003  04  04
004  38 SIN
005  86 STF
006  04  04
007  03  3
008  02  2
009  42 STO
010  30  30
011  01  1
012  02  2
013  42 STO
014  00  00
015  76 LBL
016  70 RAD
017  73 RC*
018  00  00
019  55  ÷
020  43 RCL
021  01  01
022  95  =
023  72 ST*
024  30  30
025  91 R/S
026  01  1
027  44 SUM
028  00  00
029  44 SUM
030  30  30
031  61 GTO
032  70 RAD
033  76 LBL
034  38 SIN
035  03  3
036  02  2
037  42 STO
038  00  00
039  76 LBL
040  80 GRD
041  73 RC*
042  00  00
043  91 R/S
044  01  1
045  44 SUM
046  00  00
047  61 GTO
048  80 GRD
049  76 LBL
050  12  B
051  87 IFF
052  03  03
053  39 COS
054  86 STF
055  03  03
056  43 RCL
057  02  02
058  85  +
059  01  1
060  02  2
061  95  =
062  42 STO
063  00  00
064  85  +
065  02  2
066  00  0
067  95  =
068  42 STO
069  30  30
070  76 LBL
071  60 DEG
072  73 RC*
073  00  00
074  55  ÷
075  43 RCL
076  02  02
077  95  =
078  72 ST*
079  30  30
```

```
080  91 R/S
081  01  1
082  44 SUM
083  00  00
084  44 SUM
085  30  30
086  61 GTO
087  60 DEG
088  76 LBL
089  39 COS
090  43 RCL
091  02  02
092  85  +
093  03  3
094  02  2
095  95  =
096  42 STO
097  00  00
098  76 LBL
099  49 PRD
100  73 RC*
101  00  00
102  91 R/S
103  01  1
104  44 SUM
105  00  00
106  61 GTO
107  49 PRD
108  76 LBL
109  13  C
110  87 IFF
111  05  05
112  58 FIX
113  36 PGM
114  16  16
115  11  A
116  36 PGM
117  16  16
118  12  B
119  36 PGM
120  16  16
121  17  B'
122  91 R/S
123  43 RCL
124  01  01
125  75  -
126  01  1
127  95  =
128  42 STO
129  11  11
130  42 STO
131  58  58
132  91 R/S
133  35 1/X
134  65  ×
135  43 RCL
136  10  10
137  95  =
138  91 R/S
139  43 RCL
140  09  09
141  91 R/S
142  43 RCL
143  02  02
144  75  -
145  01  1
146  95  =
147  42 STO
148  59  59
149  49 PRD
150  11  11
151  91 R/S
152  35 1/X
153  65  ×
154  43 RCL
155  09  09
156  95  =
157  76 LBL
158  48 EXC
159  91 R/S
160  43 RCL
161  09  09
162  85  +
163  43 RCL
164  10  10
165  75  -
166  43 RCL
167  08  08
168  95  =
169  94 +/-
170  91 R/S
171  55  ÷
172  43 RCL
173  11  11
174  91 R/S
175  95  =
176  42 STO
177  48  48
178  91 R/S
179  43 RCL
180  08  08
181  91 R/S
182  43 RCL
183  03  03
184  75  -
185  01  1
186  95  =
187  91 R/S
188  87 IFF
189  05  05
190  57 ENG
191  36 PGM
192  16  16
193  18  C'
194  42 STO
195  12  12
196  36 PGM
197  16  16
198  18  C'
199  42 STO
200  13  13
201  86 STF
202  05  05
203  76 LBL
204  57 ENG
205  43 RCL
206  58  58
207  36 PGM
208  22  22
209  11  A
210  43 RCL
211  11  11
212  36 PGM
213  22  22
214  12  B
215  43 RCL
216  12  12
217  91 R/S
218  36 PGM
219  22  22
220  13  C
221  91 R/S
222  43 RCL
223  59  59
224  36 PGM
225  22  22
226  11  A
227  43 RCL
228  11  11
229  36 PGM
230  22  22
231  12  B
232  43 RCL
233  13  13
234  91 R/S
235  36 PGM
236  22  22
237  13  C
238  92 RTN
239  76 LBL
240  15  E
241  00  0
242  42 STO
243  53  53
244  42 STO
245  50  50
246  42 STO
247  54  54
248  76 LBL
249  34 √X
250  91 R/S
251  42 STO
252  51  51
253  33 X²
254  44 SUM
255  53  53
256  91 R/S
257  42 STO
258  52  52
259  65  ×
260  43 RCL
261  51  51
262  95  =
263  44 SUM
264  50  50
265  01  1
266  44 SUM
267  54  54
268  43 RCL
269  54  54
270  61 GTO
271  34 √X
272  76 LBL
273  14  D
274  65  ×
275  43 RCL
276  58  58
277  65  ×
278  43 RCL
279  48  48
280  65  ×
281  43 RCL
282  53  53
283  55  ÷
284  53  (
285  43 RCL
286  59  59
287  85  +
288  01  1
289  54  )
290  95  =
291  76 LBL
292  52 EE
293  34 √X
294  42 STO
295  49  49
296  85  +
297  43 RCL
298  50  50
299  95  =
300  32 X⇌T
301  43 RCL
302  50  50
303  75  -
304  43 RCL
305  49  49
306  95  =
307  92 RTN
308  76 LBL
309  19  D'
310  65  ×
311  43 RCL
312  59  59
313  65  ×
314  43 RCL
315  48  48
316  65  ×
317  43 RCL
318  53  53
319  55  ÷
```

```
320  53  (
321  43 RCL
322  58  58
323  85  +
324  01  1
325  54  )
326  95  =
327  61 GTO
328  52 EE
329  76 LBL
330  11  A
331  36 PGM
332  15  15
333  11  A
334  42 STO
335  30  30
336  43 RCL
337  08  08
338  91 R/S
339  43 RCL
340  14  14
341  91 R/S
342  43 RCL
343  15  15
344  91 R/S
345  43 RCL
346  13  13
347  91 R/S
348  43 RCL
349  16  16
350  91 R/S
351  43 RCL
352  17  17
353  42 STO
354  48  48
355  91 R/S
356  43 RCL
357  12  12
358  91 R/S
359  43 RCL
360  09  09
361  75  -
362  01  1
363  95  =
364  91 R/S
365  43 RCL
366  30  30
367  91 R/S
368  43 RCL
369  14  14
370  36 PGM
371  22  22
372  11  A
373  43 RCL
374  16  16
375  36 PGM
376  22  22
377  12  B
378  43 RCL
379  30  30
380  36 PGM
381  22  22
382  13  C
383  42 STO
384  31  31
385  92 RTN
386  76 LBL
387  10 E'
388  00  0
389  42 STO
390  53  53
391  42 STO
392  50  50
393  42 STO
394  54  54
395  76 LBL
396  33 X²
397  91 R/S
398  42 STO
399  51  51
400  91 R/S
401  42 STO
402  52  52
403  91 R/S
404  35 1/X
405  65  ×
406  43 RCL
407  51  51
408  33 X²
409  95  =
410  44 SUM
411  53  53
412  43 RCL
413  51  51
414  65  ×
415  43 RCL
416  52  52
417  95  =
418  44 SUM
419  50  50
420  01  1
421  44 SUM
422  54  54
423  43 RCL
424  54  54
425  61 GTO
426  33 X²
427  76 LBL
428  16 A'
429  65  ×
430  43 RCL
431  14  14
432  65  ×
433  43 RCL
434  48  48
435  65  ×
436  43 RCL
437  53  53
438  95  =
439  34 √X
440  42 STO
441  49  49
442  85  +
443  43 RCL
444  50  50
445  95  =
446  32 X:T
447  43 RCL
448  50  50
449  75  -
450  43 RCL
451  49  49
452  95  =
453  92 RTN
454  76 LBL
455  58 FIX
456  43 RCL
457  10  10
458  91 R/S
459  55  ÷
460  43 RCL
461  58  58
462  91 R/S
463  95  =
464  91 R/S
465  43 RCL
466  09  09
467  91 R/S
468  55  ÷
469  43 RCL
470  59  59
471  91 R/S
472  95  =
473  61 GTO
474  48 EXC
```

ZS–9: LABELS AND LOCATIONS

```
001  14  D
007  10 E'
091  11  A
105  23 LNX
119  12  B
135  13  C
139  28 LOG
155  18 C'
168  70 RAD
193  65  ×
203  16 A'
213  38 SIN
239  85  +
252  58 FIX
262  75  -
267  17 B'
278  15  E
303  19 D'
381  43 RCL
```

```
000  76  LBL
001  14   D
002  32  X:T
003  91  R/S
004  78  Σ+
005  91  R/S
006  76  LBL
007  10  E'
008  43  RCL
009  03   03
010  75   -
011  02   2
012  95   =
013  42  STO
014  15   15
015  69  OP
016  11   11
017  65   ×
018  43  RCL
019  03   03
020  95   =
021  42  STO
022  07   07
023  32  X:T
024  65   ×
025  43  RCL
026  03   03
027  95   =
028  42  STO
029  08   08
030  65   ×
031  43  RCL
032  07   07
033  95   =
034  34  √X
035  65   ×
036  69  OP
037  13   13
038  42  STO
039  44   44
040  65   ×
041  69  OP
042  12   12
043  42  STO
044  46   46
045  32  X:T
046  42  STO
047  45   45
048  94  +/-
049  85   +
050  43  RCL
051  07   07
052  95   =
053  55   ÷
054  43  RCL
055  15   15
056  95   =
057  34  √X
058  42  STO
059  33   33
060  65   ×
061  53   (
062  43  RCL
063  05   05
064  55   ÷
065  43  RCL
066  03   03
067  55   ÷
068  43  RCL
069  08   08
070  54   )
071  34  √X
072  95   =
073  42  STO
074  36   36
075  43  RCL
076  33   33
077  55   ÷
078  43  RCL
079  08   08
080  34  √X
081  95   =
082  42  STO
083  37   37
084  43  RCL
085  15   15
086  91  R/S
087  42  STO
088  31   31
089  92  RTN
090  76  LBL
091  11   A
092  43  RCL
093  31   31
094  65   ×
095  43  RCL
096  37   37
097  95   =
098  42  STO
099  39   39
100  43  RCL
101  45   45
102  42  STO
103  40   40
104  76  LBL
105  23  LNX
106  85   +
107  43  RCL
108  39   39
109  95   =
110  32  X:T
111  43  RCL
112  40   40
113  75   -
114  43  RCL
115  39   39
116  95   =
117  91  R/S
118  76  LBL
119  12   B
120  43  RCL
121  31   31
122  65   ×
123  43  RCL
124  36   36
125  95   =
126  42  STO
127  39   39
128  43  RCL
129  46   46
130  42  STO
131  40   40
132  61  GTO
133  23  LNX
134  76  LBL
135  13   C
136  71  SBR
137  70  RAD
138  76  LBL
139  28  LOG
140  65   ×
141  43  RCL
142  31   31
143  95   =
144  42  STO
145  39   39
146  43  RCL
147  00   00
148  69  OP
149  14   14
150  42  STO
151  40   40
152  61  GTO
153  23  LNX
154  76  LBL
155  18  C'
156  71  SBR
157  70  RAD
158  33  X²
159  85   +
160  43  RCL
161  33   33
162  33  X²
163  95   =
164  34  √X
165  61  GTO
166  28  LOG
167  76  LBL
168  70  RAD
169  42  STO
170  00   00
171  75   -
172  79   x̄
173  32  X:T
174  95   =
175  33  X²
176  55   ÷
177  43  RCL
178  08   08
179  85   +
180  43  RCL
181  03   03
182  35  1/X
183  95   =
184  34  √X
185  65   ×
186  43  RCL
187  33   33
188  95   =
189  42  STO
190  38   38
191  92  RTN
192  76  LBL
193  65   ×
194  42  STO
195  28   28
196  91  R/S
197  94  +/-
198  85   +
199  69  OP
200  12   12
201  92  RTN
202  76  LBL
203  16  A'
204  71  SBR
205  65   ×
206  32  X:T
207  95   =
208  55   ÷
209  43  RCL
210  37   37
211  95   =
212  76  LBL
213  38  SIN
214  42  STO
215  30   30
216  36  PGM
217  21   21
218  15   E
219  42  STO
220  29   29
221  43  RCL
222  28   28
223  32  X:T
224  00   0
225  67  EQ
226  85   +
227  01   1
228  94  +/-
229  67  EQ
230  75   -
231  43  RCL
232  29   29
233  94  +/-
234  85   +
235  01   1
236  95   =
237  91  R/S
238  76  LBL
239  85   +
```

```
240  93   .
241  05   5
242  32  X⇄T
243  43  RCL
244  29   29
245  77   GE
246  58  FIX
247  65   ×
248  02   2
249  95   =
250  91  R/S
251  76  LBL
252  58  FIX
253  94  +/-
254  85   +
255  01   1
256  95   =
257  65   ×
258  02   2
259  95   =
260  91  R/S
261  76  LBL
262  75   -
263  43  RCL
264  29   29
265  91  R/S
266  76  LBL
267  17  B'
268  71  SBR
269  65   ×
270  95   =
271  55   ÷
272  43  RCL
273  36   36
274  95   =
275  61  GTO
276  38  SIN
277  76  LBL
278  15   E
279  42  STO
280  28   28
281  43  RCL
282  44   44
283  42  STO
284  40   40
285  33  X²
286  94  +/-
287  85   +
288  01   1
289  95   =
290  34  √X
291  35  1/X
292  65   ×
293  43  RCL
294  15   15
295  34  √X
296  65   ×
297  43  RCL
298  40   40
299  95   =
300  61  GTO
301  38  SIN
302  76  LBL
303  19  D'
304  75   -
305  79   x̄
306  95   =
307  42  STO
308  43   43
309  33  X²
310  55   ÷
311  43  RCL
312  08   08
313  95   =
314  85   +
315  53   (
316  53   (
317  53   (
318  43  RCL
319  33   33
320  65   ×
321  43  RCL
322  31   31
323  54   )
324  33  X²
325  55   ÷
326  43  RCL
327  08   08
328  54   )
329  94  +/-
330  85   +
331  43  RCL
332  45   45
333  33  X²
334  54   )
335  42  STO
336  42   42
337  65   ×
338  53   (
339  43  RCL
340  03   03
341  35  1/X
342  85   +
343  01   1
344  54   )
345  95   =
346  29  CP
347  22  INV
348  77   GE
349  43  RCL
350  34  √X
351  65   ×
352  43  RCL
353  31   31
354  65   ×
355  43  RCL
356  33   33
357  55   ÷
358  43  RCL
359  42   42
360  95   =
361  42  STO
362  39   39
363  43  RCL
364  45   45
365  65   ×
366  43  RCL
367  43   43
368  55   ÷
369  43  RCL
370  42   42
371  95   =
372  85   +
373  79   x̄
374  32  X⇄T
375  95   =
376  42  STO
377  40   40
378  61  GTO
379  23  LNX
380  76  LBL
381  43  RCL
382  00   0
383  35  1/X
384  92  RTN
```

ZS–10: LABELS AND LOCATIONS

```
001  10  E'
195  11   A
209  23  LNX
223  12   B
239  13   C
255  16  A'
269  80  GRD
290  57  ENG
302  65   ×
310  75   -
317  17  B'
333  18  C'
349  15   E
372  14   D
```

```
000  76  LBL
001  10  E'
002  43  RCL
003  03   03
004  65   ×
005  43  RCL
006  05   05
007  65   ×
008  43  RCL
009  02   02
010  95   =
011  42  STO
012  27   27
013  02   2
014  65   ×
015  43  RCL
016  04   04
017  65   ×
018  43  RCL
019  01   01
020  65   ×
021  43  RCL
022  06   06
023  95   =
024  44  SUM
025  27   27
026  43  RCL
027  03   03
028  65   ×
029  43  RCL
030  06   06
031  33  X²
032  95   =
033  22  INV
034  44  SUM
035  27   27
036  43  RCL
037  04   04
038  33  X²
039  65   ×
040  43  RCL
041  02   02
042  95   =
043  22  INV
044  44  SUM
045  27   27
046  43  RCL
047  01   01
048  33  X²
049  65   ×
050  43  RCL
051  05   05
052  95   =
053  22  INV
054  44  SUM
055  27   27
056  43  RCL
057  13   13
058  42  STO
059  33   33
060  36  PGM
061  18   18
062  11   A
063  42  STO
064  41   41
065  65   ×
066  43  RCL
067  07   07
068  95   =
069  22  INV
070  44  SUM
071  33   33
072  36  PGM
073  18   18
074  12   B
075  42  STO
076  42   42
077  65   ×
078  43  RCL
079  11   11
080  95   =
081  22  INV
082  44  SUM
083  33   33
084  36  PGM
085  18   18
086  13   C
087  42  STO
088  43   43
089  65   ×
090  43  RCL
091  12   12
092  95   =
093  22  INV
094  44  SUM
095  33   33
096  00   0
097  36  PGM
098  18   18
099  14   D
100  42  STO
101  44   44
102  43  RCL
103  33   33
104  55   ÷
105  53   (
106  43  RCL
107  03   03
108  75   -
109  03   3
110  54   )
111  42  STO
112  15   15
113  95   =
114  34  ΓX
115  42  STO
116  33   33
117  43  RCL
118  05   05
119  65   ×
120  43  RCL
121  02   02
122  75   -
123  43  RCL
124  06   06
125  33  X²
126  95   =
127  55   ÷
128  43  RCL
129  27   27
130  95   =
131  34  ΓX
132  65   ×
133  43  RCL
134  33   33
135  95   =
136  42  STO
137  36   36
138  43  RCL
139  03   03
140  65   ×
141  43  RCL
142  02   02
143  75   -
144  43  RCL
145  01   01
146  33  X²
147  95   =
148  55   ÷
149  43  RCL
150  27   27
151  95   =
152  34  ΓX
153  65   ×
154  43  RCL
155  33   33
156  95   =
157  42  STO
158  37   37
159  43  RCL
160  03   03
161  65   ×
162  43  RCL
163  05   05
164  75   -
165  43  RCL
166  04   04
167  33  X²
168  95   =
169  55   ÷
170  43  RCL
171  27   27
172  95   =
173  34  ΓX
174  65   ×
175  43  RCL
176  33   33
177  95   =
178  42  STO
179  38   38
180  43  RCL
181  07   07
182  55   ÷
183  43  RCL
184  03   03
185  95   =
186  42  STO
187  08   08
188  43  RCL
189  15   15
190  91  R/S
191  42  STO
192  31   31
193  92  RTN
194  76  LBL
195  11   A
196  43  RCL
197  41   41
198  42  STO
199  40   40
200  43  RCL
201  31   31
202  65   ×
203  43  RCL
204  36   36
205  95   =
206  42  STO
207  39   39
208  76  LBL
209  23  LNX
210  85   +
211  43  RCL
212  40   40
213  95   =
214  32  X:T
215  43  RCL
216  40   40
217  75   -
218  43  RCL
219  39   39
220  95   =
221  92  RTN
222  76  LBL
223  12   B
224  43  RCL
225  42   42
226  42  STO
227  40   40
228  43  RCL
229  31   31
230  65   ×
231  43  RCL
232  37   37
233  95   =
234  42  STO
235  39   39
236  61  GTO
237  23  LNX
238  76  LBL
239  13   C
```

```
240  43 RCL
241  43  43
242  42 STO
243  40  40
244  43 RCL
245  31  31
246  65  ×
247  43 RCL
248  38  38
249  95  =
250  42 STO
251  39  39
252  61 GTO
253  23 LNX
254  76 LBL
255  16 A'
256  42 STO
257  28  28
258  91 R/S
259  94 +/-
260  85  +
261  43 RCL
262  41  41
263  95  =
264  55  ÷
265  43 RCL
266  36  36
267  95  =
268  76 LBL
269  80 GRD
270  42 STO
271  30  30
272  36 PGM
273  21  21
274  15  E
275  42 STO
276  29  29
277  43 RCL
278  28  28
279  29 CP
280  22 INV
281  77  GE
282  57 ENG
283  22 INV
284  67  EQ
285  65  ×
286  43 RCL
287  29  29
288  85  +
289  76 LBL
290  57 ENG
291  43 RCL
292  29  29
293  95  =
294  32 X:T
295  01  1
296  22 INV
297  77  GE
298  75  -
299  32 X:T
300  92 RTN
301  76 LBL
302  65  ×
303  01  1
304  75  -
305  43 RCL
306  29  29
307  95  =
308  92 RTN
309  76 LBL
310  75  -
311  02  2
312  75  -
313  32 X:T
314  95  =
315  92 RTN
316  76 LBL
317  17 B'
318  42 STO
319  28  28
320  91 R/S
321  94 +/-
322  85  +
323  43 RCL
324  42  42
325  95  =
326  55  ÷
327  43 RCL
328  37  37
329  95  =
330  61 GTO
331  80 GRD
332  76 LBL
333  18 C'
334  42 STO
335  28  28
336  91 R/S
337  94 +/-
338  85  +
339  43 RCL
340  43  43
341  95  =
342  55  ÷
343  43 RCL
344  38  38
345  95  =
346  61 GTO
347  80 GRD
348  76 LBL
349  15  E
350  65  ×
351  43 RCL
352  42  42
353  95  =
354  42 STO
355  00  00
356  00  0
357  91 R/S
358  65  ×
359  43 RCL
360  43  43
361  95  =
362  44 SUM
363  00  00
364  43 RCL
365  41  41
366  44 SUM
367  00  00
368  43 RCL
369  00  00
370  92 RTN
371  76 LBL
372  19 D'
373  43 RCL
374  41  41
375  91 R/S
376  43 RCL
377  42  42
378  91 R/S
379  43 RCL
380  43  43
381  92 RTN
```

Index